Responsibility and Healthcare

Responsibility and Healthcare

Edited by
BENJAMIN DAVIES, GABRIEL DE MARCO,
NEIL LEVY, JULIAN SAVULESCU

OXFORD
UNIVERSITY PRESS

OXFORD
UNIVERSITY PRESS

Great Clarendon Street, Oxford, OX2 6DP,
United Kingdom

Oxford University Press is a department of the University of Oxford.
It furthers the University's objective of excellence in research, scholarship,
and education by publishing worldwide. Oxford is a registered trade mark of
Oxford University Press in the UK and in certain other countries

© Edited by Benjamin Davies, Gabriel De Marco, Neil Levy, Julian Savulescu 2024

The moral rights of the authors have been asserted

Published in the United States of America by Oxford University Press
198 Madison Avenue, New York, NY 10016, United States of America

British Library Cataloguing in Publication Data
Data available

Library of Congress Control Number: 2023944890

ISBN 978–0–19–287223–4

DOI: 10.1093/oso/9780192872234.001.0001

Printed and bound by
CPI Group (UK) Ltd, Croydon, CR0 4YY

Links to third party websites are provided by Oxford in good faith and
for information only. Oxford disclaims any responsibility for the materials
contained in any third party website referenced in this work.

MIX
Paper | Supporting
responsible forestry
FSC® C013604

Contents

PART I THE JUSTIFICATION FOR RESPONSIBILITY-SENSITIVE HEALTH CARE POLICIES

PART II THE NATURE OF RESPONSIBILITY-SENSITIVE HEALTH CARE POLICIES

PART III RESPONSIBILITY FOR HEALTH

List of Figures and Tables

Figures

Tables

Notes on Contributors

Jana Schaich Borg is an Associate Research Professor at Duke's Social Science Research Institute, co-Director of Duke's Moral Attitudes and Decision-Making Lab, co-Director of Duke's Moral Artificial Intelligence Lab, and was formerly the Director of the Duke's Master in Interdisciplinary Data Science program. Dr. Schaich Borg is also an affiliate of Duke's Center for Cognitive Neuroscience and Duke's Institute for Brain Science, and has taught at Duke's Master of Quantitative Management at Fuqua School of Business. She authored the book "Moral AI and How We Get There" (Penguin 2024) with Walter Sinnott-Armstrong and Vincent Conitzer.

Rebecca Brown is a Senior Research Fellow at the Oxford Uehiro Centre for Practical Ethics and a research associate at Wadham College. She works largely on topics in public health ethics, including the ethics of public health communication and health promoting interventions to reduce chronic disease.

Lok Chan received his PhD in Philosophy at Duke University and held a postdoctoral research position in the Social Science Research Institute and the Department of Population Health Sciences at Duke University. At Duke's Moral AI Lab, Chan investigated issues pertaining to the ethics of artificial intelligence. He also was part of an NIH-funded project that examines the role of risk perception in the decision-making processes of cancer patients. Currently, Chan serves as an engineer at Curi Advisory, where he engages in data-centric projects that intersect health tech and risk management.

Vincent Conitzer is the Kimberly J. Jenkins Distinguished University Professor of New Technologies and Professor of Computer Science, Professor of Economics, and Professor of Philosophy at Duke University. He is also Head of Technical AI Engagement at the Institute for Ethics in AI, and Professor of Computer Science and Philosophy, at the University of Oxford. He received PhD (2006) and MS (2003) degrees in Computer Science from Carnegie Mellon University, and an AB (2001) degree in Applied Mathematics from Harvard University. Conitzer works on artificial intelligence (AI). Much of his work has focused on AI and game theory, for example designing algorithms for the optimal strategic placement of defensive resources. More recently, he has started to work on AI and ethics: how should we determine the objectives that AI systems pursue, when these objectives have complex effects on various stakeholders?

Ben Davies is a Lecturer in Political Philosophy at the University of Sheffield. He was previously a Wellcome Trust Research Fellow at the Oxford Uehiro Centre for Practical Ethics and Wellcome Centre for Ethics and Humanities, leading a project on the role of sufficiency in health care ethics and policy.

Gabriel De Marco is a Research Fellow at the Oxford Uehiro Centre for Practical Ethics. He received his PhD in Philosophy at Florida State University (2018). His research focuses on free will, moral responsibility, and their relevance to practical issues.

Nir Eyal is the Henry Rutgers Professor of Bioethics in the School of Public Health, the Director of the Center for Population-level Ethics, and an Associate Faculty Member of the Department of Philosophy, Rutgers University. He has published widely in medical ethics and population-level bioethics. He is also the co-editor of the OUP series *Population-Level Bioethics*. He has co-edited *Inequalities in Health* (Oxford University Press, 2013), *Identified vs. Statistical Persons* (Oxford University Press, 2015), and *Measuring the Global Burden of Disease: Philosophical Dimensions* (Oxford University Press, 2020).

Nadira Faber is a Full Professor of Social and Economic Psychology at the University of Bremen and an Associated Faculty Member in Philosophy at the University of Oxford. She is an experimental social psychologist by training, and works interdisciplinarily on the intersection of psychology and philosophy.

Zoë Fritz is a Wellcome fellow in Society and Ethics, and a Consultant Physician in acute medicine at Addenbrooke's Hospital. She is also Director of Studies in Clinical Medicine and Fellow at Gonville and Caius College, University of Cambridge. Her research is focused on identifying areas of clinical practice that raise ethical questions and applying rigorous empirical and ethical analysis to explore the issues and find effective solutions.

Richard Holton is a professor of philosophy at the University of Cambridge, and a fellow of Peterhouse.. He has published widely in moral psychology, ethics, and the philosophy of law. He is the author of *Willing, Wanting, Waiting* (Oxford University Press, 2009).

Andreas Kappes is a Senior Lecturer in the Psychology Department at City, University of London, and the Principal Investigator in the Other Lab (City, UL). His research focuses on how the unavoidable uncertainty in our social decisions affects our behaviour, how and when we are influenced by others, and how we learn about ourselves and others.

Jeanette Kennett is a professor of philosophy in the Department of Philosophy at Macquarie University. She has published widely in moral cognition, moral motivation, moral and criminal responsibility, addiction, and mental disorder. She is the author of *Agency and Responsibility: A Common-Sense Moral Psychology* (Oxford University Press, 2001).

Neil Levy is a professor of philosophy in the Department of Philosophy at Macquarie University, Sydney, and a Senior Research Fellow at the Oxford Uehiro Centre for Practical Ethics. He has published widely in neuroethics, applied ethics, philosophy of mind, and free will and moral responsibility. His most recent book is *Bad Beliefs* (Oxford University Press, 2021).

Peter Marber, PhD is a pioneering global investor, professor, and writer. He currently teaches at Harvard and Johns Hopkins, and he has published eight books including his latest *Augmented Education in the Global Age: Artificial Intelligence and the Future of Learning and Work* (2023, Routledge) and *Quid Periculum? Measuring and Managing Political Risk in the 21st Century* (2021, PRS). Marber is a Fellow of both the Royal Astronomical Society and Royal Society of Arts, and is a member of Chatham House. He serves and/or has served on boards for institutions including the Emerging Markets Trade

Association, St. John's College, Geolinks, University of Virginia, World Policy Institute, and Columbia's School of International & Public Affairs.

Daniel J. Miller is an Assistant Professor in the Department of Philosophy at West Virginia University and a Lecturer in the West Virginia University School of Medicine. His research focuses on topics in moral responsibility, the ethics of blame and forgiveness, and medical ethics.

Rekha Nath is an Associate Professor of Philosophy at the University of Alabama, specializing in social and political philosophy. She has published on global justice, equality, and responsibility. Her book *Why It's OK to be Fat* is forthcoming with Routledge Press.

Dana Kay Nelkin is a Professor of Philosophy at the University of California, San Diego. Her areas of research include moral psychology, ethics, bioethics, and philosophy of law. She is the author of *Making Sense of Freedom and Responsibility* (Oxford University Press, 2011), and a number of articles on a variety of topics, including self-deception, friendship, the lottery paradox, psychopathy, forgiveness, moral luck, and praise and blame. She is also a co-editor of the *The Ethics and Law of Omissions* (2017), *The Oxford Handbook of Moral Responsibility* (2022), and *Forgiveness and its Moral Dimensions* (2021). Her work in moral psychology includes participation in an interdisciplinary research collaboration of philosophers and psychologists, *The Moral Judgements Project*, which brings together normative and descriptive enquiries about the use of moral principles such as the Doctrine of Doing and Allowing and the Doctrine of Double Effect.

Anne-Marie Nussberger is a postdoctoral researcher at the Center for Humans and Machines of the Max-Planck-Institute for Human Development, Berlin. Her doctoral work in Experimental Psychology at the University of Oxford explored human decision-making under uncertainty in contexts such as health care and infectious disease. More recently, she investigates how uncertainty shapes interactions between humans and artificially intelligent agents.

Joshua Parker is a General Practitioner. He is also a PhD candidate at Lancaster University funded by the Wellcome Trust. His PhD research explores the questions of distributive justice at the intersection of health, health care and climate change.

Julian Savulescu is the Uehiro Chair of Practical Ethics and Director of the Oxford Uehiro Centre for Practical Ethics and Co-Director Wellcome Centre for Ethics and Humanities, University of Oxford. He holds a Wellcome Senior Investigator Award on Responsibility and health care.

Shlomi Segall is Professor of Political Philosophy in the Program in Politics, Philosophy, and Economics at the Hebrew University of Jerusalem. He is the author of numerous articles as well as three books, including one on the topic of the proposed collection: *Health, Luck, and Justice* (Princeton University Press, 2010), *Equality and Opportunity* (OUP, 2013), and *Why Inequality Matters* (CUP, 2016).

Elizabeth Shaw is a Senior Lecturer in the law school at Aberdeen University. She received her LLM by research in 2008 and LLB in 2010 from Aberdeen University, and received her PhD from Edinburgh University in 2014. She is also a co-director of the Justice Without

Retribution Network. Her research interests are interdisciplinary, involving criminal law, philosophy, and neuroscience. She is a co-editor of *Free Will Skepticism in Law and Society* (Cambridge University Press, 2019), and *The Routledge Handbook of Philosophy and Science of Punishment* (Routledge, 2021).

Walter Sinnott-Armstrong is Chauncey Stillman Professor of Practical Ethics in the Department of Philosophy and the Kenan Institute for Ethics at Duke University. He has secondary appointments in Law and in Psychology and Neuroscience and is core faculty in the Duke Center for Cognitive Neuroscience. He has published widely in ethics, moral psychology and neuroscience, epistemology, philosophy of law, and philosophy of religion. His most recent books include *Think Again: How to Reason and Argue* (Oxford University Press, 2018) and *Clean Hands: Philosophical Lessons from Scrupulosity* (Oxford University Press, 2019).

Acknowledgments

This volume originated in a conference on the same topic that, due to COVID-19, was held virtually on April 28–30, 2021. Many contributors also participated in that conference, and we are grateful to them for the fruitful discussion. That conference was made possible by generous funding from the Wellcome Trust, and Julian Savulescu's Wellcome-funded grant, Responsibility and Health Care (WT104848) and the Wellcome Centre for Ethics and Humanities (WT203132). For invaluable assistance in putting that conference together, we would like to thank Rachel Gaminiratne and Miriam Wood. The project is part of the Oxford Uehiro Centre for Practical Ethics, and we thank The Uehiro Foundation for Ethics and Education for their generous support to the Centre and the Uehiro Chair in Practical Ethics. We also owe a great debt of gratitude to all of our contributors, for their excellent contributions to this volume.

Introduction: Responsibility and Health Care

An Overview

Ben Davies, Gabriel De Marco, Neil Levy, and Julian Savulescu

0.1 Introduction

Many illnesses that risk death or serious harm are at least partly the result of lifestyle behaviours such as smoking, lack of exercise, or extreme sports. The WHO notes that the global prevalence of preventable, non-communicable diseases is rising, and accounts for a large proportion of deaths worldwide. According to some, insofar as such illnesses are related to exercises of agency, individuals with such illnesses may be partly, or wholly, responsible for them. One relevant question, then, concerns whether health care providers, be they state-run systems or private entities, would be justified in treating patients differently on the basis of health-related behaviours. Could a health care system, for example, be justified in offering discounts to those who engage in healthy behaviours? Or, when there are not enough resources to treat everyone who is ill, would a health care system be justified in holding patients accountable for their harmful lifestyles—for instance, by taxing tobacco products, charging higher premiums, or giving such patients lower priority for certain treatments?

Consider, for instance, the following case. Two patients require a heart transplant. They are alike in the respects often considered in such circumstances; for example, they have similar chances of successful transplantation and recovery, have similar chances of death without the transplant, and so on. However, one of these patients needs a transplant partly because of lifestyle choices they could have made differently, whereas the other needs it due to factors outside her control. There is only one heart available. Should we take the first patient's partial responsibility for needing the transplant into account when deciding who gets the heart? If people think the answer is "yes", then they would seem to think that responsibility for health-related lifestyles, and perhaps outcomes, can permissibly be used as a factor in making health care decisions.

Benjamin Davies, Gabriel De Marco, Neil Levy, and Julian Savulescu, *Introduction: Responsibility and Health Care—An Overview* In: *Responsibility and Healthcare*. Edited by: Benjamin Davies, Gabriel De Marco, Neil Levy, Julian Savulescu, Oxford University Press. © Edited by Benjamin Davies, Gabriel De Marco, Neil Levy, Julian Savulescu 2024.
DOI: 10.1093/oso/9780192872234.003.0001

In fact, as the chapter by Lok Chan, Walter Sinnott-Armstrong, Jana Schaich Borg, and Vincent Conitzer (Chapter 1) suggests, people's responses to such cases may be more complex than a simple yes or no. Chan et al. conducted experiments related to vignettes like the one above to test people's attitudes towards responsibility and its relationship to prioritising kidney transplants. Their work uncovered intriguing, and perhaps unexpected, relationships between people's attitudes towards patients' responsibility for health-related behaviours and their outcomes on the one hand, and prioritising treatment access on the other. For instance, particular importance seems to be placed on a patient's decision to stop an unhealthy behaviour upon diagnosis of a related health condition, independent of responsibility for earlier behaviour or the resulting condition. And, in contrast to some widespread philosophical assumptions, respondents seemed to think that a patient's *actual* knowledge is most relevant to blameworthiness for a health outcome, but that whether knowledge was easily accessible was relevant to kidney allocation.

Answering these questions requires delving into a variety of issues in political philosophy and ethical theory: When is a distribution of resources just or fair? Should individuals have to meet all the costs of their choices, or is there a limit on how much any individual can be expected to bear? What are the responsibilities of institutions, governments, and society for individuals' health-related lifestyle choices? It also raises questions in philosophy of action and moral psychology: What does it mean for agents to be responsible for their actions, their lifestyles, and their health-related outcomes? And responsibility for health involves cognitive science: What are we actually like as agents? When we make health-affecting decisions such as having a cigarette or failing to wear a seatbelt, do we meet the conditions for responsibility laid out by our best philosophical theories? When these questions are addressed, they are often addressed independently, yet the answers are interrelated. For instance, whether a policy that holds patients accountable is justified depends not only on what amounts to a fair distribution of goods, but also on whether patients' health and health-related behaviour is sufficiently under their control to justify holding them accountable for it. And the proper design of policies that help to improve our agency with respect to these behaviours and outcomes will depend on what we are like as agents. Similarly, the details of such policies, if they are to be justified, will depend on what sorts of things we can be responsible for, and to what extent we can be responsible at all.

Questions of responsibility in health care do not end with those receiving treatment. Health care professionals, including doctors, also have various responsibilities, particularly to their patients. The doctor–patient relationship has been conceived of in various ways with different understandings of the responsibilities each party holds. It also raises important questions: given the inherent uncertainty in medical decision-making, how should doctors and patients jointly approach health-related choices, and who is responsible when a mutually agreed decision goes wrong? What are the limits to doctors' responsibilities, given the critical nature of their profession? Further, the state, and perhaps society as a whole, have responsibilities to individual citizens, residents, and members. How should we think about those responsibilities, both at a general level and in extraordinary situations such as pandemics?

This collection brings together work by leading theorists in these topics, in order to push the debate forward by elucidating our understanding of these questions, their possible answers, and how they are related. In this Introduction, we provide a big-picture view of the debate. Our aim is to help the reader understand the debate as a whole, and to place the individual contributions in this volume within this broader context.

We begin by introducing different ways of understanding responsibility, and move on to consider how different notions of responsibility might be relevant in health care contexts. However, we wish to point out that our purpose here is to provide a brief overview of the debate, and we wish to remain as neutral as possible on substantive positions.

0.2 Responsibility and Health: Some Basics

0.2.1 Responsibility

For a volume concerning the use of responsibility in health care contexts, it is important to elaborate the various ways "responsibility" can be understood. This term is used in different ways in the various debates relevant to the volume. For instance, some speak of individual responsibility, others of personal responsibility, others of moral responsibility, and others still of responsibility, unqualified. There is common agreement that the term "responsibility" has multiple senses, and can be used to mean different things. There is less agreement on how best to categorise or understand these different senses.[1] Here, we aim to only make some rough distinctions.

An initial distinction that is important to make is between moral and other types of responsibility (as well as associated ideas such as praiseworthiness and blameworthiness). While these terms may naturally conjure up moralised inter-pretations, we may also speak of people being *epistemically* or *prudentially* responsible, praiseworthy, or blameworthy in ways that are relevant to practical decisions in health care. This is particularly important in thinking about one possible criticism of holding individuals responsible for their health, which is that we should not "moralise" health care. As we outline in the next section, however,

---

[1] See, for instance (Clarke et al. 2015: 3–12; Eshleman 2001; Hart and Gardner 2008, chapter 9; Talbert 2019; Vincent 2011; Zimmerman 2015).

holding people responsible—or describing them as blameworthy—need not involve a moral judgement at all. Thus, in the rest of this section, when we use terms such as "responsible", "blameworthy", or "praiseworthy", they are intended in a *general* sense, covering both moralised and non-moralised interpretations, unless otherwise specified.

One way of understanding responsibility concerns evaluations of people's actions, or failures to act, and their results. This sort of responsibility is associated with concepts like culpability, blameworthiness, praiseworthiness, and credit-worthiness. For instance, we may understand responsibility such that an agent's being responsible for something implies that she *deserves* some form of blame or punishment, or some form of praise, credit, or reward.[2] Or we might understand this in a broader fashion, such that an agent's being, say, praiseworthy or blame-worthy, for an action has implications for the strength of the duties we have towards them, or the strength of their rights, or the weight that we should give their interests in our deliberations. On some views of responsibility, it is intrin-sically good that the responsible agent gets the positive or negative response, and we may even have defeasible obligations to respond in these ways. On other views, whether we should take someone to be blameworthy or praiseworthy is partly determined by the forward-looking implications,[3] either of responding to a particular instance of responsible behaviour, or of adopting a system which licences or demands responding in these ways.

Most generally, these sorts of responsibility are similar in assuming that an agent's being responsible for something can make it appropriate to respond to the agent in some way or other. And, in order to be culpable, blameworthy, praise-worthy, or creditworthy for something, the agent needs to be responsible, in this sense, for that thing. If they are culpable or blameworthy, the appropriate response is a negative one; if they are praiseworthy or creditworthy, the appropriate response is positive. We will say that this family of views of responsibility are concerned with *Accountability-Responsibility*.[4]

In Chapter 6, Dana Kay Nelkin elaborates on the details of accounts of accountability-responsible agency and on the debate concerning the appropriate conditions for valid informed consent. Drawing parallels between these two debates, Nelkin argues that they have much to learn from each other. In this

---

[2] Or, slightly more broadly, that she is an apt candidate for such responses.

[3] Sometimes, this seems to be what people have in mind when they mention "forward-looking" responsibility (see, for instance, a recent issue of *The Monist* on *Forward-Looking Accounts of Responsibility*, 104: 4 (Jefferson and Robichaud, 2021)). Roughly put, this is a forward-looking form of responsibility insofar as the justification for responding in some way or other to an agent's behaviour is, at least partly, justified in terms of forward-looking considerations. Yet other times, the phrase seems to be used for something different, more akin to what we call the obligation-sense below.

[4] On some ways of categorising such views in the debate on *moral* responsibility, the term "accountability" is just one form among others; e.g., attributability, answerability, and appraisability, (for some discussion, see Jeppsson 2022; Clarke et al. 2015: 1–15). For lack of a better word, on our use of "accountability-responsibility", all of these are forms of accountability-responsibility.

chapter, Nelkin details how the notions of reasons-responsiveness and quality of opportunities for decision-making can be shared across these two domains. Further, she highlights where the conditions on responsible agency, on one hand, and valid informed consent, on the other, might come apart. In particular, since the basis for and the quality of our reasons-responsive capacities comes in degrees, there is a question of when a threshold is met so that a person can be said to have a capacity, or a sufficiently high-quality capacity, which is a difficult question to answer in both realms. In this chapter, Nelkin assesses possible answers and considers how constrained those answers should be by independent reasons to treat responsible agency and eligibility for informed consent similarly.

This sort of responsibility is to be contrasted with another, which we will call *Obligation-Responsibility*. Consider the claim that we have a responsibility to be honest. This claim seems to be about some sort of obligation or duty, perhaps a defeasible one, that we have concerning honesty.[5] Or consider the claim that Juanita is responsible for staying healthy. This suggests that, in some sense, it is on her to stay healthy, she has some sort of obligation to do so. Contrast this with the claim that John is responsible for failing to stop at the red light. Here, "responsibility" is not being used to refer to some obligation that John had; he certainly did not have an obligation to fail to stop at the red light. Here, we mean that John had the sort of control over his behaviour, and awareness of (or capacity to be aware of) the relevant features, such that responding to his failure in a certain way is now appropriate. The claims about the doctor and Juanita are about obligation-responsibility, the claim about John is about accountability-responsibility.

Although we make this rough distinction, this is not to say that these two sorts of responsibility are unrelated. In fact, it is common to think that they are importantly related. For instance, a plausible necessary condition on being blameworthy for something is that one had a responsibility *not* to do that thing, or, at least, that one thought one had such a responsibility. Thus, without being obligation-responsible for A-ing, or without thinking that one had such an obligation, one cannot be blameworthy for failing to A.

Finally, some philosophers use *Role-Responsibility* to refer to some obligations we have in virtue of particular roles that we occupy (Hart 1968: 212–14; Dworkin 1981; Goodin 1987; Cane 2016). Consider the question, uttered by a parent coming home to find the wreckage left by a teenage party, "who is responsible for this mess?" On an accountability form of responsibility, the answer is that the teenage child and her friends are responsible. But now consider a different way of understanding the question: who ought to deal with cleaning up the mess. Again,

---

[5] Sometimes, when people speak of "forward-looking" responsibility, or "prospective responsibility" it seems that it is something like obligation-responsibility that they have in mind, since it concerns something about the future; i.e., one has an obligation to do, or not do, something in the future. Given the ambiguity in "forward-looking responsibility" (see n. 3 above), and given that our rough distinction seems to clarify what the term is sometimes used for, we prefer not to use the term.

we might think that the teenage child and her friends ought to clean up the mess, and this is in virtue of their being blameworthy for creating it. However, we might also think that at least some obligation-responsibility is held by the parents: not due to their past blameworthy behaviour, but because of their *role* as a parent: this is just the sort of thing that parents are responsible for.

Health care is full of roles, some more well-defined than others. Doctors, nurses, and other medical professionals have social and professional roles, while the role of "patient" may come with a set of obligation-responsibilities too. Thus, while the primary focus of responsibility in health care settings has been on that derived from voluntary, avoidable behaviour, questions of role responsibility are of potential relevance too. In the rest of this section, we focus on accountability-responsibility and obligation-responsibility more generally; but when we come to consider broader issues of responsibility for health, such as the responsibilities of medical professionals or the state, we consider role-responsibility again.

0.2.2 Agency and Health

When it comes to the health of individuals, responsibility is intimately tied to agency. What we might hold people accountable for are certain exercises of agency, or their consequences. And what obligations people might have are limited to those things that they have agential capacities to do. This includes health-related behaviours, habits, or lifestyles, and potentially the outcomes of those things. For instance, mountain climbing may be a health-related behaviour in virtue of the risks of injury involved, while smoking and excessive drinking are health-related habits in virtue of the negative health outcomes they can bring about; the same can be said for a sedentary lifestyle.

Consider three sorts of health-related items that an individual might be responsible for: (1) individual actions and omissions, (2) habits, lifestyles, or patterns of behaviour, and (3) health outcomes. These are not identical. Smoking a single cigarette (1) does not significantly increase risk of lung cancer, yet the habit of smoking (2) does. And the habit of smoking (2), even for a lifetime, does not always result in lung cancer (3). Similar points can be made about excessive drinking and sedentary behaviour: one instance of excessive drinking (1) does not make a habit (2), and the habit does not always lead to liver disease (3); spending one day on the couch (1) does not make for a sedentary lifestyle (2), and a sedentary lifestyle does not always result in heart disease (3).[6]

---

[6] These are simple examples, but things can get more complicated. For instance, some outcomes may be the result of multiple habits. And individual habits can result in a variety of health outcomes. The risk of having *some* negative health outcome as a result of being a lifelong smoker is higher than the risk of developing lung cancer as a result of being a lifelong smoker.

The distinction between these three tracks some relevant differences when it comes to responsibility. First, there are important differences between individual actions and/or omissions on the one hand, and habits and lifestyles on the other. Some individual actions or omissions are not significantly associated with unhealthy outcomes, whereas habits or lifestyles involving a series of such actions or omissions may be. Smoking a cigar to celebrate the birth of one's child does not significantly increase the risk of a negative health outcome, yet smoking a daily cigar does. It may be true that, in some sense, we should not adopt the habit, and would be in some sense blameworthy for doing so. At the same time, it would seem false that in the same sense, we should not have a cigar to celebrate the birth of one's child, or that we would be blameworthy for doing so.

Second, since not all individuals with unhealthy lifestyles end up with associated negative health outcomes, there is luck involved in the relationship between lifestyles and outcomes; whether the negative outcome occurs is importantly influenced by non-agential factors. For example, the difference between those smokers that do develop lung cancer and those that do not is often due to factors outside smokers' awareness and control. Thus, there are some luck-related difficulties that arise with holding individuals responsible for their negative health *outcomes* that do not arise with holding them responsible for unhealthy *lifestyles*.

A further complication arises from the fact that some (perhaps most) negative health outcomes associated with a particular lifestyle can also come about via other processes; one need not be a smoker or have a sedentary lifestyle in order to develop heart disease. Thus, even if an individual practices a health-related behaviour, and has developed the negative outcome associated with this behaviour, determining whether the outcome is due to the behaviour may still not be straightforward. And since it is often thought that, in order to be accountability-responsible for an outcome, one must be responsible for the behaviour leading to it, it may be difficult to determine responsibility for the outcome itself.

0.3 Why Responsibility in Health Care?

We will shortly discuss some ways that considerations of responsibility might be incorporated into health care decision-making, and consider some objections to doing so. But first, we will outline why some may think this a worthwhile exercise. If there are no good reasons for considering responsibility in health care, then it is not worth thinking about the mechanics of doing so.

There are four broad types of reasons that may favour holding people responsible in some contexts. We will briefly consider whether and how each may apply in health care.

The first reason is worth considering only to explicitly distinguish other approaches from it. This reason is a retributive one, which says that by engaging in certain kinds of behaviour, people deserve to suffer some loss or burden. Such a view is sometimes expressed popularly in the context of criminal punishment, though it is controversial even there. A retributive stance on health care would say that some people ought to suffer poor health even if it would cost us nothing to help them. We know of nobody who has defended this view.

Turn now to a second reason supporting responsibility, discussed by Scanlon (1986, 2019). Scanlon frames the idea of responsibility in terms of the way in which our wills shape our lives via our choices. As Scanlon (2019)[7] sees it, we can in some contexts (including health care contexts) justify making "eligibility for benefits depend on individuals' choices because of the instrumental value for people of having the choice...between retaining entitlement to state-supported care by living carefully or foregoing it in order to have a different life style". Choice is important, Scanlon suggests, for several reasons: it allows us to shape the course of our lives, and express our personality and wills in our behaviour; and it is a default assumption that competent adults *should* have a choice, and a mark of disrespect when others take choice away from them or assume that they cannot choose well.

Such a justification, though, seems incomplete. To see why, consider the binary set of options Scanlon offers: living carefully and retaining entitlement to state support, or losing this entitlement by living dangerously. This raises a question: why shouldn't people also have the option of living *riskily* but still retaining entitlement to state support? Scanlon's appeal to the value of choice is incomplete here. Thus, a third potential justification for responsibility policies in health care is to appeal to fairness; we might think that a person who decides to take risks and retain their entitlement to state support chooses in the way that is most valuable for them, but which is unfair on others who have to bear the costs of their risky choices.

Health care requires the use of various kinds of resource: money, medicines, and equipment; beds; medical professionals' time; and so on. In many cases, a patient using these resources generates costs for others: either because of an increase in taxation or insurance premiums, or because other patients are thereby unable to use that resource. Fairness-based reasons for involving responsibility in health care claim that in *some* cases, individual patients generate unfair costs for others (e.g., Laverty and Harris 2018; Davies and Savulescu 2019). It is important in this context to distinguish between whether certain patients generate *higher than average* costs, and whether they generate *unjustified* costs. A patient's care could cost more than average, and yet all of those costs could be justified, for example, because none were generated by choices for which the patient can be held responsible. On the other hand, a patient's care could cost less than average, and yet be unjustified due to the reasons for their care needs.

One obvious version of a fairness-based reason for health care responsibility is a *luck egalitarian* view. Roughly speaking, luck egalitarians believe that inequalities

---

[7] The text of Scanlon's lecture was previously available online, but has now been removed in anticipation of a forthcoming volume. Quoted text is from the online version.

between different individuals are unjustified *unless* those inequalities derive from individual choices under suitable circumstances (e.g., Arneson 2006; Segall 2009; Albertsen and Knight 2015; Lippert-Rasmussen 2016). This could apply within health care: thus, if someone's health need created an inequality between them and others, luck egalitarians will say that this should be either corrected or compensated for, unless it resulted from a choice for which the individual can be properly held responsible.

However, while the basic idea underlying luck egalitarianism may seem to support holding people responsible, many actual luck egalitarians have proven reluctant to support relevant policies in practice. For there are questions about when people are (sufficiently) responsible for their health needs. Some luck egalitarians (Arneson 1989; Roemer 1993; Stemplowska 2017; Segall 2013: 179) defend the idea that we should only appeal to responsibility when people have had equal opportunities, something that does not apply in any real health care system.

While luck egalitarianism is the most widespread, any view of justice can have responsibility-sensitive versions that appeal to the unfairness of expecting others to bear (all of) the costs of our choices. For instance, Herlitz (2019) suggests that the importance of holding people responsible supports a *sufficientarian* approach to health care, where we draw a particular threshold below which individuals' claims take on particular moral importance. Relatedly, Davies and Savulescu (2020) argue that people need only have had *sufficient* opportunity, not equal opportunity, to warrant being held responsible. More recently, Andreas Schmidt (2022) has argued that "relational egalitarians"—who see the value of equality as residing in allowing people to relate to one another as equals, rather than in the equal distribution of some good—should embrace responsibility in some contexts. Schmidt suggests that the kinds of relationships that relational egalitarians value can only occur when people are disposed to hold one another responsible. However, like many luck egalitarians, Schmidt sounds a note of caution on the question of whether to make health care dependent on people's choices, since allowing significant differences in health may exacerbate relational inequalities.

Finally, some justifications of holding people responsible appeal to the positive future effects of doing so. For instance, in the literature on moral responsibility, some authors have focused on the way that holding people explicitly *morally* responsible for their behaviour can help them to develop their moral capacities (e.g., Vargas 2013; McGeer 2019). As we mentioned previously, responsibility in health care need not be moralised in this way; following Scanlon, we might think that a person who chooses to do certain things such as smoking or downhill skiing does nothing immoral, but should bear (some of) the costs of their choices.

Still, a correlate to this view might be the claim that holding people responsible will improve *prudential* reasoning or behaviour (see Sharkey and Gillam 2010: 62 for discussion) in the future. This view comes in two forms. On the first view, "individual motivational change", holding a particular patient responsible for their health-affecting choices is justified insofar as it will help *that same patient* make better choices. Thus, some people may think that if smokers have to pay higher insurance premiums, that will make them more likely to give up smoking. Daniel Callahan (2013) advocated a (widely criticised, e.g., Tomiyama and Mann 2013; Gostin 2013) policy of "social pressure" on individuals who were judged to be overweight, that is, "fat-shaming", drawing on his own experiences as a former smoker. Alternatively, one might think that holding a particular patient responsible for their health-affecting choices is justified because of its effect on *other* people, that is, on "social motivational change"; if we refuse liver transplants to heavy drinkers, it may be too late for those individuals to change their behaviour, but not too late for others who see this policy being enacted.

Since they depend on their predicted causal effects, such justifications are open to empirical challenge: that they either will not have the desired effect, or that they will make things worse (e.g., Friesen 2018). Even if these effects are well supported, there is a further question of proportionality, particularly for social motivational change. Do the costs to the individual justify the benefit that either they stand to gain from better choices in the future, or that others stand to gain? Such considerations may serve to narrow the scope of occasions when holding an individual patient responsible can be justified by appeal to motivational effects, but may not undermine the basic idea that *if* such effects can be expected, then responsibility-sensitive policies can be justified.

0.4 Responsibility and Health Care

How might a health care system, or a health care policy, be influenced by considerations of patients' responsibility concerning any (or all) of these items? If we are concerned with accountability-responsibility, this will be a matter of modifying how the system, or policy, is different with respect to those who are responsible for some item or other, when compared to those who are not.

Here, we briefly introduce some of the types of policies that have been suggested in the bioethics literature. Then, we consider some of the standard criticisms of responsibility-sensitive policies, while discussing how they may differ with respect to the different types of policies.

0.4.1 Taxation Policies

One policy involves taxing individuals when they purchase the goods relevant to the lifestyle.[8] For instance, a government might tax cigarettes or alcohol at the

---

[8] For some discussion of such policies, see (Bærøe and Cappelen 2015; Cappelen and Norheim 2005; Davies and Savulescu 2019; Le Grand 2013; Vansteenkiste et al. 2014).

point of purchase. By doing this, those with the unhealthy lifestyle pay more into the system, and the increased payments can help to account for the increased risk these individuals take on; individualised risks which, across a large enough group, translate into a certainty of higher demand by *some* individuals with the lifestyle.[9]

Some important benefits from this sort of policy, in isolation,[10] come from the fact that, in practice, it holds people responsible at the point of action. The tax holds people responsible at the point of purchase, and those who purchase more of these items (e.g., who purchase more cigarettes), end up paying more. Thus, it provides a reasonable proxy for holding responsible at the point of action— purchasing a pack of cigarettes correlates well, we suspect, with the action of smoking cigarettes—and scales with the extent to which the individual has the habit—in general, the more packs one purchases, the more cigarettes one smokes. Given that this is the only place at which such a policy holds agents responsible, it does not run into the issues of outcome luck mentioned above; two smokers who smoke an equal amount are held responsible to the same extent, regardless of whether one develops lung cancer and the other does not.

There are, however, some issues that come along with such a proposal. First, although the policy seems to work well with stock examples—for example, taxes on cigarettes, alcohol, and sugary drinks—there are other unhealthy lifestyles that cannot clearly be taxed; namely, those that are not tightly associated with the purchase of products. For instance, it is not clear how one would tax a sedentary lifestyle. Thus, if the goal is to provide a system that can incorporate responsibility for unhealthy lifestyles, this will leave many such lifestyles out (Bærøe and Cappelen 2015: 839). Relatedly, Albertsen (2016a) notes that it is unclear how a taxation policy could help with organ scarcity.

Second, although the policy discriminates between lifestyles that are tightly related to purchases and those that are not, it may not do enough to discriminate between those who perform occasional, individual actions that are not especially unhealthy, and those who have the unhealthy lifestyles constituted by performing those actions repeatedly. If taxation is understood as holding responsible, then the policy will end up holding responsible many individuals who are not blameworthy for anything, for example, the person who buys a cigar to celebrate the birth of their child, or the person who buys a Coca-Cola on their annual trip to the beach.

Third, as this policy is often understood, the revenue from the tax goes towards the health care system, in order to make up for the increased costs incurred by (some) individuals with the unhealthy lifestyles. This, it seems, may be somewhat

---

[9] However, and as some have argued, there is some reason to think that those with unhealthy lifestyles cost the health care system, and the state in general, *less* over their lifetimes. (See, for example, van Baal et al. 2008; Barendregt et al. 1997; Friesen 2018: 55; Ho 2008: 83; Persaud 1995: 284; Wilkinson 1999: 257–8.)

[10] Although we do not know of someone who defends this in print, it is possible to endorse a taxation policy which *also* holds people responsible in some of the other ways that we mention below.

straightforward for societies in which the general entity doing the taxing (the state), is also the entity in charge of the health care system (or the universal insurance scheme), as is the case in various European nations. However, this is not as straightforward in some other societies, in which the entity doing the taxing is not the entity in charge of providing health care (or the universal insurance scheme); for example, the USA. In such cases, both the health care system and individual policy providers form their own agreements with individuals (or their employers), adding complications to the determination of where the increased tax revenues should go, as well as raising questions about the standing of the taxing agency to hold individuals responsible in this way.

0.4.2 Treatment Policies

Another way that a system or policy might take accountability-responsibility into account is at the point of making decisions about treatment and/or prevention. A commonly discussed case of this is the one introduced in the Introduction, about which patient should receive a heart transplant. Taking responsibility into account here becomes complicated, both because we can vary treatments in different ways, and because there are various other relevant considerations.

Consider how we can vary treatments. One way of holding agents responsible at the level of treatment decisions is by denying treatment on the basis of their responsibility; if a patient needs heart surgery as a result of a very unhealthy diet, health care providers might deny them the surgery. A less severe response might be to offer treatment to the individual, though in some reduced form, for instance, a treatment that is less expensive, and perhaps less effective.

A different way of holding people responsible at this level would involve giving the responsible patient lower priority when compared to otherwise similar patients who are not responsible for their health conditions. If the responsible patient is treated they might end up receiving the same treatment as others, but there might be a lower chance that they receive treatment, or they may have to wait longer. Different policies might vary in terms of how much responsibility affects priority. At one end of this spectrum, we might only use responsibility as a tie-breaker, between patients who are the same in other relevant respects.[11]

By holding individuals responsible at the point of treatment decisions, rather than the point of action, such a policy avoids the practical certainty of holding blameless individuals responsible that taxation policies face; it does not hold

---

[11] Even this might not be straightforward. Segall (2009: 71) considers the possibility of using a weighted lottery in such a case. The blameworthy individual, in this case, would still have a chance of receiving the organ, yet their chances would be lower than those of the other individual. Policies making use of this can differ in how weighted the lottery is.

responsible the new parent who buys a cigar. However, there is a tradeoff, in that it takes on issues with luck in outcomes. The lifelong smoker who does not develop lung cancer may not get held responsible, whereas the one who does develop lung cancer may.[12]

Such a policy also invites a host of other concerns insofar as it requires that someone in the health care system know that responsible individuals are, in fact, responsible. This would both require knowledge that the individual had the relevant lifestyle, and that they are accountable for it. Acquiring such knowledge may be impractical and intrusive. Further, consider a method of acquiring this knowledge that relies on cooperation from individuals with health-related life-styles; individuals who know that there may be a negative response in virtue of their responsibility for health-related outcomes. This gives these individuals an incentive to either refrain from sharing some of this information, or to lie about it. This might damage the doctor–patient relationship by undermining trust. But further, it can work against the goal of improving health to the extent that some of the withheld information may be relevant to determining how best to treat the individual.[13]

0.4.3 Premium Policies

A third way in which individuals might be held responsible for their health-related behaviours is at the point of paying for health care itself. Patients responsible for health-related behaviours might have higher or lower premiums, depending on whether their behaviours are positively or negatively related to health. The lifelong smoker might have higher premiums than the non-smoker, who may have higher premiums than the regular jogger.

Premium policies are in some ways similar to taxation policies: they hold people responsible for their behaviour, and can be implemented before any negative health outcome occurs. Thus, premium policies might avoid the issues with outcome luck that treatment policies are saddled with. And whereas taxation policies cannot hold agents responsible for health-related lifestyles that are not easily taxed, premium policies may seem to avoid this problem. With knowledge of who has the relevant health-related lifestyles, premium policies can be more discriminate than taxation policies. Whereas taxation policies identify a product or service associated with a health-related lifestyle and apply a general tax,

---

[12] Again, this is a simplification (see n. 6). For example, the lifelong smoker who does not develop lung cancer may still be held responsible if they need treatment for some other smoking-related illness.
[13] For discussion of these sorts of worries, see (Bærøe and Cappelen 2015; Brown 2013; Brown et al. 2019b, 2019a; Buyx 2008; Cappelen and Norheim 2005; Cohen and Benjamin 1991; Daniels 2007; Davies and Savulescu 2019; Feiring 2008; Friesen 2018; Ho 2008; Levy 2019b; Schmidt 2009; Sharkey and Gillam 2010; Vathorst and Alvarez-Dardet 2000; Véliz 2020; Davies 2022).

premium policies might be able to identify individuals with the relevant lifestyles, and apply a change in premiums in individual cases. Further, if changes in premiums can be made on the basis of accurate attributions of habits and lifestyles, rather than every individual action, premium policies could avoid holding responsible blameless individuals who perform actions that are perfectly fine and do not have the corresponding habits or lifestyles, like the new parent who buys a cigar.

However, as with treatment policies, the implementation of premium policies requires those setting premiums to have accurate beliefs regarding individuals' lifestyles. This means that, in order to gain this benefit over the tax policies, premium policies will face similar objections as did treatment policies around acquiring knowledge about individuals' lifestyles.

0.4.4 Opportunity Policies

A final set of policies is what we will call opportunity policies.[14] On opportunity policies, individuals are only held responsible for their health-related behaviour—and, perhaps, its outcome—after they have been offered an opportunity to change their behaviour and have failed to do so. For instance, if a smoker has not had the right sort of opportunity to quit smoking, then he is not held responsible.[15] Opportunity policies are consistent with different mechanisms for holding patients responsible; they can do so at the point of paying premiums, or at the point of treatment-decisions, with corresponding objections potentially applying.

However, adding the opportunity component brings *some* benefits. First, by offering the opportunity to change one's health-related behaviour, this policy does more than merely hold patients responsible; assuming the opportunity is robust and the individual well-informed, it can empower the individual to make health-improving changes. Second, insofar as this opportunity is robust enough to increase individuals' abilities to make a change, and the opportunity is provided by a medical professional, some of the worries about acquiring knowledge concerning individuals' responsibility are mitigated: the medical professional who makes the offer is in a better position to know that the individual had the opportunity to make a change.[16] Third, it can provide for a more forgiving policy,

---

[14] For versions of such policies and related discussion, see (Bærøe and Cappelen 2015; Brown and Savulescu 2019a, 2019b; Davies and Savulescu 2019; De Marco et al. 2021; Feiring 2008; Savulescu 2018).

[15] Sometimes, "smoking" is taken to also include "vaping". However, on some versions of opportunity policies, making a change to vaping, from smoking cigarettes, counts as a successful lifestyle change (De Marco et al. 2021; Savulescu 2018).

[16] Although opportunity policies can mitigate some of these worries, they will not eliminate them all. Importantly, to make the offer, the medical professional might need to know that the individual has the lifestyle in question; and to hold such individuals responsible, someone in the health care system needs to know that the individual did not change their behaviour after the offer.

insofar as, by offering the opportunity, many individuals can be given a second chance before being held responsible. Finally, by informing the patient of the outcomes associated with their lifestyles, as well as the ways that the health care system might respond, when making the offer, such a policy can reduce worries arising from the possibility that individuals might not clearly know that they are obligation-responsible for various aspects of their health.

0.5 General Objections

In describing some forms that a responsibility-sensitive policy might take, we mentioned some benefits and drawbacks various policies might have. In this section, we consider more general criticisms, which pose problems for responsibility-sensitive policies in general.

0.5.1 Universalisation

One general issue faced by responsibility-sensitive policies is the Universalisation Objection. Proponents of this objection argue, roughly, that holding people responsible for unhealthy lifestyles would require that we do so for many more lifestyles than are usually considered, and that this would lead to undesirable results.[17] For instance, if we are to hold smokers responsible for smoking then, to be consistent, we should also hold responsible a physician who contracts Ebola while participating in a Doctors Without Borders project in an area where Ebola is prevalent; after all, both the smoker and the physician knowingly and willingly engaged in behaviour that incurred risks to their health. But holding the physician responsible, say, by increasing the proportion of treatment costs that she is required to bear, would clearly be the wrong result. The challenge is to delineate those health-related lifestyles which it would be appropriate to hold people responsible for in a way that avoids these implications.

One way of avoiding this issue would be to argue that although the physician is responsible, and in fact *praiseworthy*, for participating in the programme, she is not *blameworthy* for contracting Ebola; and this is so even if she is, in some sense, responsible for contracting Ebola. In contrast, smokers may be both responsible *and* blameworthy for their lifestyles, and, perhaps, their health outcomes.

This response, however, may simply change the question: how do we distinguish between those health-related behaviours, and their outcomes, that agents are blameworthy for, and those that they are not? Different authors provide different

---

[17] For discussion, see (Buyx 2008; Cappelen and Norheim 2005; Friesen 2018; Sharkey and Gillam 2010; Shiu 1993).

answers. For instance, Savulescu has argued that one major factor will be whether the risks incurred by the lifestyle are unreasonable (Savulescu 2018; Davies and Savulescu 2019), while Stemplowska (2017) suggests that we should consider whether the costs to an individual of avoiding a disadvantage (such as a health need) would have been greater than compensating them for that disadvantage. The physician who contracts Ebola did not incur an unreasonable risk, whereas the smoker does incur an unreasonable risk by continuing to smoke.[18] Relatedly, Shlomi Segall argues that a relevant factor will be whether it would have been reasonable for society to expect the individual to avoid the negative health outcome (Segall 2009).

In contrast, Nir Eyal's chapter (Chapter 4) argues that it is agents who make *un*reasonable (or "pointlessly imprudent") decisions who have the strongest claim not to be held responsible, since their poor choices are very likely indicative of some deficit—for instance, in decision-making or will-power—for which they are not blameworthy. Eyal's proposal is a response to an influential argument from John Roemer (e.g., 1993, 1996), which offers a different pragmatic way to decide whether a particular choice was autonomous. Roemer's solution is to appeal to demography: if we can fix a number of demographic characteristics (such as sex, race, and socioeconomic background) as outside individual control, we can then form social "types", look to the *average* behaviours for each type, and judge responsibility according to a person's type-average. Imagine that the typical male steelworker smokes for twenty-five years in their life. If a particular worker has smoked for twenty-five years, we may not be able to identify *what* has caused him to smoke; but since he is the average for his type, we should conclude that there is some set of causes, common to men working in the steel industry, that results in people smoking for twenty-five years. Thus, we can skip difficult epistemic questions about how to know for sure whether a choice was genuinely autonomous. Eyal, though, is doubtful, and offers examples that undermine the connection between social typicality and lack of autonomy, and social atypicality and autonomy; thus, he suggests, considering the prudential value of a choice is a better heuristic for whether it is autonomous.

An alternative distinction that might mitigate worries about universalisation appeals to the social value of certain activities. Consider firefighters. Firefighting is a professional choice that nobody is forced to make, and which raises your risk of certain kinds of serious health problems. In one sense, then, firefighters might be seen as responsible when they are badly burned, or suffer from trauma related to what they have seen (at least, they meet the standard requirements on

---

[18] Savulescu's discussion is in the context of developing what we called an opportunity policy above (Brown and Savulescu 2019a, 2019b; Davies and Savulescu 2019; De Marco et al. 2021; Savulescu 2018). Thus, the smoker would be blameworthy *after* he has been presented with an appropriate opportunity to quit.

responsibility). But if we held firefighters substantively responsible for these health needs, we might see far fewer firefighters in society. Thus, we might collectively agree that even if firefighters are responsible for their health needs, we are happy to pay the associated costs. This differs from a reasonability criterion: there may be a number of perfectly reasonable choices whose costs we are *not* collectively happy to bear; and we might be happy to bear the costs of firefighters' health needs even if we personally regard it as a prudentially unreasonable line of work to choose.

0.5.2 Stigmatisation

Whereas the universalisation objection is worried about applying responsibility too widely, others have worried that in practice, any responsibility-sensitive policy in health care would not be applied widely enough. For instance, Daniel Wikler (2002) suggests that many proposals for health care responsibility focus on behaviours that are moralised as "sins", or on activities (such as drug addiction) that are associated with marginalised groups (see also Ubel 1997; Friesen 2018). It may even be that for the very same behaviours, members of marginalised groups are more likely to be held responsible than other individuals.[19]

In addition, one might worry that responsibility-sensitive policies will increase or create stigmatisation. This is perhaps most obvious with policies such as Callahan's proposal around weight, which explicitly advocated stigmatisation. But even policies which do not explicitly aim to stigmatise may be experienced in this way by patients, especially where this involves making exceptions of patients with a health care service.

0.5.3 Harshness

Another objection to responsibility-sensitive policies is the "harshness" objection. This objection can take different forms, but roughly, the idea is that to hold people responsible for their health-related behaviour, or its outcomes, is overly harsh.[20] A paradigm example is that of a motorcyclist who decides to ride his motorcycle without wearing a helmet. While speeding down the road, he slips and is involved in an accident, resulting in severe injuries which likely could have been avoided had he been wearing a helmet. On a strict responsibility-sensitive policy, he should

---

[19] For discussion of a related issue in criminal sentencing, see Vagins and McCurdy (2006).

[20] For some discussion, see (Albertsen 2016b, 2020; Albertsen and Nielsen 2020; Anderson 1999; Bærøe and Cappelen 2015; Cappelen and Norheim 2006; Daniels 2007: 76–7; Duus-Otterström 2012; Fleurbaey 1995; Segall 2009, chapters 4–5; Vansteenkiste et al. 2014; Voigt 2007, 2013).

be denied health care in virtue of the fact that he is responsible for his injuries. This would be to abandon the driver, and strikes most people as overly harsh.

Whether a proposed policy faces this objection will depend on its details. Not all responsibility-sensitive policies need to recommend abandoning the motor-cyclist. However, some versions of this objection are not easy to avoid. Recall, again, the question of which patient should receive a heart. In such a case, someone *will* go without the heart; automatically denying the blameworthy patient the heart might therefore constitute abandonment. Still, as some defenders of responsibilisa-tion point out (e.g., Segall 2009: 70; Albertsen and Knight 2015), these difficult cases are such that someone must receive "harsh" treatment; and so deciding at random rather than on the basis of responsibility may be no better in this respect.

Insofar as one sees the objection as relying on the claim that a particular response to the blameworthy individual would be overly harsh, this style of objection can apply to any of the responsibility-sensitive policies we mentioned above; a tax can be overly harsh, as can an increase in premiums. However, the seemingly most powerful versions of this objection, which involve some sort of abandonment, seem to apply mostly to policies which hold responsible at the point of making treatment-decisions. Gabriel De Marco's chapter (Chapter 5) is dedicated to this objection. De Marco considers different features that a responsibility-sensitive health care policy might have, delineating different pos-sible sources of harshness in a policy, as well as discussing different potential responses to the objection.

0.5.4 Responsibility Scepticism

Another source of opposition to responsibility-sensitive policies comes from those who think that either we are not accountability-responsible, or that we cannot form justified beliefs about whether individuals are accountability-responsible. If we are not accountability-responsible, then we should not adopt a health care policy that holds us accountable for them.

One way of reaching this result is by arguing for a wholesale rejection of responsibility, and adopting the claim that we are not responsible for *anything*.[21] If we are not responsible for anything, then we are not responsible for health-related behaviours and outcomes.

Alternatively, one might argue that we are not responsible for health-related behaviours (and their outcomes) without thereby arguing that we are not respon-sible for anything. One might argue that, with respect to some, or all, health-

---

[21] For some prominent defences of a wholesale responsibility-scepticism, at least for some sorts of responsibility, see (Caruso 2021; Double 1991; Levy 2011; Pereboom 2014; Strawson 2010; Waller 2011).

related behaviours, we either fail to meet the control condition on responsibility,[22] or we fail to meet the epistemic condition on responsibility.[23]

For instance, socioeconomic factors, many of which are outside of our control, can have a significant impact on our health-related behaviour.[24] A single mother of three working two jobs may not have the time to cook nutritious meals every day. Or one might not have the time to have a consistent exercise routine, the money to buy the often more expensive healthy food, or access to quality food in one's local grocery stores. Others point out that some lifestyles, such as smoking or drug-use, involve addiction, and thus threaten to undermine the control condition on responsibility for such behaviours.[25] Yet others point out that genetic factors, which are out of our control, can have a significant impact, both in terms of underlying a predisposition for certain behaviours—such as addictive behaviours—and for certain outcomes—such as obesity.[26] Each of these factors, one might argue, risks either mitigating, or fully undermining, responsibility for health-related behaviour and their outcomes.

Two chapters in this volume consider variations of this more local version of the objection. Rekha Nath's chapter (Chapter 8) develops a version of this objection for obesity and obesity-related behaviours. Nath argues that obesity does not sufficiently track unhealthy choices, and that our eating habits are too strongly shaped by social structures and forces for us to have significant responsibility over them, or for policies penalising obesity to have positive motivational effects. Thus, she says, obesity-penalising policies cannot be justified.

Taking her cue from the fact that many of our health-related behaviours are relatively automatic and habitual, Rebecca Brown's chapter (Chapter 9) engages with the question of how it is that agents can be responsible for habitual behaviour. Discussing both theoretical work on moral responsibility as well as empirical work on the mechanisms behind habitual action, Brown concludes that for much of our habitual behaviour, we either do not meet conditions on responsibility, or we meet them to a much lesser extent than we might have thought, due to their

---

[22] For discussion, see (Bærøe and Cappelen 2015; Brown 2013; Brown et al. 2019a, 2019b; Brown and Savulescu 2019a; Buyx 2008; Cavallero 2019; Davies and Savulescu 2019; Friesen 2018; Ho 2008; Levy 2019a, 2019b; MacKay 2019; Resnik 2007; Sharkey and Gillam 2010; Thornton 2009; Vathorst and Alvarez-Dardet 2000).

[23] For discussion, see (Brown et al. 2019b; Brown and Savulescu 2019b; Davies and Savulescu 2019; Resnik 2007; Sharkey and Gillam 2010).

[24] For discussion, see (Bærøe and Cappelen 2015; Brown 2013; Brown et al. 2019b, 2019a; Buyx 2008; Cappelen and Norheim 2005; Cavallero 2019; Daniels 2007; Davies and Savulescu 2019; De Marco et al. 2021; Feiring 2008; Ho 2008; Levy 2019a, 2019b; MacKay 2019; Resnik 2007; Schwan 2021; Sharkey and Gillam 2010; Thornton 2009; Vathorst and Alvarez-Dardet 2000).

[25] For discussion, see (Bærøe and Cappelen 2015; Brown 2013; Brown and Savulescu 2019a; Buyx 2008; Cavallero 2019; Friesen 2018; Levy 2019a, 2019b; Resnik 2007; Sharkey and Gillam 2010; Thornton 2009; Vathorst and Alvarez-Dardet 2000; Véliz 2020).

[26] For discussion, see (Bærøe and Cappelen 2015; Buyx 2008; Cavallero 2019; Davies and Savulescu 2019; Friesen 2018; Ho 2008; MacKay 2019; Sharkey and Gillam 2010; Thornton 2009; Vathorst and Alvarez-Dardet 2000; Véliz 2020).

often-automatic nature, the potential for unintentionally acquiring habits, and the limited control we exercise over environments which may drive both these factors. This can have significant implications for the extent to which we are responsible for much of our health-related behaviour in general.

A different, yet related, form of the objection does not deny—nor need it accept—that at least some individuals are responsible for their health-related behaviours, habits, and outcomes. Rather, it merely denies that we can know, in particular instances, whether individuals are responsible for some item, or, more specifically, for their health-related behaviour, habits, and outcomes.[27] If so, then we should not implement a responsibility-sensitive policy, which requires knowledge that we cannot achieve.

Consider, again, how various factors outside of our control—for example, genetic or socioeconomic factors—can significantly influence our health-related behaviour, habits, and outcomes. One might think that although such factors can mitigate, or even preclude, responsibility for our behaviours, habits, and outcomes, they need not clearly do so for everyone. However, given their prevalence, and the difficulty in coming to know whether an individual has faced such obstacles, we may not be justified in believing that any particular individual did not face such obstacles. As with the objection that we are not accountability-responsible, this objection can come in several different forms, and two chapters in this volume present versions of the objection.

First, one might argue for this as a global claim, and from a theoretical level. Elizabeth Shaw's chapter (Chapter 3) explores this possibility, adapting an argument from the debate on criminal responsibility. In the context of criminal responsibility, some have argued that we should not implement practices that rely on a notion of deserved punishment; such punishment produces significant harm, and in order to be justified in inflicting such harm our beliefs justifying these actions should meet a high standard. Yet the question of whether we deserve *anything* is yet to be resolved, and there are some reasonable arguments for the claim that we do not. Consequently, the claim that we deserve such punishment does not meet that high standard, and we are not justified in inflicting harm through criminal punishment.[28] Shaw's chapter extends this style of argument to the context of holding individuals responsible for their health-related behaviours, habits, and outcomes.

Neil Levy's chapter (Chapter 7) takes a different, though related, approach. Levy argues that even if people are responsible for their health-related behaviour, *and* we are well justified in believing this as a general claim, we might still not be in a position to know whether a particular agent is responsible for his or her health-related behaviour, habits, or outcome(s). Levy introduces conditions on responsibility

---

[27] For a prominent version of this objection to our *moral* responsibility practices, see (Rosen 2004).
[28] For some discussion, see (Caruso 2020; Corrado 2017; Jeppsson 2021; Vilhauer 2009, 2012, 2015).

over time, which is the sort that is most relevant to our health-related habits and patterns of behaviour. Levy discusses different possible standards for assessing whether individuals—and in some cases, classes of individuals—meet these conditions, concluding that we will rarely, if ever, be able to come to solid conclusions when applying said standards, and that this applies even to some of the weaker forms of taking responsibility into account, such as using responsibility as a tie-breaker.

In defence of such policies, one might note that all policies we consider face some problems, but this alone isn't enough to reject a policy (let the policy that is without sin cast the first stone).

0.6 Beyond Patient Responsibility

So far, we have focused on policies that hold individuals responsible for their health or health-related behaviours. But responsibility is potentially relevant to health in various other contexts, and for various other actors.

Questions of responsibility in health also apply to health professionals and to policymakers. While the responsibility of these non-patient actors has been touched on by earlier chapters—for instance, the claim that our health-affecting choices are shaped by our social environment may lend itself to the further claim that government should take responsibility for poor health outcomes—the final five chapters of this volume take on this broader focus more explicitly.

0.6.1 Medical Professionals

Responsibility has been a central issue in the evolving conception of the relationship between doctors and patients. According to the standard story (though see McCullough 2011), early iterations of this relationship were characterised by a paternalistic monopoly of decision-making authority, and thus responsibility, in physicians. Today, things might be said to have moved in the opposite direction: while physicians' medical expertise means they retain responsibility for suggesting realistic treatment options, ultimate decision-making authority, and perhaps responsibility, lies with the patients (at least those who have decision-making capacity). Other models of this relationship have been suggested, most prominently the "shared decision-making" approach (Emanuel and Emanuel 1992), but still give the patient a highly active role. Richard Holton and Zoë Fritz's chapter (Chapter 10) engages with the under-explored issue of *uncertainty* in the doctor–patient relationship. Whereas earlier chapters have focused on responsibility for choices that occur prior to the patient's interaction with the health care system, such as the choice whether to smoke or to ride a motorcycle without a helmet,

Holton and Fritz consider responsibility for choices made within the health care system. They note that when doctors communicate options to their patients, they will rarely give them the whole story; for instance, the doctor may omit potential treatments that she deems unlikely to succeed. Holton and Fritz argue that this involves doctors taking responsibility for uncertainty, and that this presents challenges to standard approaches to central medico-ethical concepts, including informed consent and shared decision-making. They conclude that what is required in such cases is a new notion, of "informed trust".

Holton and Fritz's chapter focuses on a direct way that doctors can be responsible for their patients' health outcomes. In modern health care services, there are also a number of indirect ways that doctors and other health care professionals might be held responsible. In part spurred on by the Covid-19 pandemic of the early 2020s, governments and health institutions have shown concern about the mental and physical health of health care professionals, as increasing numbers suffer from "burnout" and other associated workplace issues. However, Joshua Parker and Ben Davies's chapter (Chapter 11) argues that this concern has been problematically expressed for two reasons. First, concern for health care professionals' well-being has largely taken an instrumental form, focusing on professionals' role responsibilities towards their patients, and thus the implications of burnout and other mental health issues for patient care, rather than the intrinsic importance of individuals' well-being. Second, proposed solutions have largely framed the issue as one of personal responsibility, with a focus on self-care practices such as mindfulness or exercise. Parker and Davies argue that this obscures the important role that governments and health care institutions play in damaging medical professionals' well-being, and that it unreasonably instrumentalises and overloads individuals.

0.6.2 Responsibility and Institutions

It is also important to consider the responsibilities of governments and other institutions when it comes to health care. For instance, even if one rejects the kinds of policies discussed in previous sections, institutions involved in promoting and protecting health still need to recognise the role that agency can play in many illnesses. Insofar as agency can play a significant role in leading to, or preventing, the development of various illnesses, health care policies intended to help individuals exercise such agency and improve their behaviour, as it relates to health, may do a better job at preventing illnesses, or preventing them from getting worse, across a population. For instance, opportunity policies need not hold people responsible for failing to change their lifestyles, but might instead offer individuals information about how their behaviour can affect their health, and provide them with opportunities to change their lifestyles. Such a policy could empower individuals to improve their health. Relatedly, when communicating health care information to individuals, health care systems and professionals can make use of current knowledge arising from behavioural economics and psychology which can help to improve knowledge acquisition, and perhaps improve uptake of various health benefits.

A well-known example of this latter type of approach is "nudging". "Nudging" refers to interventions into people's environment aimed at improving their behaviour, either to achieve better outcomes for those who are nudged or to achieve better outcomes for society (Thaler and Sunstein 2008). Many nudges are aimed at improving agents' health. For example, there is evidence that making healthy food choices more salient to people—by, for example, positioning them at eye height— increases the proportion of such foods consumed, relative to less healthy foods (Bucher et al. 2016). In principle, nudges can target a wide range of health-related behaviours. For example, offering individuals the chance to prepay for gym membership might serve as a commitment device that encourages attendance.

In much of the literature, nudging and responsibility are held to be opposed. On the one hand, nudging behaviour fills the gap left if agents cannot, or cannot reasonably be expected to, take responsibility for their own health. That is, if we cannot reasonably expect agents to autonomously make good choices, then we should nudge them into such choices. On this view, nudging is the least bad option we have available, since the alternatives are coercion or a decline in welfare (without any corresponding gain in autonomy). On the other hand, those theorists who highly value individual responsibility (and autonomy, which while not identical to responsibility is implicated in it) are suspicious of nudges because they are held to treat agents as lacking responsibility. Some theorists even maintain that nudging brings it about that agents are not responsible.

A number of theorists have held that nudges bypass autonomous behaviour, the kind of behaviour for which agents are responsible, because they bypass reasoning. Agents are responsible for their behaviour when they are capable of recognising reasons and modulating their actions in response to these reasons (Fischer and Ravizza 1998). But nudges subvert the process of reasoning. They don't present agents with reasons for their behaviour (*that the fresh fruit is at eye level* is not a reason for preferring fruit to crisps). Instead, they take advantage of our non-rational cognitive processes. Because they bypass reasoning, some theorists maintain, and such bypassing is disrespectful of agency and corrosive of responsibility, we ought to minimise or eliminate the use of nudges (Bovens 2008; Wilkinson 2013).

Proponents of nudges largely accept that nudges bypass rational decision-making. They maintain that it is psychologically unrealistic to expect people to guide these kinds of routine decisions by reasons. We make multiple health-related decisions a day, and we can expect to be tired or distracted for many of them. Rational decision-making, nudge proponents often argue, is a limited resource, a resource we must save for major decisions. We need the support of nudges to make good routine decisions (Thaler and Sunstein 2008).

Moreover, since non-rational processes will guide our routine decisions whether we like it or not, a failure to nudge makes it likely that we will not choose well. This is particularly so given that advertisers can be expected to fill any gap we leave. Some interventions might plausibly be conceived of as aiming at countering the nudges of others. For example, plain paper packaging of cigarettes, now required by law in Australia, might be aimed at countering the attractive packaging that nudged buying and using tobacco products.

One of us has recently argued that at least some nudges do not bypass rational reasoning (Levy 2021). Rather, they present agents with genuine reasons for their choices. While it is true that they do not point to good-making features of those choices in the way that arguments usually do, they provide *recommendations* to agents, and recommendations are genuine reasons. Placing the food at eye level recommends it to agents, and those people who do not have strong antecedent preferences between their options rationally take these recommendations into account. How broadly this framework extends remains unclear. If it is true that (some) nudges provide recommendations, then we do not need to choose between nudging agents and treating them as responsible agents: we may do both at once.

It is important to note that there has been a recent debate over whether nudges work at all. The debate was sparked off by a meta-analysis reporting a moderate effect size for nudges (Mertens et al. 2022). Many psychologists found the reported effect size larger than was credible, and a number of limitations with the paper were identified (and acknowledged by the authors). Meta-analyses attempt to measure the true size of an effect by analysing the entire published literature, usually after eliminating low quality work. A good meta-analysis must control for *publication bias*, that is, the fact that reports of positive findings, supportive of the hypothesis being tested, can be expected to be over-represented in the published literature. Unsuccessful experiments are intrinsically difficult to interpret and it is hard to get them published at all. Publication bias is smaller than it has been in the past (because researchers are much more likely to make their unsuccessful experiments available through repositories, and because journals are less unwilling to publish them), but it persists, and is especially problematic for meta-analyses that include many older studies.

Mertens et al. attempted to control for publication bias, but in their reply to that paper, Maier et al. (2022) used a different analytic technique and concluded that there is no good evidence that the effect size of nudges is greater than zero. In the absence of further evidence, they conclude, nudges are hype.

This debate better illustrates the limitations of meta-analyses than the limits of nudging, though it does have important implications for philosophers and ethicists. While it is possible to control for publication bias to some degree, there are multiple problems with the published literature that renders it uninterpretable as whole. It is simply not possible to estimate the effectiveness of nudges on the basis of this literature. They may be too heterogeneous for any such measure to be sensible. Moreover, their effectiveness is context-dependent: a nudge that measurably changes the behaviour of one group might have no effect on another, if the groups differ in important ways. This follows, it should be noted, from Levy's suggestion that nudges often function as recommendations. These recommendations will alter the behaviour only of those who have no strong antecedent preferences and no reason to discount the recommendations. Plain paper packaging of cigarette products will have a much more powerful effect when the behaviour of a group containing a large number of people who might or might not take up or continue to smoke. It will have a much smaller effect among those who are committed to smoking or committed to avoiding smoking.

To know whether nudges work, we must replace estimates of publication bias with preregistered studies: that is, with studies in which researchers commit beforehand to plan of data collection and analysis and to reporting their findings. Preregistration not only avoids publication bias, by ensuring that failed experiments are reported, but also the problem of p-hacking, where researchers slice data in inventive ways to "find" an effect and then pretend they were looking for that effect all along (unfortunately, the meta-analysis by Mertens et al. includes a number of papers that were certainly p-hacked). Fortunately, preregistration is becoming standard in psychology. From work that is preregistered (or used another means of ensuring that all studies are reported fully) we know that some nudges work.

For example, DellaVigna and Linos (2022) were given access to all trials conducted by two large US nudge units: 126 randomised controlled trials, covering 23 million individuals. As expected, effect sizes across trials are heterogeneous, but on average significant. Default effects seem especially robust. Given the heterogeneity, it is perhaps best if philosophers and ethicists avoid talk of nudges without qualification. The category may not pick out a kind. Instead, we should talk of particular kinds of interventions, and be aware that their efficacy is relative to the population studied. We can be confident that some nudges are sufficiently effective that the issues concerning responsibility remain live ones.

The chapter by Daniel Miller, Anne-Marie Nussberger, Nadira Faber, and Andreas Kappes (Chapter 12) considers precisely this possibility. Making use of philosophical accounts of responsibility-relevant capacities and empirical research on how we reason, and form attitudes about vaccination, this chapter sheds light on the extent to which we have such capacities when it comes to making decisions concerning vaccination, as well as some of the obstacles to the exercise of such capacities. With this framework in place, Miller et al. provide recommendations on how to communicate information about vaccines which can help to improve the quality of our exercises of these capacities, and increase vaccine uptake.

As well as thinking about ways in which institutions can engage with individuals to generate more responsible behaviour, it is also important to think about the responsibilities of institutions, including governments, themselves. Many major health problems require significant coordination and guarantees of general compliance with health-protecting norms. Institutions have an important role to play in these contexts, and doing so gives rise to an important set of institutional responsibilities.

Peter Marber and Julian Savulescu's chapter (Chapter 13) focuses on a major, topical health-related responsibility of governments, related to their handling of pandemics. Taking the Covid-19 pandemic as their starting point, Marber and Savulescu suggest that while individual citizens must take responsibility in a pandemic, this is only a reasonable demand if governments take responsibility as well (see also Davies and Savulescu 2022). They outline a "progressive reciprocal responsibility framework", delineating key responsibilities of governments, including compensation for businesses and individuals who lose out due to lockdowns, adjustments in taxation policies, and support for those whose education or childcare arrangements are affected.

Governments have responsibilities to their current citizens. But they may also have responsibilities to people who do not currently exist. We have already discussed the luck egalitarian view, which says that as a society we have a responsibility to prevent or compensate for all inequalities between individuals that are not the result of autonomous choices. For instance, as Shlomi Segall's chapter (Chapter 14) says, luck egalitarians will regard it as unfair that there are differences in life expectancy depending on which region of the UK a person is born in. Such inequalities between actual people are unchosen, and thus cannot be something for which the individuals affected can be held responsible. Thus, on a standard luck egalitarian approach, these inequalities should be addressed, either by being eliminated or through compensation for those who are worse off. Luck egalitarians could go further than this, however, and insist that not only inequalities between actual individuals, but also those between *prospective* individuals, are cause for concern (Voorhoeve 2021). Prospective individuals are those who *might* come into existence, but also might not, depending on our choices. Segall argues that luck egalitarians should not be concerned by inequalities between prospective individuals, thus outlining a further exception to the general luck egalitarian principle that all unchosen inequalities are ethically problematic.

The volume is organised as follows. Part One—with chapters by Chan and colleagues; Kennett; and Shaw—considers questions relating to the justification of responsibility-sensitive health care policies. Part Two—with chapters by Eyal and De Marco—is concerned with questions relating to the nature of such policies, and how they may (or may not) be able to surmount particular difficulties. Part Three—with chapters by Nelkin, Levy, Nath, and Brown—focuses on the nature of individual responsibility, detailing the conditions on responsibility for behaviour and/or outcomes, and how these apply in the context of responsibility for health. Part Four—with chapters by Holton and Fritz, and Parker and Davies—considers role-responsibilities that physicians may or may not have. The final Part Five—with chapters by Miller and colleagues; Savulescu and Marber; and Segall—concerns health-related responsibilities held by the state or society at large.

References

Albertsen, A. (2016a), "Taxing unhealthy choices: The complex idea of liberal egalitarianism in health", in *Health Policy* 120/5: 561–6.

Albertsen, A. (2016b), "Drinking in the last chance saloon: Luck egalitarianism, alcohol consumption, and the organ transplant waiting list", in *Medicine, Health Care and Philosophy* 19/2: 325–38.

Albertsen, A. (2020), "Personal responsibility in health and health care: Luck egalitarianism as a plausible and flexible approach to health", in *Political Research Quarterly* 73/3: 583–95.

Albertsen, A., and Knight C. (2015), "A framework for luck egalitarianism in health and health care", in *Journal of Medical Ethics* 41/2: 165–9.

Albertsen, A., and Nielsen, L. (2020), "What is the point of the harshness objection?", in *Utilitas* 32: 427–43.

Anderson, E.S. (1999), "What is the point of equality?", in *Ethics* 109/2: 287–337.

Arneson, R. (1989), "Equality and equal opportunity for welfare", in *Philosophical Studies* 56/1: 77–93.

Arneson, R. (2006), "Luck egalitarianism interpreted and defended", in *Philosophical Topics* 32/1–2: 1–20.

Barendregt, J.J., Bonneux, L., and van der Maas, P.J. (1997), "The health care costs of smoking", in *New England Journal of Medicine* 337: 1052–7.

Bærøe, K., and Cappelen, C. (2015), "Phase-dependent justification: The role of personal responsibility in fair health care", in *Journal of Medical Ethics* 41/10: 836–40.

Bovens, L. (2008), "The ethics of nudge", in M.J. Hansson and T. Grüne-Yanoff (ed.), *Preference Change: Approaches from Philosophy, Economics and Psychology* (Springer). 207–20.

Brown, R.C.H. (2013), "Moral responsibility for (un)healthy behaviour", in *Journal of Medical Ethics* 39/11: 695–8.

Brown, R.C.H., Maslen, H., and Savulescu, J. (2019a), "Against moral responsibilisation of health: Prudential responsibility and health promotion", in *Public Health Ethics* 12/2: 114–29.

Brown, R.C.H., Maslen, H., and Savulescu, J. (2019b), "Responsibility, prudence and health promotion", in *Journal of Public Health* 41/3: 561–5.

Brown, R.C.H., and Savulescu, J. (2019a), "Responsibility in health care across time and agents", in *Journal of Medical Ethics* 45/10: 636–44.

Brown, R.C.H., and Savulescu, J. (2019b), "Response to commentaries on 'Responsibility in health care across time and agents'", in *Journal of Medical Ethics* 45/10: 652–3.

Bucher, T., Collins, C., Rollo, M E., et al. (2016), "Nudging consumers towards healthier choices: A systematic review of positional influences on food choice", in *British Journal of Nutrition* 115/12: 2252–63.

Buyx, A.M. (2008), "Personal responsibility for health as a rationing criterion: Why we don't like it and why maybe we should", in *Journal of Medical Ethics* 34/12: 871–4.

Callahan, D. (2013), "Obesity: Chasing an elusive epidemic", in *The Hastings Center Report* 43/1: 34–40.

Cane, P. (2016), "Role responsibility", in *Journal of Ethics* 20: 279–98.

Cappelen, A.W., and Norheim, O.F. (2005), "Responsibility in health care: A liberal egalitarian approach", in *Journal of Medical Ethics* 31/8: 476–80.

Cappelen, A.W., and Norheim, O.F. (2006), "Responsibility, fairness and rationing in health care", in *Health Policy* 76/3: 312–9.

Caruso, G.D. (2020), "Justice without retribution: An epistemic argument against retributive criminal punishment", in *Neuroethics* 13/1: 13–28.

Caruso, G.D. (2021), *Rejecting Retributivism* (Cambridge University Press).

Cavallero, E. (2019), "Opportunity and responsibility for Health", in *Journal of Ethics* 23/4: 369–86.

Clarke, R., McKenna, M., and Smith, A.M. (ed.). (2015), *The Nature of Moral Responsibility: New Essays* (Oxford University Press).

Cohen, C., and Benjamin, M. (1991), "Alcoholics and liver transplantation", *JAMA* 265/10: 1299–301.

Corrado, M.L. (2017), "Punishment and the burden of proof", *SSRN Scholarly Paper No. 2997654*, Social Science Research Network.

Daniels, N. (2007), *Just Health: Meeting Health Needs Fairly*, 1st edn (Cambridge University Press).

Davies, B. (2022), "Responsibility and the recursion problem", in *Ratio* 35: 112–22.

Davies, B., and Savulescu, J. (2019), "Solidarity and responsibility in health care", in *Public Health Ethics* 12/2: 133–44.

Davies, B., and Savulescu, J. (2020), "From sufficient health to sufficient responsibility", in *Journal of Bioethical Inquiry* 17: 423–33.

Davies, B., and Savulescu, J. (2022), "Institutional responsibility is prior to personal responsibility in a pandemic", in *Journal of Value Inquiry* doi:10.1007/s10790-021-09876-0

DellaVigna, S., and Linos, E. (2022), "RCTs to scale: Comprehensive evidence from two nudge units", in *Econometrica* 90/1: 81–116.

De Marco, G., Douglas, T., and Savulescu, J. (2021), "Health care, responsibility and golden opportunities", in *Ethical Theory and Moral Practice* 24/3: 817–31.

Double, R. (1991), *The Non-Reality of Free Will* (Oxford University Press).

Duus-Otterström, G. (2012), "Weak and strong luck egalitarianism", in *Contemporary Political Theory* 11/2: 153–71.

Dworkin, G. (1981), "Taking risks, taking responsibility", in *The Hastings Centre Report* 11/5: 26–31.

Emmanuel, E.J., and Emmanuel, L.L. (1992), "Four models of the physician–patient relationship", in *Journal of the American Medical Association* 267: 2221–6.

Eshleman, A. (2001), "Moral responsibility", in E.N. Zalta (ed.), *Stanford Encyclopedia of Philosophy* https://plato.stanford.edu/archives/fall2019/entries/moral-responsibility/ (accessed 17 July 2022).

Feiring, E. (2008), "Lifestyle, responsibility and justice", in *Journal of Medical Ethics* 34/1: 33–6.

Fischer, J.M., and Ravizza, M. (1998), *Responsibility and Control: An Essay on Moral Responsibility* (Cambridge University Press).

Fleurbaey, M. (1995), "Equal opportunity or equal social outcome?", in *Economics and Philosophy* 11/1: 25–55.

Friesen, P. (2018), "Personal responsibility within health policy: Unethical and ineffective", *Journal of Medical Ethics* 44/1: 53–8.

Goodin, R. (1987), "Apportioning responsibilities", in *Law and Philosophy* 6: 167–85.

Gostin, L.O. (2013), " 'Enhanced, edgier': A euphemism for 'shame and embarrassment'?", in *Hastings Center Report* 43/3: 3–4.

Hart, H.L.A. (1968), *Punishment and Responsibility* (Oxford University Press).

Hart, H.L.A., and Gardner, J. (2008), *Punishment and Responsibility: Essays in the Philosophy of Law*, 2nd edn (Oxford University Press).

Herlitz, A. (2019), "The indispensability of sufficientarianism", in *Critical Review of International Social and Political Philosophy* 22/7: 929–42.

Ho, D. (2008), "When good organs go to bad people", in *Bioethics* 22/2: 77–83.

Jefferson, A., and Robichaud, P. (ed.). (2021). "Forward-looking accounts of responsibility" (Special Issue), *The Monist* 104: 4

Jeppsson, S.M.I. (2021), "Retributivism, justification and credence: The epistemic argument revisited", in *Neuroethics* 14/2: 177–90.

Jeppsson, S.M.I. (2022), "Attributability, answerability, and accountability: On different kinds of moral responsibility", in D. Nelkin and D. Pereboom (ed.), *Oxford Handbook of Moral Responsibility* (Oxford University Press), 73–90.

Laverty, L., and Harris, R. (2018), "Can conditional health policies be justified? A policy analysis of the new NHS dental contract reforms", in *Social Science and Medicine* 207: 46–54.

Le Grand, J. (2013), "Individual responsibility, health, and health care", in N. Eyal, S. Hurst, O.F. Norheim, and D. Wikler (ed.), *Inequalities in Health: Concepts, Measures, and Ethics* (Oxford University Press): 299–306.

Levy, N. (2011), *Hard Luck: How Luck Undermines Free Will and Moral Responsibility* (Oxford University Press).

Levy, N. (2019a), "Applying Brown and Savulescu: The diachronic condition as excuse", in *Journal of Medical Ethics* 45/10: 646–7.

Levy, N. (2019b), "Taking responsibility for responsibility", in *Public Health Ethics* 12/2: 103–13.

Levy, N. (2021), *Bad Beliefs* (Oxford University Press).

Lippert-Rasmussen, K. (2016), *Luck Egalitarianism* (Bloomsbury).

MacKay, K. (2019), "Reflections on responsibility and the prospect of a long life", in *Public Health Ethics* 12/2: 130–2.

Maier, M., Bartoš, F., Stanley, T.D., et al. (2022), "No evidence for nudging after adjusting for publication bias", in *Proceedings of the National Academy of Sciences of the United States of America* 119/31: e2200300119.

McCullough, L.B. (2011), "Was bioethics founded on historical and conceptual mistakes about medical paternalism?", in *Bioethics* 25: 66–74.

McGeer, V. (2019), "Scaffolding agency: A proleptic account of the reactive attitudes", in *European Journal of Philosophy* 27: 301–23.

Mertens, S., Herberz, M., Hahnel, U.J.J, and Brosch, T. (2022), "The effectiveness of nudging: A meta-analysis of choice architecture interventions across behavioral domains", in *Proceedings of the National Academy of Sciences of the United States of America* 119/1: e2107346118.

Pereboom, D. (2014), *Free Will, Agency, and Meaning in Life* (Oxford University Press).

Persaud, R. (1995), "Smokers' rights to health care", in *Journal of Medical Ethics* 21/5: 281–7.

Resnik, D.B. (2007), "Responsibility for health: personal, social, and environmental", in *Journal of Medical Ethics* 33/8: 444–5.

Roemer, J.E. (1993), "A pragmatic theory of responsibility for the egalitarian planner", in *Philosophy and Public Affairs* 22/2: 146–66.

Roemer, J.E. (1996), *Theories of Distributive Justice* (Harvard University Press).

Rosen, G. (2004), "Skepticism about moral responsibility", in *Philosophical Perspectives* 18: 295–313.

Savulescu, J. (2018), "Golden opportunity, reasonable risk and personal responsibility for health", in *Journal of Medical Ethics* 44/1: 59–61.

Scanlon, T.M. (1986), "The significance of choice", in *The Tanner Lectures on Human Values* 7: 149–216.

Scanlon, T.M. (2019), "Responsibility for health and the value of choice", in *Lanson Lecture*. Previously available at http://bioethics.med.cuhk.edu.hk/assets/imgs/userupload/Scanlon,%20Lanson%20Lecture,rev.pdf (accessed 17 July 2022).

Schmidt, A. (2022), "From relational equality to personal responsibility", in *Philosophical Studies* 179/4: 1373–99.

Schmidt, H. (2009), "Just health responsibility", in *Journal of Medical Ethics* 35/1: 21–6.

Schwan, B. (2021), "Responsibility amid the social determinants of health", in *Bioethics* 35/1: 6–14.

Segall, S. (2009), *Health, Luck, and Justice* (Princeton University Press).

Segall, S. (2013), *Equality and Opportunity* (Oxford University Press).

Sharkey, K., and Gillam, L. (2010), "Should patients with self-inflicted illness receive lower priority in access to health care resources? Mapping out the debate", in *Journal of Medical Ethics* 36/11: 661–5.

Shiu, M. (1993), "Refusing to treat smokers is unethical and a dangerous precedent", in *BMJ: British Medical Journal* 306/6884: 1048–9.

Stemplowska, Z. (2017), "Rarely harsh and always fair: Luck egalitarianism and unhealthy choices", in S.M. Liao (ed.), *Current Controversies in Bioethics* (Routledge), 149–59.

Strawson, G. (2010), *Freedom and Belief* (Oxford University Press).

Talbert, M. (2019), "Moral responsibility", in E.N. Zalta (ed.), *The Stanford Encyclopedia of Philosophy* (Winter 2019). Metaphysics Research Lab, Stanford University. https://plato.stanford.edu/archives/win2019/entries/moral-responsibility/ (accessed 3 March 2022).

Thaler, R.H., and Sunstein, C.R. (2008), *Nudge: Improving Decisions about Health, Wealth and Happiness* (Yale University Press).

Thornton, V. (2009), "Who gets the liver transplant? The use of responsibility as the tie breaker", in *Journal of Medical Ethics* 35/12: 739–42.

Tomiyama, A.J., and Mann, T. (2013), "If shaming reduced obesity, there would be no fat people", in *Hastings Center Report* 43(3): 4–5.

Ubel, P.A. (1997), "Transplantation in alcoholics: Separating prognosis and responsibility from social biases", in *Liver Transplantation and Surgery* 3: 343–6.

Vagins, D.J., and McCurdy, J. (2006), *Cracks in the System: 20 Years of the Unjust Federal Crack Cocaine Law* (American Civil Liberties Union). Available at: https://www.aclu.org/other/cracks-system-20-years-unjust-federal-crack-cocaine-law (accessed 17 July 2022).

van Baal P.H.M., Polder J.J., de Wit G.A., et al. (2008), "Lifetime medical costs of obesity: Prevention no cure for increasing health expenditure", *PLOS Medicine* 5/2: e29.

Vansteenkiste, S., Devooght, K., and Schokkaert, E. (2014), "Beyond individual responsibility for lifestyle: Granting a fresh and fair start to the regretful", *Public Health Ethics* 7/1: 67–77.

Vargas, M. (2013), *Building Better Beings* (Oxford University Press).

Vathorst, S.V.D., and Alvarez-Dardet, C. (2000), "Doctors as judges: The verdict on responsibility for health", in *Journal of Epidemiology and Community Health* 54/3: 162–4.

Véliz, C. (2020), "Not the doctor's business: Privacy, personal responsibility and data rights in medical settings", in *Bioethics* 34/7: 712–8.

Vilhauer, B. (2009), "Free will and reasonable doubt", in *American Philosophical Quarterly* 46/2: 131–40.

Vilhauer, B. (2012), "Taking free will skepticism seriously", in *Philosophical Quarterly* 62/249: 833–52.

Vilhauer, B. (2015), "Free will and the asymmetrical justifiability of holding morally responsible", in *Philosophical Quarterly* 65/261: 772–89.

Vincent, N.A. (2011), "A structured taxonomy of responsibility concepts", in N.A. Vincent, I. van de Poel, and J. Hoven (ed.), *Moral Responsibility: Beyond Free Will and Determinism* (Springer): 15–35.

Voigt, K. (2007), "The harshness objection: Is luck egalitarianism too harsh on the victims of option luck?", in *Ethical Theory and Moral Practice* 10/4: 389–407.

Voigt, K. (2013), "Appeals to individual responsibility for health: Reconsidering the luck egalitarian perspective", in *Cambridge Quarterly of Health Care Ethics* 22/2: 146–58.

Voorhoeve, A. (2021), "Equality for orospective people: A novel statement and defense", in *Utilitas* 33/3: 304–20.

Waller, B.N. (2011), *Against Moral Responsibility* (MIT).

Wikler, D. (2002), "Personal and social responsibility for health", in *Ethics and International Affairs* 16/2: 47–55.

Wilkinson S. (1999), "Smokers' rights to health care: Why the 'restoration argument' is a moralising wolf in a liberal sheep's clothing", in *Journal of Applied Philosophy* 16/3: 255–69.

Wilkinson, T.M. (2013), "Thinking harder about nudges", in *Journal of Medical Ethics* 39/8: 486.

Zimmerman, M.J. (2015), "Varieties of moral responsibility", in R. Clarke, M. McKenna, and A.M. Smith (ed.), *The Nature of Moral Responsibility* (Oxford University Press): 45–64.

PART I

THE JUSTIFICATION FOR RESPONSIBILITY-SENSITIVE HEALTH CARE POLICIES

1
Should Responsibility Affect Who Gets a Kidney?

*Lok Chan, Walter Sinnott-Armstrong, Jana Schaich Borg,
and Vincent Conitzer*

1.1 Introduction

Most people have two kidneys and can live comfortably with only one, if it functions well enough. If both kidneys completely fail, however, patients will die quickly unless they receive kidney dialysis or transplant. Dialysis has severe costs in money, time and discomfort, and patients face approximately a 40 percent chance of mortality after five years on dialysis (USRDS 2020: chapter 5). For these reasons, many dialysis patients eventually need a kidney transplant, after which they have a survival rate of 90 percent after 3–5 years (Briggs, 2001). Unfortunately, there are about 98,000 people in the USA waiting for a kidney transplant, but only around 20,000 kidneys become available each year (OPTN, n.d.).

This scarcity of medical resources forces doctors to make difficult decisions about which patient gets a kidney when only one is available for many patients who urgently need a transplant. One moral issue regarding these difficult decisions is which characteristics of patients should determine which patient gets a kidney transplant. In the USA, kidney allocation decisions are regulated by the Organ Procurement and Transplantation Network, which considers primarily biological features of the patient (such as age and levels of antigens) or administrative matters (such as time on the wait list and proximity to the donor).[1] In contrast, we have found that many people believe that other factors should also be considered in kidney allocations, such as violent criminal record and number of dependents, including children, disabled or elderly (Freedman et al., 2020). Whether a patient has dependents affects how many people will be helped by giving a kidney to the patient, so this feature is forward-looking into the future. In contrast, a patient's violent criminal record is for past acts, so this feature is backward-looking.

---

[1] https://optn.transplant.hrsa.gov/media/1200/optn_policies.pdf

Lok Chan, Walter Sinnott-Armstrong, Jana Schaich Borg, and Vincent Conitzer, *Should Responsibility Affect Who Gets a Kidney?* In: *Responsibility and Healthcare*. Edited by: Benjamin Davies, Gabriel De Marco, Neil Levy, Julian Savulescu, Oxford University Press. © Edited by Benjamin Davies, Gabriel De Marco, Neil Levy, Julian Savulescu 2024.
DOI: 10.1093/oso/9780192872234.003.0002

Another backward-looking feature that many people see as morally relevant to who should get a kidney when not enough are available is responsibility for getting kidney disease and then needing a kidney transplant. Doctors have found several risk factors that increase the chances of kidney disease, including taking drugs, drinking alcohol, smoking tobacco and eating unhealthy foods (high in fat, sugar and salt). When people know or should know that such behaviors are risk factors for medical problems, including kidney disease, and yet they engage in such behaviors anyway, they are sometimes viewed as responsible for their own health needs and outcomes. Their responsibility might then be seen as a reason to give the kidney to someone else who also needs it if that other person is not responsible (or perhaps not as responsible).

The point is not that such people should not get a kidney transplant. If a kidney is available and nobody else needs it, then almost everyone (and we) would agree that a person who engaged in such behaviors should receive the kidney. To let them die when treatment is available and not needed by others seems too harsh (see Gabriel DeMarco's chapter in this volume). Nonetheless, when one must choose between a person who is responsible for their own medical problems and another who is less responsible or not responsible at all, then some people might give priority to the person who is less responsible and thereby deprive the person who is more responsible of a needed kidney, sometimes leading to death.

To explore these issues, we conducted two experiments. We asked participants to make decisions and judgments about responsibility, fairness and kidney allocation in various scenarios where two patients—one with a history of unhealthy behavior and one without—are competing for a single kidney. We found that our participants were favorable to patients who chose to stop the unhealthy behavior after diagnosis. We also found that, while *not knowing* the risks of one's behavior makes one seem less at fault, people's judgments on who should get the kidney are more reliably predicted by whether the relevant health information was *accessible* to the patient.

1.2 Experiment 1

1.2.1 Introduction: Responsible for What?

In order to determine whether people take responsibility to be morally relevant to kidney allocation, we first need to determine when patients are seen as responsible and what they are seen as responsible for. One could see the patient as responsible for the behavior itself—for abusing drugs, drinking alcohol, smoking tobacco or eating unhealthy foods—but not responsible for contracting kidney disease if they had no way of knowing that their behavior could cause kidney disease (or any other health problem). Moreover, one could see a patient as responsible for their behaviors and also for the resulting disease but then, if they are denied a kidney, one still might not say that it was the patient's own fault that they did not get—or were denied or deprived of—a kidney for transplant. One might hold that, instead of the patient, the doctor, the policy or the health system is responsible for this patient not getting a kidney transplant, perhaps because the doctor, policy or system is seen as unfair to the patient.

Because responsibility can have at least these three targets, we asked separately whether a patient who engaged in a certain behavior is responsible for that behavior, responsible for the disease that it causes or responsible for not getting a kidney. We expected people to hold different views on these issues, so our first step was to survey variations among nonexperts. It would also be interesting to explore the opinions of medical experts, including those who decide kidney allocations, but that project had to be left for future research.

To determine nonexpert views on responsibility, we constructed a survey that contained a set of scenarios in which only one kidney was to be distributed between two patients, one of whom always lost out due to past unhealthy behaviors. Participants were prompted to indicate the extent to which they thought that this patient was responsible for, respectively, their behavior, their kidney disease and their being deprived of a kidney.

We also embedded a within-subject study in the survey: each participant was given two versions of the scenarios. In one version, the patient with past unhealthy behavior stopped their behavior immediately after receiving the diagnosis. In the other version, that patient continued the same behavior after the diagnosis. We hypothesized that participants would judge patients who stopped the behavior after diagnosis as less responsible for being deprived of a kidney.

1.2.2 Participants, Design and Procedure

Two-hundred participants were recruited using Amazon's Mechanical Turk (40 percent female; 69 percent white; average age 36). As an attention check, participants were asked to report whether they took the tasks seriously and paid attention. After accounting for twelve attention check failures, the final sample size was 188.

Participants were first provided information about kidney transplants and donations. They were then presented with eight scenarios in which only one kidney is available for transplant for two hypothetical patients, called A and B. Participants were also told that whoever receives the kidney will undergo an operation that is almost always successful, and whoever does not will remain on dialysis and is likely to die within a year.

Two patients were stated to be identical in all relevant aspects except for a certain negative attribute. The attributes that distinguish A from B were one of following: drug abuse, heavy alcohol consumption, unhealthy eating habits and smoking cigarettes. Participants were informed that drug abuse, heavy alcohol consumption, unhealthy eating and smoking habits can negatively impact kidney function and sometimes lead to kidney failure.

In each scenario, B was always the patient with the negative attribute. For instance, in the case of drinking, B was said to abuse alcohol before they were diagnosed with kidney failure. Furthermore, in all scenarios, the kidney was given to A instead of B on the basis of B's negative attribute. In accordance with our within-subject experimental design, each participant saw two versions for each of the drinking, drug abuse, smoking and eating habit scenarios. In one version B was said to *continue* the unhealthy behavior after being told the diagnosis. In another, B *stopped* the behavior after receiving the diagnosis. In order to keep the experimental conditions as close to each other as possible, it was not specified that B was deprived of a kidney *because* of having continued these behaviors. These two sets of four scenarios then constituted our two experimental conditions: Stopped and Continued.

For each scenario, a participant was asked three questions:

1. *Responsibility for behavior*: Is Patient B responsible for their own [drug abuse, heavy alcohol consumption, smoking, unhealthy eating]?
2. *Responsibility for disease*: Is Patient B responsible for their own organ failure?
3. *Responsibility for deprivation*: Is Patient B responsible for not being selected to receive a transplant?

For each of the three questions, participants responded on a 5-point Likert scale, from "definitely no" to "neutral" to "definitely yes."

1.2.3 Analyses

We hypothesized that whether B stopped the unhealthy behavior would influence a participant's perception of whether B is responsible for not being given a kidney. To test this hypothesis, we calculated two responsibility-for-deprivation scores: (1) the mean of responses to the responsibility for deprivation question for the four Continued scenarios and (2) the means of the same responses for the Stopped scenarios. We then determined whether participants attributed significantly different degrees of responsibility for not receiving a kidney between patients who continued and stopped their unhealthy behaviors.

We also investigated whether participants responded differently regarding the three kinds of responsibility—for behavior, for disease and for deprivation. We first took the average of each responsibility over all behaviors. We then labeled each participant's averaged responses as either Responsible (≥ 3.5), Neutral (< 3.5

and >2.5) or Not Responsible (≤2.5). This allowed us to determine whether or not the proportions of participants who were categorized as Responsible, Neutral or Not Responsible were different between the Continued scenarios and Stopped scenarios. We also explored whether the participants' judgments of someone being responsible for not receiving medical resources was mediated by their judgment of someone being responsible for their behaviors and disease. The goal of the mediation analysis was to quantify the degree to which the observed effect on participants' judgments of responsibility for deprivation can be attributed *directly* to the stopping of unhealthy behaviors and the degree to which this effect was *indirect*: a participant might have thought that patients who stopped were less responsible for being deprived of a kidney, not because stopping directly led one to be less responsible for not getting a kidney, but because stopping made the patient less responsible for, say, having kidney failure, which in turn influenced the perception of how responsible the patient was for not getting a kidney. Technical details of the statistical tests and mediation analysis for experiment are described in Sections 1.5.1 and 1.5.2, respectively.

1.2.4 Results

1.2.4.1 Targets of Responsibility
Our first question was whether the continuation or cessation of unhealthy behaviors made a difference to whether participants attributed responsibilities for our three targets. We did find a difference in the case of responsibility for deprivation. When averaged across all behaviors, participants' responses to the question "Is Patient B responsible for not being selected to receive a transplant?" were 0.79 lower when B stopped their unhealthy behavior compared to when B continued. This result was highly statistically significant ($p < 0.001$) (Table 1.1).

In contrast, there was no statistically significant difference in the cases of responsibility for behavior and responsibility for disease. Thus, our experimental condition (stopping versus not stopping) affected attributions of one target of responsibility (deprivation) but not the others (behavior and disease) (Table 1.2).

1.2.4.2 Mediation Analysis
Our mediation analysis suggested that the mean difference of 0.79, as reported in 1.2.4.1, was largely due to whether the patient continued or stopped the behavior. We found that roughly 75 percent of the total effect was due to the change of behavior, and 25 percent was due to indirect effect through responsibility for behavior and responsibility for disease. The most prominent indirect effect that stopping behaviors had on responsibility for deprivation was through responsibility for disease, which accounts for 16 percent of the total effect. Our mediation model is described in detail in Section 1.5.2.

Table 1.1 Means (standard deviations) in Experiment 1

	Behavior Continued				Behavior Stopped			
	Alcohol	Drugs	Eating	Smoking	Alcohol	Drugs	Eating	Smoking
Responsible for...								
Unhealthy behavior	4.34 (0.93)	4.55 (0.83)	4.55 (0.82)	4.42(0.88)	4.54(0.83)	4.38 (0.93)	4.42 (0.83)	4.19 (1.02)
Kidney disease	4.36 (0.85)	4.25 (0.89)	3.85 (1.02)	3.96 (1.12)	4.00 (0.73)	3.90 (0.86)	3.46 (1.04)	3.73 (10.3)
Deprived of a kidney	4.14(1.20)	4.01(1.30)	3.70 (1.18)	3.97 (1.20)	3.32(1.23)	3.18 (1.30)	2.97 (1.20)	3.20 (1.20)

Table 1.2 Categorical variables in Experiment 1

Responsible for…	Behavior		Disease		Deprivation	
	Continued	Stopped	Continued	Stopped	Continued	Stopped
Responsible ($X \geq 3.5$)	167	162	157	141	135	88
Neutral ($2.5 < X < 3.5$)	12	17	16	30	25	43
Not Responsible ($X \leq 2.5$)	3	3	9	11	22	51

1.2.5 Discussion

According to our mediation model, stopping a behavior exerted only very minor influence on the judgment of responsibility for behavior or disease. This is not surprising, because the behavior and disease occurred *before* the patient stopped, so stopping cannot affect the earlier behavior or disease. In contrast, whether the patient got the needed kidney was determined *after* the decision to stop, so stopping the behavior could affect the later consequence of being deprived of the kidney. Thus, it makes sense that a patient's failure to stop unhealthy behavior after being diagnosed with kidney disease can make it their own fault that they did not get a kidney. Many participants seemed to think this way.

The effect of stopping unhealthy behaviors on responsibility for deprivation can also be interpreted in light of how many more participants said that the patient was responsible for being deprived of a kidney when the patient did stop as opposed to when the patient continued. In the case of heavy alcohol consumption, for example, the Responsible/Neutral/No-Responsible split for responsibility for deprivation for the Continued scenario is 75 percent/13 percent/12 percent, but we see a shift for the Stopped scenario to 52 percent/21 percent/27 percent. Thus, around twice as many participants thought that B was not responsible for being deprived of a kidney when B stopped heavy alcohol consumption than when B did not stop heavy alcohol consumption. The difference might have been even higher if participants had been told in the Continued scenarios that B was deprived of a kidney *because* B continued the behavior. A substantial percentage of participants then seem to hold that it was B's own fault that B did not get a kidney transplant when B did not stop, but this outcome was not B's fault when B did stop.

1.3 Experiment 2

1.3.1 Introduction: Knowledge and Responsibility

Our first study investigated *what* people hold kidney patients responsible for. A separate question is *when* (or *why*) they hold patients responsible to various degrees. Our second study explores this new question.

One factor that is commonly assumed to affect responsibility is knowledge (cf. Sher, 2009). If a person knows that an act will cause or risk causing harm to another person, but they do it anyway, then they are typically seen as more responsible than someone who harms another without knowing that the act will cause or risk causing the harm. Still, even if the agent did not in fact know that the act would or could be harmful, sometimes they should have known better, and then they are more responsible than an agent who could not have known that the act might cause harm. For example, if I choose to eat peanuts when I know that

I am allergic to them, then I am more responsible for the harm to myself than if I eat peanuts only knowing that peanut allergies are common but without any knowledge I personally have this allergy. I am even less responsible if I had no idea that anyone could have nut allergies.

These distinctions are encoded into categories of *mens rea* in the Model Penal Code of the American Law Institute (1962). Laws based on this model distinguish causing death *knowingly* (when the defendant knew that the act would cause death), *recklessly* (when the defendant did not know that death would result but did know that the act created a high risk of death), *negligently* (when the defendant did not know the risk but should have known, because reasonable people would have known), and not even negligently (when the defendant had no access to the information that the act created a risk, so a reasonable person in the defendant's situation would not have known the risk). Notice that reckless and negligent agents both lack knowledge that their acts *will* actually cause harm, but reckless agents do know the *risk*, and negligent agents do not know the risk (but should know it). The defendant's level of knowledge or ability to know the result or risk then affects the degree of murder or of other crimes and, hence, the length of sentence. Recent surveys have found that nonlawyers often distinguish these levels of responsibility in relation to knowledge (Ginther et al., 2018).

These criminal cases involve harm to other people, but it is less clear whether the same distinctions apply to the responsibility of medical patients for causing harm to themselves. After all, criminal and medical triage situations differ in important ways (to be discussed in Section 1.4.2). Do people hold patients more responsible when they knew or should have known that their behavior would cause or increase the risk of their own medical condition for which they now need a scarce treatment, such as a kidney for transplant? Which degree of responsibility is seen as sufficient or necessary for justifying giving priority to another patient who is not responsible or is less responsible on account of not knowing or not having access to information about the risks of their behaviors? These and related questions are addressed in our second study.

1.3.2 Participants, Design and Procedure

Five-hundred-and-twenty participants were recruited using mTurk (48 percent female; 75 percent white; average age 40). After a brief introduction about kidney failures and transplants, participants were given three scenarios in random order. In each of the scenarios, participants were asked to consider two hypothetical patients (A and B), both of whom need a kidney transplant. These patients are said to be the same in all relevant aspects except that B engaged in some unhealthy behavior specified in each scenario. The three scenarios are: alcohol consumption, drug abuse and smoking.

There are several important differences from Experiment 1. First, A was explicitly described as having never engaged in the unhealthy behavior before or after their diagnosis, while B was described as having frequently engaged in said behavior. Second, in all scenarios, participants were told that B *stopped* the unhealthy behavior altogether after being told that continuing to do so would make the disease worse. Third, we specified that according to B's doctor, relapse to unhealthy behavior after transplant was unlikely, and that, due to the change in behavior, B was expected to have the same prospect of recovery and life expectancy as A. This third change was intended to ensure that participants would focus on backward-looking fault or responsibility instead of predicting different future prospects for patients who engaged in the past behaviors. Fourth, eating habit was removed as a scenario to reduce cognitive load.

For Experiment 2, we also switched to a between-subjects design: to determine whether someone's knowledge of the consequences of their unhealthy behaviors would influence a participant's opinion on medical resource distribution, participants were randomly assigned to three groups. For all scenarios X, where X is either heavy alcohol consumption, drug abuse or smoking, participants were given one of the following messages about the patients' epistemic state, depending on their group assignment:

1. *Knowledge*: A and B *knew* that X creates a risk of kidney failure when they chose whether or not to X.
2. *Access*: A and B *did not know* that X creates a risk of kidney failure when they chose whether or not to X. However, A and B *had easy access to this information* when they chose whether or not to X, and most people in their community *know* that X creates a risk of kidney failure.
3. *No-Access*: A and B *did not know* that X creates a risk of kidney failure when they chose whether or not to X. Moreover, A and B *had no easy access to this information* when they chose whether or not to X, and most people in their community *did not know* that X creates a risk of kidney failure.

In addition, we no longer stipulated that the kidney was given to B. Instead, this choice was presented to the participant. For each scenario, the participant was asked to choose among three options:

- I would give the kidney to Patient A.
- I would give the kidney to Patient B.
- I would decide randomly which patient gets the kidney.

Participants were also asked to indicate their agreement with the following statements in a 7-point Likert scale from "strongly disagree" to "neutral" to "strongly agree":

1. *Unfairness*: It would be unfair for the doctor to decide which patient gets the kidney on the basis of patient B's behavior X.
2. *Responsibility for behavior*: Patient B was responsible for their own X.
3. *Responsibility for disease*: Patient B was responsible for their own kidney failure.
4. *Fault for deprivation*: If the kidney is given to A because of B's X, then it is B's own fault that B did not get the kidney.

The statement about not getting a kidney was phrased in terms of being at fault (in Experiment 2) instead of being responsible (in Experiment 1). This change was due to the concern that lay participants might not uniformly understand the idea of "being responsible for not receiving the kidney." To alleviate this concern, we explicitly invoked the concept of fault by asking whether this outcome would have been B's own fault.

Two additional questions were given to participants at the end of the survey in order to account for the potential influence of the participants' background beliefs about the unhealthy behaviors in question. First, participants were asked to indicate the percentage of people in their community they thought would know that heavy drinking, drug abuse and heavy smoking can cause kidney problems. Second, we also asked participants to indicate how much voluntary control they believe patients have over their alcohol consumption, drug abuse and smoking habits on a 5-point scale, from "none at all" to "a great deal."

Three attention checks were randomly distributed throughout the survey. Participants were asked to indicate whether Patient B stopped their unhealthy behavior after diagnosis. They were also asked whether A and B knew that the unhealthy behavior creates a risk of kidney failure, and the correct answer would correspond to the treatment group to which they belong. They were also asked to indicate that they were paying attention by choosing a particular answer. Only participants who answered all three attention-check questions correctly were included in the analytic cohort. Seventy-five participants failed at least one of the attention checks, leaving n = 149 for the Knowledge group, 151 for the Access group and 145 for the No-Access group.

1.3.3 Analyses

As in Experiment 1, participants' responses to unfairness, responsibility for behavior, responsibility for disease and fault for deprivation for all scenarios were averaged in order to generate an overall score for each measure. To account for effects from background beliefs, we conducted a regression analysis between how much they thought patient B was at fault for being deprived of a kidney, the extent to which they think their community would know certain unhealthy habits can cause kidney problems and how much control they think people have over these habits. Another similar regression analysis was carried out on the averaged judgments of unfairness and of responsibility for behavior. We also conducted exploratory mediation analyses to examine the possible relationships among the hypothetical patients' access to information, participants' judgments of fault for deprivation and decisions of who gets the kidney. For technical details of our analyses, see Section 1.5.3.

1.3.4 Results

1.3.4.1 Between-Group Differences
Does the satisfaction of the epistemic condition affect people's opinions on fault for deprivation? On average, participants in the Knowledge group gave a score for 5.13 out of 7, which indicated that they "somewhat agree" that B is at fault for not getting a kidney. Responses by the Knowledge group were significantly higher than those by both Access group (4.77) and No-Access group (4.57) (Table 1.3). Details of our results are reported in Section 1.5.3.

1.3.4.2 Background Beliefs
Our analysis showed that participants' beliefs about how much patients had control over their unhealthy behaviors and how commonly known these behaviors' impacts on kidney functions were *positively* associated with their judgments of fault for deprivation. In other words, the more a participant thought that people have control over their unhealthy behaviors and the more a participant thought these behaviors' effects on kidney function are commonly known in their communities, the more strongly a participant agreed with the statement that it was Patient B's own fault that they were not given a kidney. Nonetheless, our model indicates that the between-group differences (in 1.3.4.1) remained statistically significant even after accounting for the effects of community knowledge and voluntariness. Consider, for example, two participants who held similar beliefs about voluntariness and community awareness, but one was assigned to Knowledge group and another to No-Access group. The participant in the Knowledge group will be predicted to agree more strongly than the one in the No-Access group with the claim that it was B's own fault that B was deprived of a kidney. Thus, our conclusion about between-group differences in Section 1.3.4.1 do not depend on the background beliefs that we checked.

When a similar analysis was done on judgments of unfairness, we found that the *only* reliable predictor of judgments of unfairness was how much voluntary control a participant believed that patients have over these unhealthy behaviors. The more they believed in such control, the *less* they thought it would be unfair to consider those behaviors as relevant. Epistemic situations (Knowledge, Access,

Table 1.3 Descriptive statistics by group in Experiment 2

Who Gets the Kidney, percent of Participants

	Knowledge	Access	No-Access
Drinking			
A	70%	68%	62%
B	6.7%	4%	0.7%
Decide Randomly	23.5%	28%	37%
Drug Abuse			
A	74.5%	73.5%	63.4%
B	5.4%	2%	0.7%
Decide Randomly	20.1%	24,5%	35.9%
Smoking			
A	68.5%	67.5%	61%
B	6%	1.3%	2%
Decide Randomly	25.5%	31.1%	37%
Responsibility for Behavior, M(SD)			
Drinking	5.93(1.18)	5.94(1.13)	5.91(1.12)
Drug Abuse	5.86(1.26)	5.88(1.19)	5.85(1.23)
Smoking	6.05(1.21)	6.06(1.21)	6.04(1.10)
Responsibility for Disease, M(SD)			
Drinking	5.55 (1.17)	5.19(1.38)	5.00(1.40)
Drug Abuse	5.38(1.26)	5.26(1.44)	5.08(1.52)
Smoking	5.23(1.46)	5.01(1.41)	4.86(1.61)
Unfair to Decide Based on Behavior, M(SD)			
Drinking	3.82(1.90)	3.97(1.80)	4.06(1.63)
Drug Abuse	3.74(1.87)	3.80(1.83)	3.82(1.75)
Smoking	3.88(1.93)	3.99(1.85)	3.99(1.80)
Fault for Deprivation, M(SD)			
Drinking	5.22(1.54)	4.74(1.64)	4.56(1.72)
Drug Abuse	5.21(1.50)	4.90(1.62)	4.55(1.81)
Smoking	5.09(1.53)	4.66(1.70)	4.59(1.76)

No-Access) did not make any statistical difference. As expected, we found the *opposite* effect for responsibility for behavior: the belief in voluntary control of these behaviors is again the only reliable predictor here, but now the more participants believed in such control, the *more* they thought that patients are responsible for their own behavior. The attribution of responsibility of behavior is stable across groups (Knowledge, Access, No-Access). This is consistent with the results of the first experiment and suggests that participants believed that an agent can be responsible for a behavior (taking drugs, for example), even if the agent does not and cannot know the risks of that behavior.

1.3.4.3 Exploratory and Mediation Analyses on Decision-Making
Here we are primarily interested in whether having knowledge or access to knowledge made a difference in how participants distributed kidneys. What was

the underlying process that led a participant to (1) give A the kidney, as opposed to (2) distributing it randomly or (in rare cases) giving it to B? First, we found that participants' decision-making was not significantly influenced by the type of behavior done by B. In other words, the probability that the kidney was given to A was not changed whether B was a heavy drinker, drug abuser or a heavy smoker. Second, the only fact that made a statistical difference to whether A got the kidney was whether or not B had access to the information about the impact these unhealthy behaviors had on kidney functions. Based on this analysis, we constructed a mediation analysis to explore whether the effect of access to information on the participant's likeliness to give A the kidney was direct or indirect. We found that this effect was almost completely indirect: the effect of access to information on kidney allocation was almost entirely due to the effect it had on participants' judgments on fault for deprivation. Section 1.5.4 describes technical details of our methods and results of our analyses.

1.3.5 Discussion

Our most striking finding was how the epistemic conditions differently influenced judgments of fairness, fault and decisions regarding kidney allocation. First, there was no statistically significant difference among the Knowledge, Access and No-Access groups in thinking that it would be unfair to consider past behaviors. Second, unlike fairness, we found a detectable difference in the attributions of fault for deprivation between participants in the Knowledge group and participants in Access and No-Access groups. Our participants saw patients with or without access to the information (but without knowledge) as similar and less at fault for not receiving a kidney than patients who had knowledge of the risks of their actions. With regard to fault for deprivation, then, patients with access but without knowledge were grouped with patients who had no access to information about the risks of their behavior. Thus, what mattered to our participants in ascribing fault for deprivation was actual knowledge of risks rather than access to knowledge about risks.

In contrast, patients with access but without knowledge of risks were instead grouped with patients who had knowledge of risks when the question was to whom they would give the kidney. In decisions about kidney allocation, participants in the No-Access group were more in favor of a random decision procedure than participants in the Knowledge and Access groups. Specifically, there is only 1 percent difference in how often participants in the Knowledge and Access groups would give the kidney to A (71 percent vs 70 percent), but the No-Access group would give the kidney to A only 62 percent of the time on average.

Putting these results together, patients with access but no knowledge would receive a kidney less often than patients without access, even though these two kinds of patients were seen as equally at fault for not getting a kidney. Thus, some participants seem to use different factors to determine who is at fault for not getting a kidney than they use for determining whom to give the kidney.

1.4 General Discussion

1.4.1 Scientific Discussion

The results of Experiment 1 show that participants tended to assign less responsibility for being deprived of a kidney to patients who stopped their behaviors, independent of how responsible they thought those patients were for their earlier behaviors and diseases, and even without knowing how these patients came to develop these behaviors. This result makes sense in light of common views about what makes a person responsible. When a patient receives a diagnosis of kidney disease and is told by a doctor that continuing certain behaviors—such as abusing drugs or alcohol, smoking tobacco and eating unhealthy foods—will exacerbate their kidney disease, then the patient learns that the behavior is not only risky in the abstract but harmful to them in concrete ways. The diagnosis also prompts the patient to think at a higher level of generality about their lasting pattern of behavior instead of only whether to engage in the behavior on a particular occasion. These effects of the diagnosis give the patient a new kind of opportunity to stop the harmful behavior. If the patient does not take advantage of this opportunity, then observers often see the patient as responsible for later harms caused by not stopping, such as being deprived of a kidney.

Experiment 2 explored the epistemic conditions of responsibility as well as its implications for the allocation of kidneys to patients in need. In broad strokes, participants considered two patients who did not know the risks of their unhealthy behaviors to be the similarly less at fault for deprivation than someone who knew, even if one had easy access to this information whereas the other did not. However, when it came to making decisions about which patient to give the kidney, access rather than knowledge became the determining factor, as participants were much more likely to give Patient B a chance of getting a kidney when B lacked easy access than when B had easy access, even if B lacked knowledge with or without easy access.

One interpretation of these results is that participants were sensitive to the distinction—present in the Model Penal Code (American Law Institute, 1962)—between negligent acts whose agents should have known the risk of harm versus non-negligent acts of which it is not true that their agents should have known the risk (FitzPatrick, 2008). In both the Knowledge and Access conditions, Patient B either already knew about the risks of their behaviors or could have easily found out if they wanted. Given the importance of these risks, participants might have assumed that an agent who could have known the risks should have taken steps to find out about them. On this assumption, Patient B should have known better and therefore was seen as more at fault for the consequence of their actions. On the other hand, in the No-Access condition where the risks of the unhealthy behaviors were not easily accessible to B, participants might have thought that B *could not* have known the risks and, therefore, one could not say either that they ought to have known better or that they ought to pay a price for their mistake.

This interpretation allows us to explain why our participants thought that patients with and without access to knowledge of risks were similarly at fault when they both lacked knowledge of risks, but our participants were more still more likely to give the kidney to the patient without access to information about risks than to the patient with access to that information, even though both lacked knowledge. The two patients' shared lack of knowledge of the harm might have put them on a similar or equal footing in terms of being at fault, but participants seem to have considered the fact that a patient *could not* have reasonably known the risks of their behavior as a reason for giving them just as much chance to get a kidney as someone who did not engage in the unhealthy behavior at all.

The results of Experiment 2 are relevant to the debates about whether or when it is possible to reliably attribute responsibility for behavior to people (Brown & Savulescu, 2019; Friesen, 2018). We found that people do use a patient's knowledge in judging responsibility. As a result, they need to look beyond mere behaviors to the mental states behind those behaviors in order to judge responsibility. However, it is often hard to determine whether or not someone possessed knowledge of the health risks of their behaviors. This difficulty might explain why people do not use whether someone actually possesses knowledge to decide who gets a scarce medical resource. For this life-affecting decision, they instead use a consideration that might be easier to determine, namely, whether the patient had easy access, so a reasonable person would have known, and the patient ought to have known. Our results show that people *do* use these kinds of considerations to make decisions regarding the distribution of scarce medical resources.

1.4.2 Concluding Unscientific Postscript

What, if anything, do these popular opinions tell us about who *should* get a kidney? Can we draw normative conclusions from surveys? No. The fact that many people base their moral judgments about kidney allocation on their judgments about responsibility does not prove that it is correct to do so.

Many philosophers will reject these popular opinions. Act consequentialists (including act utilitarians) claim that what one ought to do depends only on what will make the world better in the future (Sinnott-Armstrong, 2019). However, whether someone is responsible for needing a kidney or for not getting one depends on facts about past behaviors, such as abusing drugs, drinking alcohol, smoking tobacco or eating unhealthy foods. Therefore, act consequentialists deny that whether one ought to give a kidney to one patient instead of another depends on past acts that make one of the patients responsible. Past behavior can still be evidence of future consequences, but it is not what makes a decision right or wrong.

In contrast, rule consequentialists (including rule utilitarians) can say that responsibility matters indirectly to who should get the kidney. A widely known rule that gives priority to patients who are not responsible for their own medical needs might provide additional incentives not to engage in unhealthy behaviors with bad consequences. If so, and if one ought to follow rules that are beneficial when widely known, as many rule consequentialists argue, then whether one ought to give a kidney to a patient can depend indirectly on whether potential recipients' past acts make them responsible.

Thus, act consequentialism does conflict with, but rule consequentialism does not conflict with, the judgments made by our survey participants who give priority to patients who are not responsible, at least if our participants assumed that rules that base allocation on responsibility have beneficial consequences. Of course, these findings do not show that act consequentialism is incorrect or that rule consequentialism is correct (or that some nonconsequentialist theory is correct). These popular opinions still might be inaccurate. Nonetheless, our surveys do place some burden on act consequentialists both to show why the popular opinions are incorrect and also to explain how so many people become confused about the importance of responsibility in medical triage. This point should not be surprising, because act consequentialists need to shoulder similar burdens in many other cases where act consequentialism conflicts with popular opinions.

Our findings also become relevant to normative issues in a more practical way if we assume that hospital policies should not conflict too much with local moral standards. This assumption might be supported by reference to some kind of right. At least when hospitals are funded by public monies, people who contribute to public funds might be seen as having a right to determine or limit what those hospitals do. The same assumption could also be supported by consequentialist considerations. Hospital policies that conflict too much with local moral standards might reduce trust in those hospitals, and distrust can undermine the ability of hospitals to serve their communities. If so, then act consequentialists can agree with rights-theorists who think that hospitals should not stray too far from public judgments about morality.

Critics have objected (in conversation) that doctors do not know enough about their patients to determine whether a patient is responsible and also that patients should be able to defend themselves before they are deprived of a kidney on grounds of responsibility. It might seem unjust not to hold a legal hearing or trial before punishing a patient by denying them a kidney that they need in order to stay alive.

In response, it is crucial to realize how triage differs from punishment (see Neil Levy's chapter in this volume). In a criminal trial, the decision is whether or not the defendant goes to jail. The issue is not whether this person as opposed to another person goes to jail. In contrast, when a doctor decides not to give a kidney to a patient in need, that decision is made only because another patient needs that kidney. Nobody thinks that a person who is responsible for their own kidney disease should not receive a kidney that nobody else needs. The relevant question is which person in need gets a kidney and not whether this person gets a kidney at all. That makes triage very different from asking whether or not a criminal defendant gets punished. Moreover, when someone is sentenced to punishment in a criminal trial, that punishment expresses blame and condemnation of the criminal. In contrast, a doctor does not need to condemn, punish or blame the patient who does not get the kidney (though see Jeanette Kennett's chapter in this volume). It is very important to separate blame and condemnation from the judgment that a person is responsible, not only in triage but also in other contexts (such as addiction; see Pickard, 2018). Furthermore, punishment involves actively inflicting harm on the criminal, whereas giving a kidney to another person in need is simply failing to help the person who is deprived of that kidney (but who still might get another kidney later). The facts that triage—in contrast with punishment—does not involve active harm or blame and does help another person in need explain why a trial is not required in order to determine responsibility before a triage decision. And it is a good thing, too, because a trial that allowed both parties an adequate chance to present evidence would take much too long, given the time pressure of triage decisions.

Nonetheless, a practical problem remains. How can doctors determine which patients are responsible for their misbehaviors or for their own kidney disease? Our surveys did not mention the particular circumstances that led patients to use drugs, abuse alcohol, smoke or eat unhealthy foods. Perhaps some patients—even if only a few—had no available option or no way of knowing that these behaviors were risky (or risky to their kidneys in particular). Nonetheless, general policies for kidney triage can be made with a default of responsibility and a chance for patients to argue for exceptions to the policy when special conditions are met. These policies will not be easy to formulate precisely, but they can be developed by experts with general information that is not readily available to all doctors. The admitted difficulty in determining responsibility in individual cases is, thus, not an insuperable obstacle to considering patient responsibility in allocating kidneys and perhaps some other scarce medical resources.

Our conclusion is not that surveys reveal what really is morally right or wrong. Nonetheless, surveys can reveal both burdens on theories as well as practical costs of ignoring popular opinions, including the common view that responsibility should affect who gets the kidney when there are not enough kidneys for all in need.

The same point applies to other situations of scarcity. Imagine Abe knows that COVID-19 is highly contagious in crowded rooms but still voluntarily goes to an inside party where nobody (including Abe) has been vaccinated or is wearing a mask or socially distancing. As a result, Abe contracts COVID-19, experiences severe breathing problems, goes to a hospital and needs to be placed on a ventilator in order to save his life. If there are not enough ventilators to go around, is the fact that Abe is responsible for his own medical problems any or adequate reason to give priority to others in need who are not so responsible? This question is more complicated in this situation because of time pressure. When a patient shows up at a hospital with breathing problems, there is often not enough time or any practical way to determine where or how they caught COVID-19. In contrast, a patient with kidney disease does not usually need a kidney transplant immediately, since they can live on dialysis for a while. Thus, each situation comes with its own complications, and we need to think through each case separately.

Surveys cannot solve these pressing moral problems. Our hope is only that they can provide a little help in understanding what is at stake and in guiding future discussions and policies.

1.5 Technical Appendix

1.5.1 Experiment 1: Statistical Analyses and Results

To determine within-subject differences between participants' responsibility for deprivation score in the Continued scenarios and those in the Stopped scenarios, we employed a paired t-test. We found a statistically significant difference of 0.79 ($p < 0.001$) with a 95 percent confidence interval of -0.66 and -0.91.

To determine differences between the proportions of participants who responded positively, neutrally and negatively regarding responsibility for behavior, disease and deprivation when B continued versus when B stopped, we used Pearson's Chi-squared test t. We found that responsibility for deprivation differed significantly between Continued and Stopped ($p < 0.001$, $X^2 = 26.2$). This difference did not reach statistical significance for either responsibility for behavior ($p = 0.63$, $X^2 = 0.94$) or responsibility for disease ($p = 0.07$, $X^2 = 5.32$).

1.5.2 Experiment 1: Mediation Analysis and Results

We constructed a mediation model by using the ordinary least squares regression approach (Hayes, 2009; Hayes & Matthes, 2009). Values for statistically significant relationships are presented in Figure 1.1.

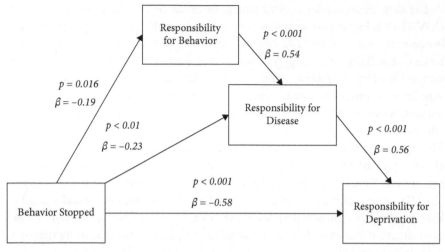

Figure 1.1 Mediation analyses in Experiment 1

According to our mediation model, stopping unhealthy behaviors leads to a *decrease* of 0.58 in attributions of responsibility for deprivation. Thus, after accounting for the effects of these mediators, the direct effect of Stopped versus Continued remains strong.

Our model suggests that the indirect effect of Stopped on responsibility for deprivation is primarily through its effect on responsibility for disease. First, Stopped negatively relates to responsibility for disease ($\beta = -0.23$, $p = 0.007$), which in turn is positively associated with responsibility for deprivation ($\beta = 0.56$, $p < 0.0001$), yielding an *indirect effect* of -0.13. In other words, stopping unhealthy behavior indirectly lowers responsibility for deprivation by lowering responsibility for disease.

Second, Stopped has a statistically significant influence on responsibility for behavior ($\beta = -0.19$, $p = 0.02$). This allows responsibility for behavior to mediate between Stopped and responsibility for disease ($\beta = 0.56$, $p < 0.0001$), and finally led to an indirect effect of -0.06 on responsibility for deprivation. There is no statistical evidence that responsibility for behavior has any direct effect on responsibility for deprivation ($\beta = 0.11$, $p = 0.19$). The relevant statistics for total effect, direct effects and all indirect effects are summarized in Table 1.4.

1.5.3   Experiment 2: Statistical Analyses and Results

We conducted a prospective power analysis which found that a sample of 150 per group (Knowledge, Access and No-Access) was needed to achieve the power of 0.82 to detect a small effect size of 0.15 to a significance level of 0.05. After the data

Table 1.4 Total, direct and indirect effects in Experiment 1

	Effect	SE	p	C.I. Lower	C.I. Upper
Behavior Stopped → Responsibility for Deprivation					
Total Effect	−0.79	0.11	< 0.0001	−1.01	−0.57
Direct Effect	−0.59	0.11	< 0.0001	−0.79	−0.37
Indirect Effects					
Total Indirect Effects	−0.21	0.06	NA	−0.32	−0.10
Behav. Stopped → Res. for Behav. →Res. for Depriv.	−0.02	0.02	NA	−0.07	0.01
Behav. Stopped→ Res. for Disease→ Res. for Depriv.	−0.13	0.05	NA	−0.23	−0.05
Behav. Stopped→ Res. for Behav. → Res. for Disease → Res. for Depriv.	−0.06	0.02	NA	−0.11	−0.01

was gathered, fault for deprivation was found to violate normality (Shapiro-Wilk test, $p < 0.0001$), so the nonparametric Kruskal–Wallis test was used to detect group differences, and the Wilcoxon rank sum test was used to make a multiple pairwise comparison, with p-values adjusted using Bonferroni correction. The Kruskal–Wallis test showed that there was indeed a statistically significant difference between the median fault for deprivation scores ($p < 0.001$). The median of the Knowledge group was statistically different from both the median of the Access group (adjusted $p = 0.04$) and the median of the No-Access group (adjusted $p < 0.01$). However, there was no statistically significant difference between the two groups (Access and No-Access) that lacked knowledge (adjusted $p = 1$). Medians and 25th and 75th percentiles are presented in Figure 1.2.

To study participants' decisions on who should get the kidney, we dichotomized each decision into giving the kidney to A versus either giving it to B or deciding randomly. The dividing line was motivated by analytic expediency and because the kidney was given to B in only roughly 3 percent of cases. Logistic regression was used to examine the relationship between the experimental conditions (Knowledge, Access and No-Access) and the probability of participants giving the kidney to Patient A.

In order to explore whether these differences between the Knowledge, Access and No-Access groups could be explained by background beliefs held by participants, we carried out three regression analyses (Table 1.5). We found that participants' background beliefs about voluntariness and community knowledge were both statistically associated with a higher fault for deprivation score ($p < 0.001$, $\beta = 0.51$ and $p < 0.001$, $\beta = 0.01$). When a similar regression was done on judgments of unfairness, voluntariness was found to be the only statistically

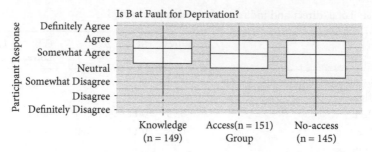

Figure 1.2 Group medians and 25/75 percentiles for fault for deprivation in Experiment 2

Table 1.5 Regression analyses in Experiment 2

Fault for Deprivation	β	SE	p
Intercept	2.67	0.31	< 0.001
Voluntary Control of Behavior	0.51	0.31	< 0.001
Community Knowledge of Consequences	0.01	0.002	< 0.01
Access Group	−0.37	0.16	0.02
No-Access Group	−0.57	0.16	< 0.001
Unfairness of Deciding Based on Unhealthy Behavior			
Intercept	6.05	0.36	< 0.001
Voluntary Control of Behavior	−0.59	0.08	< 0.001
Community Knowledge of Consequences	≈ −0.0001	0.003	0.98
Access Group	0.08	0.18	0.66
No-Access Group	0.12	0.18	0.51
Responsibility for Behavior			
Intercept	3.61	0.20	< 0.001
Voluntary Control of Behavior	0.57	0.04	< 0.001
Community Knowledge of Consequences	0.003	0.001	0.09
Access Group	0.05	0.10	0.65
No-Access Group	0.02	0.11	0.87

significant predictor ($p < 0.001$, $\beta = -0.59$). Lastly, voluntariness, again the sole statistically significant predictor, was positively associated with responsibility for behavior ($p < 0.001$, $\beta = 0.57$).

1.5.4 Experiment 2: Exploratory and Mediation Analyses and Results

We carried out a logistic regression to investigate whether the experimental conditions (Knowledge, Access and No-Access) and the types of behaviors (drinking, drug abuse or smoking) are associated with whether the kidney would be given to A (Table 1.6). We found that the type of behavior bears no significant association with the probability of the kidney being given to A. In fact, the only significant predictor of the kidney being given to A is the No-Access group ($p < 0.001$, $\beta = -0.39$). Translating the odds into probabilities, we saw that when the patients possess knowledge of the risks of the unhealthy behavior, the probability that A will receive the kidney is 0.70, but this drops to 0.61 when they do not know the consequences and lack easy access to the information.

Since the outcome variable of interest is now dichotomous, we used the Baron and Kenny method (1986) in our mediation analyses and used a method documented by Rijnhart et al. (2019) to calculate direct, indirect and total effects the access to information had on the odds of a participant giving the kidney to A. We examined whether the No-Access group's lower probability of giving A the kidney compared to the other group was mediated by judgments of fault for deprivation. Since the logistic regression analysis showed that there was no difference between the Knowledge group and the Access group in their decisions on who gets the kidney ($p = 0.70$, $\beta = -0.06$), we constructed our mediation model by treating having access to health or not as a dichotomous antecedent variable (assuming that patients in Knowledge scenarios had access to the information). Fault for deprivation was modeled as the mediator (Figure 1.3).

Table 1.6 Logistic regression model of decision to give A the kidney in Experiment 2

Odds of Giving A the Kidney

	β	SE	p
Intercept	0.45	0.13	< 0.001
Scenario: Drugs	0.18	0.15	0.22
Scenario: Smoking	−0.04	0.14	0.78
Access Group	−0.06	0.15	0.70
No-Access Group	−0.39	0.14	< 0.001

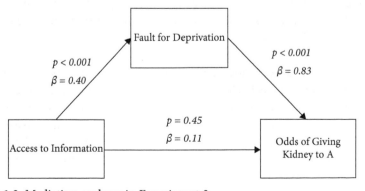

Figure 1.3 Mediation analyses in Experiment 2

Table 1.7 Total, direct and indirect effects in Experiment 2

	β	SE	p
Mediation Analysis: Accessibility to Giving A the Kidney			
Total Effect	0.36	0.12	0.003
Direct Effect Adjusted with Fault for Depriv.	0.11	0.14	0.45
Estimated Indirect Effect through Fault for Depriv.	0.33	NA	NA

We found that the effect of access to information on how often a participant gave the kidney to A was almost entirely mediated by fault for deprivation (Table 1.7). When regressing giving the kidney to A only on access to information, we saw that the latter had a statistically significant *total* effect on the former (p = 0.003, β = 0.36, Table 1.7). However, this effect became insignificant after we accounted for fault for deprivation (p = 0.45, β = 0.11), so its *direct* effect was not statistically significant. Statistically significant relationships were found between access to information and fault for deprivation (p < 0.001, β = 0.40), and between fault for deprivation and the odds of the kidney being given to Patient A (p < 0.001, β = 0.83), thus providing evidence for the presence of mediation. The indirect effect was estimated to be 0.33 (Figure 1.3).

1.5.5 Limitations and Directions for Future Research

In Study 2, the mediation analysis cannot determine what causes what. Participants might judge that it would be B's own fault if B did not get a kidney, and this judgment might then cause them to give the kidney to A. However, they might instead first decide to give the kidney to A, and this might then cause them to ascribe fault to B in order to justify their decision. To determine the causal order, future research could intervene on the order in which participants give their opinions on various issues, including responsibility for deprivation and who should get the kidney.

Study 2 also could not determine whether responsibility is affected by knowledge or access to information that a behavior is risky to kidneys in particular as opposed to being risky to health in general. Future studies could explore how specific information needs to be in order to affect judgments about whether a patient is responsible.

In both studies, our scenarios did not specify the circumstances that led patients to stop or continue the behavior. Patients might stop either because doctors or relatives put pressure on them or because they feel guilty about past behaviors or scared about their future health. These circumstances could influence participants' judgments of responsibility as well as allocations of kidneys, so one direction for future research could involve intervening on the patient's reasons or motives for stopping or continuing the unhealthy behavior.

In both experiments, all of our scenarios were also one-sided: Patient A was always blameless relative to Patient B. In Experiment 2, we also told participants that A and B were on equal footing when it came to life expectancy and prospect for recovery. More realistic scenarios could create a tension between the responsibility and benefit. For instance, consider the role of responsibility in liver allocation. Alcohol-related liver disease is the second leading cause of liver transplantation in the United States and Europe, but some have argued that alcoholics should be deprioritized relative to patients who are not at fault for developing their liver diseases (Moss & Siegler, 1991). Responsibility motivated considerations, in addition to concerns about relapse, have led many programs to require alcoholic patients to abstain from alcohol for six months. However, many patients affected by this rule will not survive this period and would benefit from an earlier transplant, so benefit and urgency seem to dictate that alcoholics should be prioritized (Mathurin et al., 2011). A promising direction for future research is to create a scenario in which participants have to trade responsibility for utility: giving an organ to someone who caused their own disease will lead to better consequences, but doing so might amount to depriving someone who developed the same condition through no fault of their own.

References

American Law Institute. (1962), *Model Penal Code*. American Law Institute.

Baron, RM., and Kenny, DA. (1986), 'The Moderator–Mediator Variable Distinction in Social Psychological Research: Conceptual, Strategic, and Statistical Considerations' in *Journal of Personality and Social Psychology* 51: 1173–82.

Briggs, JD. (2001), 'Causes of Death After Renal Transplantation', in *Nephrology, Dialysis, Transplantation: Official Publication of the European Dialysis and Transplant Association—European Renal Association*, 16/8: 1545–9.

Brown, RCH., and Savulescu, J. (2019), 'Responsibility in Healthcare Across Time and Agents' in *Journal of Medical Ethics*, 45/10: 636–44.

FitzPatrick, WK. (2008), 'Moral Responsibility and Normative Ignorance: Answering a New Skeptical Challenge'. *Ethics* 118: 589–613.

Freedman, R., Schaich Borg, J., Sinnott-Armstrong, W., Dickerson, J.P., and Conitzer, V. (2020), 'Adapting a Kidney Exchange Algorithm to Align with Human Values' in *Artificial Intelligence* 283: 103261. https://arxiv.org/abs/2005.09755

Friesen, P. (2018), 'Personal Responsibility Within Health Policy: Unethical and Ineffective' in *Journal of Medical Ethics* 44/1: 53–8.

Ginther, MR., Shen, FX., Bonnie, RJ., Hoffman, MB., Jones, OD., and Simons, KW. (2018), 'Decoding Guilty Minds: How Jurors Attribute Knowledge and Guilt' in *Vanderbilt Law Review* 71/1: 241–83.

Hayes, AF. (2009), 'Beyond Baron and Kenny: Statistical Mediation Analysis in the New Millennium' in *Communication Monographs* 76/4: 408–20.

Hayes, AF., and Matthes, J. (2009), 'Computational Procedures for Probing Interactions in OLS and Logistic Regression: SPSS and SAS Implementations' in *Behavior Research Methods* 41/3: 924–36.

Mathurin, P., Moreno, C., & Samuel, D. et al. (2011), 'Early Liver Transplantation for Severe Alcoholic Hepatitis' in *The New England Journal of Medicine* 365/19: 1790–800.

Moss, AH., and Siegler, M. (1991), 'Should Alcoholics Compete Equally for Liver Transplantation?' in *JAMA: The Journal of the American Medical Association*, 265/10: 1295–8.

Organ Procurement and Transplantation Network (OPTN). (No date.) https://optn.transplant.hrsa.gov/data/view-data-reports/national-data/

Pickard, H. (2018), 'Responsibility Without Blame'. www.responsibilitywithoutblame.org

Rijnhart JJM., Twisk JWR., Eekhout I., and Heymans, MW. (2019), 'Comparison of Logistic-Regression Based Methods for Simple Mediation Analysis with a Dichotomous Outcome Variable' in *BMC Medical Research Methodology*; 19, 19.

Sher, G. (2009), *Who Knew? Responsibility Without Awareness*. (Oxford University Press).

Sinnott-Armstrong, W. (2019), 'Consequentialism' in E. Zalta (ed.) *Stanford Encyclopedia of Philosophy*. plato.stanford.edu/entries/consequentialism/

United States Renal Data System (USRDS). (2020), *Annual Data Report* https://adr.usrds.org/2020/end-stage-renal-disease/5-mortality

2

Undeserving Patients

Against Retributivism in Health Care

Jeanette Kennett

2.1 Introduction

Societies with universal public health care systems have good reason to reduce the load on their systems by enabling and promoting healthy behaviours, such as regular exercise and a healthy diet, and discouraging unhealthy behaviours such as smoking, excessive consumption of alcohol, and the like. Given the rising cost of, and demands upon, our health care systems, this is an important public health goal.

Holding adults responsible for poor health-related choices and behaviours, by restricting or denying access to medical treatments for conditions that arise from those choices, is periodically raised as an appropriate means of rationing health care resources, and as a tool to encourage individuals to make better health choices (e.g., Donnelly 2017; O'Chee 2017; UniMed 2018). A system which rations medical treatment on the basis of *past*, health-implicating choices allegedly encourages taking responsibility in the *forward*-looking sense for one's health. Arguments for rationing medical treatment in this way often focus on smoking-related conditions, obesity-related conditions and, more recently in the context of COVID-19, illness as a result of failure to vaccinate. These are conditions for which it is argued treatment may justifiably be limited or denied if the person is deemed to have contributed to or caused the health condition by their own behaviour.

Unfortunately, such arguments are unsupported by hard evidence that prospective denial of treatment has any effect on the health-related choices of the targeted populations. If the forward-looking consequentialist argument fails, then we are left with desert-based arguments for service restriction or denial. While such arguments draw upon intuitively appealing claims about fairness, I shall claim that the denial of treatment based on past behaviour is effectively a form of retributive punishment. Punishment rightly carries a particularly heavy justificatory burden that cannot be discharged in health care settings or by health care professionals. Moreover, the division of patients into deserving and undeserving categories aggravates existing health inequities and introduces additional moral harms.

Jeanette Kennett, *Undeserving Patients: Against Retributivism in Health Care* In: *Responsibility and Healthcare*. Edited by: Benjamin Davies, Gabriel De Marco, Neil Levy, Julian Savulescu, Oxford University Press. © Edited by Benjamin Davies, Gabriel De Marco, Neil Levy, Julian Savulescu 2024. DOI: 10.1093/oso/9780192872234.003.0003

2.2 Punishment and Its Purposes

2.2.1 What Is Punishment?

Punishment involves the imposition of treatment, which is intentionally unwelcome, burdensome, or harsh in response to action or actions undertaken by the target, *for which the target is thought to be blameworthy*. Duff (1986: 41) argues that "an essential part of [punishment's] meaning and justification lies in its relation to moral blame."

Punishment is thus a moral and not merely a conventional sanction or penalty, like a parking ticket. As Feinberg (1970: 98) notes, "Punishment is a conventional device for the expression of attitudes of resentment and indignation, and of judgments of disapproval and reprobation, on either the part of the punishing authority himself or of those in whose name the punishment is inflicted. Punishment...has a symbolic significance." Strawson (1974: 23) argues "the readiness to acquiesce in the infliction of suffering on the offender *which is an essential part of punishment*" flows from the moral reactive attitudes of indignation and resentment that we experience when moral demands are disregarded (my emphasis).

Given that punishment involves the intentional infliction of suffering or deprivation, it carries a heavy justificatory burden. Punishment must, at a minimum, be deserved. Hoskins and Duff (2022) distinguish three subsidiary questions relevant to the justification of punishment:

- Who may properly be punished?
- What principles or aims should determine the allocations of punishments to individuals? (To which I would add: and who may properly decide upon and administer the punishment?)
- How should the appropriate amount of punishment be determined?

At the broad level, theories of the *purposes* and principled *justifications* of punishment have divided roughly into the retributivist and the utilitarian. Retributivism offers distinctively desert-based, and so backward-looking, justifications of punishment focused on the agent and the wrongful act or acts for which they are considered blameworthy. On any retributivist account, justified punishment has to be fair, proportionate to the offence, and deserved. Harsh treatment is seen as restoring the balance of rights disturbed by the offence and offering redress for the wrong committed. In communicative versions of the theory, punishment is a critical vehicle for the expression and communication of the community's justified moral condemnation of the offence and the offender.

Second, there are functions of punishment that are primarily utilitarian and forward looking. These are concerned with both individual and general deterrence and with protection of the community. Recently Ellis (2003) argues that the deterrence and community protection functions of punishment gain their legitimacy, not from utilitarian considerations of welfare, but from the fundamental moral right of self-defence of one's significant interests. Punishment achieves this through the deterrence provided by the credible threat of punishment.

Third, a concern with *reform* of behaviour has elements of both the utilitarian and retributive approach. Punishment is supposed to achieve reform by deterrence or by bringing the offender to a sense of their own wrongdoing.

While much discussion of punishment, its aims and justification, has focused upon the criminal justice system it should be apparent that formal and informal systems of punishment are widespread in society. There are punitive sanctions for rule-breaking and other forms of misbehaviour in workplaces, clubs, sport, and interpersonal relationships, to name a few. These may include warnings, reprimands, fines, and demotions but the most severe and final punishment available is exclusion. You may be suspended, fired, evicted, blocked, or shunned. It is my claim that denial of medical treatment, that would otherwise have been offered, *on the basis of supposed desert* is a form of punitive social exclusion. Such denial of treatment comes at considerable cost to patients, exposing them to pain, suffering, disability, and even death. Treatment denial meets the criteria of being unwelcome, harsh, burdensome, and coercively imposed. Moreover, treatment denial typically carries and communicates social judgements of disapproval and reprobation. It thus carries a heavy justificatory burden. Can this burden be discharged?

2.3 Arguments in Favour of Denial or Restriction of Health Care on the Basis of Contributory Behaviour

In overloaded public health care systems, triaging and rationing of services is a fact of life. People cannot be offered every treatment that might be beneficial to them. Questions about what criteria should govern rationing, who should make the decisions, and at what point in the system they should be made are clearly important. I leave most of those questions aside. My focus here is upon arguments that could favour including personal responsibility in the criteria for treatment restriction or denial, and an assessment of personal responsibility as part of the decision-making process.

Buyx (2008) surveys a number of justifications for the incorporation of personal responsibility into rationing criteria. I will briefly examine these justifications before addressing more closely the issue of service denial as unfairly discriminatory and whether it can properly be claimed to constitute punishment of those denied treatment on the basis of their voluntary behaviour.

On a *libertarian* view there is no right to provision of health care by the state; health is a personal responsibility, and it is up to individual citizens to decide how much of their resources to invest in health care. Wealthy individuals can buy the health services they need or desire while poorer people "only have their health-related behaviour as a 'health resource'" (Buyx 2016: 343).

Libertarian views might at first appear irrelevant to states with universal public health care systems where wealth is not a criterion for accessing public health services. However, as will become apparent, the notion of one's health-related choices as a health resource—which may be spent wisely or unwisely—is fundamental to justifications offered for personal responsibility as a rationing criterion and reintroduces into public health care systems the divisions between rich and poor that the libertarian is happy to accept.

According to Buyx, on a *communitarian* view, "The primacy of the health of a community and of the common good justifies that the state or community require individuals to contribute to this common good by showing responsible health behaviours. It also legitimises a public system that exerts pressure and withholds resources in cases where individuals do not comply" (2008: 871). This is both explicitly consequentialist in its focus on community health and the common good as justification for a requirement of personal responsibility, and explicit in its use of denial of treatment as a tool of compliance and as punishment for failure to comply. It has echoes of the self-defence justification for punishment offered by Ellis (2003) in that the targeted behaviour is seen as damaging community values and the defence of those values permits or requires a credible threat of punishment. But if denial of treatment is not effective in promoting personal responsibility can its apparently punitive character be justified?

Here the *luck egalitarian* may provide an answer. On this view, withholding treatment is *not* punitive because society has no obligation to provide health care for conditions arising from voluntary behaviour:

> inequalities warranting compensation are those resulting from factors that individuals have no choice about (so-called "brute luck") As for inequalities resulting from freely chosen behaviour, such as lifestyle choices or risky behaviour (so-called "option luck"), these do not warrant compensation ... personal responsibility is one of the most important criteria for allocation in medicine. A public healthcare system need compensate only for treatment of conditions that do not result from chosen behaviour.   (Buyx 2008: 872)

The luck egalitarian view combines the notion of health-related behaviour as a health resource which may be expended wisely or unwisely, with a fairness criterion. It is a requirement of fairness or equity to compensate/treat people for health conditions that they acquired through no fault of their own. There is no such requirement of fairness to treat people who acquired the same condition through their lifestyle choices, and we may decline to do so without wronging them. In a situation where health care resources are scarce, and we cannot provide treatment for all, fairness requires us to prioritize those whose conditions are a matter of brute luck. Personal responsibility on this view is a backward-looking criterion that allows us to decide what is fair; the view of health care as *compensation* for brute bad luck vitiates any surface appearance of treatment denial as punitive.

This view of luck and fairness is intuitively appealing—or would be if there was a clean distinction between brute luck and option luck that could underwrite the division of patients into deserving and undeserving categories. For brute luck determines and shapes options in myriad ways, and what counts as brute luck in a health context may in any case be contested. When we take into account the social factors that shape health behaviour and health options (the social determinants of health; see Marmot et al. 2008; WHO 2008; Marmot 2018) we should become less confident in our judgements of fairness, and less confident that denial of treatment on the basis of chosen behaviour is not in fact punitive.

The position favoured by Buyx is what she terms solidarity. The obligation to exercise personal responsibility in our health-related choices rests upon a moral requirement of social reciprocity. She says "members of a liberal and solidary society owe one another a reasonable degree of effort and care, which are essential to support and preserve the system and its institutions in the long run." Hence she argues that "a moderate expectation that people contribute towards this system and behave responsibly within it is justified" (Buyx 2008: 873). In her view, taking account of lifestyle choices that have led to ill health is one permissible element in a decision matrix, and could be overridden by other considerations.

The moral case for reciprocal responsibility and mutual care set out by Buyx is compelling. But as a justification for restriction or denial of service it would require a significantly more fair and equal society than is generally the case. In practice, denial of treatment on grounds of health choices and lifestyle tends to further disadvantage marginalized people and communities.

It is to this issue that I now turn.

2.4 Discrimination and Moralized Denial of Treatment

In this section I highlight some of the discrimination and denial of service that currently occurs informally in health care settings. I suggest that denial of service is often based on a view of the worthiness of the patient to receive treatment, rather than on an assessment of their medical need. These are desert-based judgements—they divide patients into deserving and undeserving categories. The effects of this division are experienced as a form of social punishment by those affected. Indeed, the harsh treatment described in some of the examples below is directly and intentionally punitive.

Lancaster et al. (2017) note that the use of stigma in public health campaigns has been defended and cite arguments that "stigmatisation or de-normalisation of alcohol, tobacco and other drugs use might serve a positive public health function, in that such approaches can change social norms" (Bayer 2008; Bayer and Fairchild 2015). Public health campaigns, along with restrictions on smoking in public places, have surely contributed to the marginalization of smoking as an acceptable activity, and to the significant reduction in smoking rates. Given these health benefits to the community, the burden of stigma on individuals who continue to engage in unhealthy behaviours might be considered justifiable on a communitarian or solidarity view of the obligation to look after one's own health to avoid imposing unnecessary health costs on the community.

But the stigmatization of groups and individuals as a public health tool must be used with great care. It is often counterproductive and can bleed into the attitudes and decisions of frontline health care workers in ways that are unfair and even dangerous for the person seeking treatment.[1]

It is notable that those who experience treatment denial or restriction in medical settings already tend to be people with stigmatized characteristics such as obesity, or people who belong to stigmatized groups including people with addiction, trans- and gender-diverse people, and members of minority ethnic and racialized groups. People living in poverty bear the brunt of discrimination since the kinds of health conditions for which people are most likely to be blamed— smoking-related illnesses, obesity-related conditions, conditions related to poor dental health—are diseases of the poor, rather than diseases of affluence. People injured pursuing extreme (and expensive) sports are not denied health care when they injure themselves. Heart disease resulting from executive stress is not frowned upon in the way that the same condition resulting from smoking might be.

It is thus worth surveying the experiences of persons with stigmatized characteristics and conditions who seek health care.

A study of the experiences of *transgender people* found that refusal of care from GPs, psychiatrists, dentists, and other medical specialists was a common experience, with some saying doctors had denied them treatment on moral or religious grounds (Stark 2015).

A national (US) study of over 6,000 transgender adults found that 28 per cent had experienced harassment in medical settings, 19 per cent were refused care, and 2 per cent experienced violence in their doctor's office (Grant et al. 2011).

---

[1] Stigma is clearly counterproductive as an approach to obesity. Weight stigma has been found to be "prospectively related to heightened mortality and other chronic diseases and conditions. Most ironically, it actually begets heightened risk of obesity through multiple obesogenic pathways" (Tomiyama et al. 2018).

"I have been refused emergency room treatment even when delivered to the hospital by ambulance with numerous broken bones and wounds."

"I have been living with excruciating pain in my ovaries because I can't find a doctor who will examine my reproductive organs" (transgender man).

(Grant et al. 2011)

People with *substance use disorders* also regularly experience discrimination in health care settings.

Of 300 injecting drug users interviewed "'ill treatment' had been experienced [at the hands of]...hospital staff (60%), doctors (57%), pharmacists (57%)... dentists (33%), methadone providers (33%), drug treatment services (33%) and community health workers (7%)" (Lancaster et al. 2017).

Persons with a history of injecting drug use are regularly denied care that would be offered were it not for their history.

"I broke my leg in three places after a motorbike accident. At the hospital they wouldn't give me anything other than Panadol, even though the bone was sticking out through the skin."   (AIVL 2011)

"Some [nurses] used to [say], 'We haven't got time for you, there are sicker people than you'."   (AIVL 2011)

"I went to the doctor for antibiotics for an infected and really badly swollen arm. He came out into the waiting room, took one look at me, told me 'I don't treat scum like you', and then picked up my bag and threw it out on the footpath."

(AIVL 2011)

"There's been a couple of situations where I've been at the hospital because I've been really sick with a bad migraine, and what-not, and because it's on my file that I used to use, they won't treat me up there, they just put me down that I'm an ice junkie, you know, and I don't even use the stuff" (Interview 10, male 38 years).

(Brener et al. 2010)

These experiences of refusal of service do seem clearly based on *negative moral assessments* of the person denied service or their condition—the judgement that they are *undeserving* of the medical attention they are seeking because of prior behaviour, group membership, or identity.

Obesity similarly attracts negative moral assessments from health care workers. A 2016 study similarly found that nurse practitioners judged obese persons to be untidy, unhealthy, and less good and successful than others (Ward-Smith and Peterson 2016). A 2015 study reports that "individuals with obesity are frequently the targets of prejudice, derogatory comments and other poor treatment in a variety of settings, including health care" and that "there is a growing body of evidence that physicians and other healthcare professionals hold strong negative opinions about people with obesity." They are seen as "lazy, undisciplined and weak-willed." Physicians have been found to spend less time with obese patients, rate encounters with them as a waste of time, overattribute presenting symptoms to obesity, and fail to refer for diagnostic testing that other patients would receive (Phelan et al. 2015).

These cases should raise concerns about any push to institutionalize a practice of holding people responsible for their health conditions, since it appears that part of what influences the selection of certain conditions for discussions of responsibility for health conditions that might warrant treatment restriction or denial, are moralized judgements, which are not about health but about character. People who smoke, people who are obese, people whose conditions result from drug or alcohol use, are seen as displaying the vices of gluttony, self-indulgence, or weakness of will, which render them *less deserving* of access to scarce health resources. Unlike the hard-driving executive or dedicated athlete, they lack self-discipline and are irresponsibly focused on short-term pleasures over long-term goods.

Moreover, when such harmful judgements of character are made, they tend to influence medical judgements: "people who use drugs become seemingly inseparable from their behaviour and are labelled as 'drug users' or 'addicts' and not, as in other health-related discourse, 'a person [who] has cancer, heart disease, or the flu'" (Link and Phelan 2001: 370).

The medical consequences of moralized refusal of treatment as in the examples provided may be profound. They include a failure to properly investigate symptoms the patient presents with, leading to untreated illness and avoidable suffering; patient avoidance of health care settings; and increased stress and trauma due to experienced stigma (Fruh et al. 2016).

In a report for the Queensland Health Commission, Lancaster et al. (2017) argued that "having been refused medical care or having being [sic] verbally abused are factors associated with people who inject drugs subsequently avoiding health care services."

> You're doing the right thing coming to a doctor, and then that happens, well, it doesn't make you want to go to the doctor and seek help, or go seek help anywhere because you think you're going to get judged everywhere else because of that one thing. So it makes it a lot harder.
>
> (Interview 15, female 27 years)

The authors also note research showing that "when health care workers believe that injecting behaviour is within a client's control or that it is a choice, health care workers *are more likely to assume that health problems are caused by a person's injecting drug use status and not any other underlying health issue*" (Lancaster et al. 2017; my emphasis).

2.5 Is Responsibility-Based Denial of Health Care Punitive?

At this point it might be acknowledged that these patterns of discrimination in health care are troubling but that they do not amount to punishment, and moreover that institutionalizing careful assessments of responsibility for health conditions could even ameliorate the discrimination faced by members of dis-favoured groups and ensure that service restrictions and denial were properly and fully justified. Such assessments could determine, for example, when a person with smoking or obesity or addiction-related illnesses was *not* to blame for their condition or when these factors were *not* causally relevant to their current illness.

On the first point, it is true that decisions to restrict or deny treatment may not be presented in moralized terms. Physicians do not represent their actions as punishing the patient for past behaviour. Rather, the patient with smoking- or obesity-related conditions may be characterized as less compliant, less likely to benefit, less likely to make a full recovery than another patient. The addict presenting with pain is viewed as drug-seeking rather than ill—and thus as not requiring treatment. The decision by an individual health care professional or treating team will usually be couched in medical, not moral, terms. Nevertheless, the evidence is that patterns of discrimination in health care are driven by moralized attitudes to persons and conditions and are experienced by those at the sharp end *as punitive*—exclusion from treatment *is* harsh, unwelcome, and coercively imposed. For even if the patient might not have benefitted as much as another patient from treatment on some metric—perhaps full recovery—this does not mean that the benefit to them would not have been considerable, and subjectively comparable to that gained by other patients. (We are not talking about futile treatment here.) Given that these patients are already disadvantaged a refusal to treat is likely to further entrench health inequalities.

Could unwarranted, moralized discrimination be ameliorated by systematic impartial assessments of contributory responsibility by a medical tribunal? Such assessments, if they are to be fair, would be highly complex and time consuming. What does it mean to be personally responsible for one's health-related condition? Consider addiction-related illnesses and medical needs. Perhaps we can and should distinguish between the innocently addicted—for example those who became involuntarily addicted to prescribed painkillers after an accident or surgery—and those who became addicted to illicit drugs which they voluntarily consumed in pursuit of pleasure. But if this is to hold water, given the high cost to the individual of treatment denial, we would need to assure ourselves that the individual's pursuit of drugs was not related to poverty, trauma, abuse, self-medication for mental illness, the normalization of drug use in their early social environment, or early induction into drug use. The roots of addiction are multiple and the allocation of personal responsibility for a particular health condition, decades into such use, is likely beyond the competence and expertise of the medical staff who make treatment decisions.

Tribunal members would also need a keen awareness of the social determinants of health (Lynch et al. 1997; Martinson 2012; Braveman et al. 2014). People living in poverty are generally at increased risk of experiencing discrimination and denial of service since stigmatized conditions intersect with the privations and stresses of poverty. Moreover, in Britain, the USA, and Australia, poverty itself is stigmatized and is widely represented as reflecting on the character of the person who lives in poverty.

Obesity is a condition that differentially affects poor communities for well-documented reasons. Cost, availability, storage, and preparation of healthy food are factors along with the cost and availability of options for recreation and exercise. Drewnowski and Specter (2004) point out that "the highest rates of obesity occur among population groups with the highest poverty rates and the least education." There is also "an inverse relation between energy density (MJ/kg) and energy cost (US dollars/MJ."

Put simply, junk food is cheaper, more readily available, and more satisfying. Hungry people with limited money understandably choose food that fills them up over expensive fresh fruit, vegetables, and lean protein. As Linda Tirado explains in her essay:

> We have learned not to try too hard to be middle-class. It never works out well and always makes you feel worse for having tried and failed yet again. Better not to try. It makes more sense to get food that you know will be palatable and cheap and that keeps well. Junk food is a pleasure that we are allowed to have; why would we give that up? We have very few of them.   (Tirado 2013)

Smoking rates are also higher in poor communities and given the high cost of smoking, the habit can exacerbate not just smoking-related illnesses, but also reduces the money available for food and other necessities, contributing to other health conditions (de Beyer et al. 2001; Reed 2021).

Surely then, the choice to smoke is both an immediately irresponsible choice and since the longer-term health risks are widely known, counts as the kind of behaviour that would justify restriction or denial of health services to the smoker. But again, the judgement is too quick. Tirado (2013) again:

> I smoke. It's expensive. It's also the best option. You see, I am always, always exhausted. It's a stimulant. When I am too tired to walk one more step, I can smoke and go for another hour. When I am enraged and beaten down and incapable of accomplishing one more thing, I can smoke and I feel a little better, just for a minute.... It is not a good decision, but it is the only one that I have access to.

On a libertarian view, Tirado must reap the consequences of spending her health-related resources unwisely. On the communitarian view she might be considered blameworthy for not considering the costs to the community of her decisions. On the luck egalitarian view, we are not obliged to compensate her if she develops a smoking-related health condition since it results from choice and not brute bad luck. But when we examine the reasoning behind Tirado's choices none of these rationales for treatment denial or restriction seem fair and would only serve to entrench the disadvantages of poverty. The solidarity view favoured by Buyx could not unequivocally blame Tirado since it could be argued that society has failed to stand in solidarity with the poor.

Poor people are responsible agents. They do not lack the relevant capacities for choice and reflection (e.g., Wallace 1994). But they choose from a severely limited set of options and most often their circumstances mean that they do not have the luxury of taking the long view of their choices or confidence in their capacity to secure diachronic goods. They understandably have little confidence in their ability to exert control over their circumstances. Poor people may bear *causal* responsibility for health-related conditions arising from their decisions but, since denial of treatment is effectively punitive—and often harshly so—it could only be justified if they were also *blameworthy* for their decisions. They must be *morally culpable*. I accept that there will be cases where people are culpable in the required sense for their health conditions. Perhaps some people who are anti-vaccination meet a threshold for culpability. But I doubt that medical professionals are equipped to make such determinations or should be required to do so.

It could be argued, reasonably enough, that courts make determinations of responsibility all the time, including in cases where people have limited options. But courts must satisfy themselves of guilt beyond reasonable doubt before they inflict punishment, and the accused person has an advocate. The infliction of punishment is also governed by strict sentencing criteria and must take account of various mitigating circumstances which may lessen the offender's moral culpability.[2] Should health services set up quasi-hearings to determine if someone is not just causally responsible, but blameworthy, for their health condition and provide for independent patient advocates? And would we do it for *all* health conditions that do not clearly result from brute luck or just some of them? This seems like an expensive, time-consuming, and cumbersome process. Moreover, such a process would not only have to decide on the level of responsibility the patient bore but also what would be a proportionate response in terms of restrictions on treatment offered. *How much* ongoing, *preventable*, ill health and suffering is warranted by the patient's past voluntary behaviour? Is, for example, the suffering

---

[2] It should also be noted that poor people and those with mental health conditions are disproportionately represented in courts and in jails. The justice system tends to replicate and perpetuate disadvantage. Advocates of a personal responsibility approach to provision of health services may achieve the same.

resulting from decisions not to offer adequate pain relief to a person with a history of drug use, a fair, proportionate, and warranted response to that history?

Given these significant practical problems, advocates of responsibility as a criterion for access to medical treatment cannot mean to recommend processes akin to an investigation or trial of the presenting individual. More likely, they envisage that considerations of responsibility will inform policy or institution-level decisions about whom and when to treat, involving something like the provision of a decision matrix to triage staff, or the delegation of the decision to an AI programme to ensure impartiality. But decision matrices and AI programmes tend to import discriminatory patterns and reproduce discriminatory outcomes (Ferrer 2021). If one of the aims of the matrix is to restrict the treatments offered to those the matrix deems causally and morally responsible for their condition it will be punitive in effect and perceived to be so.

2.6 Conclusion: Eschewing Punishment: Responsibility Without Blame?

Is there a role in health care for an emphasis on personal responsibility for one's health that can avoid the retributive attitudes and punitive outcomes outlined above? In this final section I briefly outline Hanna Pickard's responsibility-without-blame approach and suggest that this is consistent with harm minimization practices and policies that have proven successful in relation to addiction and in reducing transmission of infectious diseases (e.g., HIV/AIDS) and provides a useful non-punitive model for health care professionals seeking to assist patients to make lifestyle changes that will improve their health.

Pickard, in a discussion of addiction, acknowledges both the agency of the individual seeking treatment for (or with) stigmatized conditions and the often-difficult circumstances and constrained nature of their choices. Her position sharply distinguishes acknowledging patients' agency over their harmful choices from the "emotions and attitudes constitutive of affective blame" (2017, fn. 17). According to Pickard, the aim of good clinical practice should be "to support and empower people to make different choices" (175). In order to do this, we must "mobilize a sense of agency and empowerment" (176). But affective blame under-cuts good clinical practice. It

> undermine[s] the capacity of responsibility and accountability to enable change and empower, because of its propensity to make patients feel rejected, worthless, ashamed and uncared for, thereby rupturing the therapeutic relationship as well as damaging any sense of hope for the future they might otherwise have, and, correspondingly, any motivation or belief that they really can overcome their difficulties.   (175)

Our belief in our own moral value and our efficacy as agents is strongly affected by how others treat us, and by the political and social narratives that shape our institutions and practices. Preventative and ameliorative health measures will be most successful when they engage and respect the agency of the patient. A person may not be ready to give up injecting drug use, for example, but may be willing to discuss and implement safer, more hygienic practices. The person living in poverty might be able to make some changes to their diet consistent with their budget without going hungry, or to consider feasible methods for dealing with stress that might allow them to reduce smoking. The purpose of probing the personal and social context for the patient's clinical presentation should be to tailor best treatment not to allocate responsibility or divide patients into deserving and undeserving categories.

As Pickard argues, blame is counterproductive. And it is so, both because of how it is received by the patient and how it disposes the clinician. When clinicians blame patients for their condition they will be disposed, as we have seen, to practice differently. They will examine less, provide different and lower quality treatment, or exclude the patient from treatment altogether. Patients who are blamed, rightly experience such treatment as punitive. I have argued that restricting or denying medical treatment on the basis of judgements of the blameworthiness of the patient meets the criteria for retributive punishment and is not justified by communitarian, egalitarian, or solidarity considerations of fairness. It reinforces social stereotypes and stigma, entrenches health inequalities, and is inappropriate to good medical practice. As such it should be combated, not formalized. We should be particularly wary about entrenching existing stigmatization at a structural and institutional level through triaging policies that selectively exclude some patients for treatment that would benefit them and that they would otherwise have been offered, on the grounds that they bear responsibility for their condition.

References

AIVL. (2011), *"Why wouldn't I discriminate against all of them?" A Report on Stigma & Discrimination Towards the Injecting Drug User Community* (Australian Injecting & Illicit Drug Users League).

Braveman, P., and Gottlieb, L. (2014), "The social determinants of health: It's time to consider the causes of the causes," in *Public Health Reports*, 129/suppl 2: 19–31.

Brener, L., Von Hippel, W., Kippax, S., and Preacher, K.J. (2010), "The role of physician and nurse attitudes in the health care of injecting drug users," in *Substance Use & Misuse*, 45/7–8: 1007–18.

Buyx, A. (2008), "Personal responsibility for health as a rationing criterion: Why we don't like it and why maybe we should," in *Journal of Medical Ethics*, 34: 871–4.

Buyx, A. (2016), "Allocating healthcare resources: The role of personal responsibility," in J. Gunning, S. Holm, and I. Kenway (eds). *Ethics, Law and Society, Vol. 4* (Routledge), 341–50.

de Beyer, J., Lovelace, C., and Yürekli, A. (2001), "Poverty and tobacco," in *Tobacco Control*, 10: 210–11.

Donnelly, L. (2017), "NHS provokes fury with indefinite surgery ban for smokers and obese," in *The Telegraph*, October 17. Available at https://www.telegraph.co.uk/news/2017/10/17/nhs-provokes-fury-indefinite-surgery-ban-smokers-obese/

Drewnowski, A., and Specter, S. (2004), "Poverty and obesity: The role of energy density and energy costs," in *The American Journal of Clinical Nutrition*, 79/1: 6–16.

Duff, A. (1986), *Trials and Punishments* (Cambridge University Press).

Ellis, A. (2003), "A deterrence theory of punishment," in *The Philosophical Quarterly*, 53/212: 337–51.

Feinberg, J. (1970), *Doing and Deserving: Essays in the Theory of Responsibility* (Princeton University Press).

Ferrer, X. (2021), "Bias and discrimination in AI: A cross-disciplinary perspective: IEEE Technology and Society," in *Technology and Society*, August 7. Available at https://technologyandsociety.org/bias-and-discrimination-in-ai-a-cross-disciplinary-perspective/ [Accessed 02/09/2022].

Fruh, S., Nadglowski, J., Hall, H., Davis, S., Crook, E., and Zlomke, K. (2016), "Obesity stigma and bias," in *The Journal for Nurse Practitioners*, 12/7: 425–32.

Grant, J., Lisa, A., Mottet, J. et al. (2011), *Injustice At Every Turn: A Report of the National Transgender Discrimination Survey* (National Center for Transgender Equality and National Gay and Lesbian Task Force).

Hoskins, Z., and Duff, A. (2022), "Legal punishment," in *The Stanford Encyclopedia of Philosophy* (Summer 2022 edition), E.N. Zalta (ed.). Available at https://plato.stanford.edu/archives/sum2022/entries/legal-punishment

Lancaster, K., Seear, K., and Ritter, A. (2017), *Reducing Stigma and Discrimination for People Experiencing Problematic Alcohol and Other Drug Use: A Report for the Queensland Mental Health Commission*. DPMP Monograph No. 26 (NDARC: UNSW).

Link, B., and Phelan, J. (2001), "Conceptualizing stigma," in *Annual Review of Sociology*, 27: 363–85.

Lynch, J., Kaplan, G., and Shema, S. (1997), "Cumulative impact of sustained economic hardship on physical, cognitive, psychological, and social functioning," in *New England Journal of Medicine*, 337: 1889–95.

Marmot, M. (2018), "Health equity, cancer, and social determinants of health," in *The Lancet Global Health*, 6/S29.

Marmot, M., Friel, S., Bell, R., Houweling, T., Taylor, S., and Commission on Social Determinants of Health (2008), "Closing the gap in a generation: Health equity through action on the social determinants of health," *Lancet*, 372/9650: 1661–9.

Martinson, M. (2012), "Income inequality in health at all ages: A comparison of the United States and England," in *American Journal of Public Health,* 102: 2049–56.

O'Chee, B. (2017), "Should smokers be denied health care?," in *Sydney Morning Herald,* October 24. Available at https://www.smh.com.au/lifestyle/health-and-wel lness/should-smokers-be-denied-health-care-20171024-p4ywmi.html   [Accessed 02/09/22].

Phelan, S., Burgess, D., Yeazel, M., Hellerstedt, W., Griffin, J., and van Ryn, M. (2015), "Impact of weight bias and stigma on quality of care and outcomes for patients with obesity," in *Obesity Reviews,* 16/4: 319–26.

Pickard, H. (2017), "Responsibility without blame for addiction," in *Neuroethics,* 10: 169–80.

Reed, H. (2021), *Estimates of Poverty in the UK Adjusted for Expenditure on Tobacco: 2021 Update* (Landman Economics). Available at https://ash.org.uk/uploads/Smoking-and-poverty-July-2021.pdf [Accessed 02/09/22].

Stark, J. (2015), "'We don't look after people like you.' Transgender people refused medical care," in *Sydney Morning Herald,* October 15. Available at https://www.smh.com.au/national/we-dont-look-after-people-like-you-transgender-people-refused-medical-care-20151015-gk9ss1.html. [Accessed 5/08/22]

Strawson, P.F. (1974), *Freedom and Resentment and Other Essays.* Methuen.

Tirado, Linda (2013), "This is why poor people's bad decisions make perfect sense," *HuffPost Latest News.*

Tomiyama, A., Carr, D., Granberg, E. et al. (2018), "How and why weight stigma drives the obesity 'epidemic' and harms health," *BMC Med,* 16: 123. https://doi.org/10.1186/s12916-018-1116-5

UniMed (2018), "Debate: Should the NHS fund treatment for smokers?" https://www.uni-med.co.uk/debate-should-the-nhs-fund-treatment-for-smokers/

Wallace, R. (1994), *Responsibility and the Moral Sentiments* (Harvard University Press).

Ward-Smith, P., and Peterson, J. (2016), "Development of an instrument to assess nurse practitioner attitudes and beliefs about obesity," in *Journal of the American Association of Nurse Practitioners,* 28/3: 125–9.

WHO (2008), *Closing the Gap in a Generation: Health Equity through Action on the Social Determinants of Health—Final Report of the Commission on Social Determinants of Health* (WHO Press). Available at https://www.who.int/publications-detail-redirect/WHO-IER-CSDH-08.1 [Accessed 02/09/2022].

3

Moral Responsibility Scepticism, Epistemic Considerations and Responsibility for Health

Elizabeth Shaw

3.1 Introduction

It has become increasingly common within the United Kingdom's National Health Service (NHS) for patients who have engaged in certain types of unhealthy behaviour to face restrictions on their treatment. Such policies have attracted considerable academic attention. The idea that people should be held responsible for their bad health decisions is often associated with "luck egalitarianism" (see, e.g., Albertsen and Knight 2015). Broadly speaking, this is the idea that inequalities that are purely down to luck are unjust, but that inequalities that reflect responsible agents' choices can be justifiable (Lippert-Rasmussen 2016). This chapter will discuss some difficulties faced by luck egalitarians (and proponents of similarly responsibility-sensitive approaches) in the context of responsibility for health, highlighting the implications of moral responsibility scepticism in this context. Theorists who have discussed the practical implications of moral responsibility scepticism have focused on criminal punishment. This chapter instead examines the implications of moral responsibility scepticism for whether patients should face penalties for unhealthy lifestyle choices.[1] When discussing punishment, moral responsibility sceptics often invoke an epistemic argument, maintaining that there are at least serious doubts about whether people are morally responsible (in the sense required for retributive punishment) and that, in view of this uncertainty, retributive punishment is unjust, given the serious harm it inflicts on offenders.[2] This chapter argues that this type of reasoning also implies that we

---

[1] I am grateful to Gabriel De Marco and Ben Davies for their helpful comments on an earlier draft of this chapter.
One of the few discussions of this topic from the perspective of moral responsibility scepticism is Levy (2018).

[2] Retributivism is the view that punishment is justified because offenders deserve to suffer hardship in return for their morally wrongful acts. There are different varieties of retributivism, cf. Moore (1997) and Duff (2001). For an overview see Walen (2020).

Elizabeth Shaw, *Moral Responsibility Scepticism, Epistemic Considerations and Responsibility for Health* In: *Responsibility and Healthcare*. Edited by: Benjamin Davies, Gabriel De Marco, Neil Levy, Julian Savulescu, Oxford University Press. © Edited by Benjamin Davies, Gabriel De Marco, Neil Levy, Julian Savulescu 2024. DOI: 10.1093/oso/9780192872234.003.0004

should not take patients' responsibility for their poor health into account when deciding whether to give these patients treatments and that the health system should not impose significant penalties on individuals for harming their own health.

After first briefly describing examples of responsibility-sensitive health care policies, Section 3.3 explains the connection between such policies and luck egalitarianism and will outline some criticisms that have been made of this view. Section 3.4 argues that (prima facie) culpability-based desert stands a better chance of helping luck egalitarians (and others with related views) to justify responsibility-sensitive health policies than some alternative approaches. However, Sections 3.5, 3.6 and 3.7 cast doubt on the idea that responsibility-sensitive health care policies based on this kind of desert are justifiable. Section 3.5 indicates reasons for doubting whether people are morally responsible for harming their own health. Section 3.6 argues that there are reasons for doubting whether harming one's own health is morally wrong. Section 3.7 argues that there are reasons for doubting that significant penalties for harming one's health would be proportionate (although very minor burdens might be less open to challenge).

3.2 Examples of Responsibility-Sensitive Health Care Policies

Such "responsibilising" policies include limiting access to treatments or deprioritising treatments for patients who are judged to be at fault for damaging their health; punishing people for poor health decisions by making them contribute financially to the costs of their treatment; and communicating stigmatic messages in health promotion campaigns about people whose illnesses are judged to be due to "bad" lifestyle choices. In England, policies requiring smokers and obese patients to take part in weight loss or smoking cessation programmes prior to elective surgery were increasingly adopted by Clinical Commissioning Groups (CCGs), organisations within the National Health Service (NHS) that made decisions about the provision of services to local populations.[3] Over a third of CCGs restricted access to some form of elective surgery for smokers or obese patients until they completed a fitness programme (Pillutla et al. 2018). The types of restriction vary (for an example of one such policy see East and North Hertfordshire CCG 2017). One justification for such policies is the forward-looking idea that patients will not sufficiently benefit from surgery until they stop smoking or lose weight. That justification has been criticised elsewhere (Royal College of Surgeons 2016). The current chapter focuses on

---

[3] CCGs were established under the Health and Social Care Act 2012. There were 106 CCGs in England prior to their abolition by the Health and Social Care Act 2022, which caused them to be subsumed into Integrated Care Systems.

whether responsibilising policies could be justified with reference to the backward-looking consideration that the relevant patients were responsible for engaging in behaviour that risked harming their health. However, it will suggest that other types of policy, such as taxes on cigarettes and minimum alcohol pricing, are not as open to challenge. My main aim is to discuss the moral rationale underlying responsibility-sensitivity in health care, rather than to examine the details of specific policies. Like Levy (2018), my conclusions will therefore be largely negative, that is, identifying which policy-rationales are most open to challenge rather than setting out positive policy proposals. Discussion of these underlying moral rationales is a precondition for defensible law and policy. As Levy (2018: 460) writes, "effective policy development requires the expertise of many different kinds of people with different kinds of disciplinary backgrounds. Offering purely negative advice can be an important contribution to such development."

3.3 Responsibility-Sensitive Health Care Policies and Luck Egalitarianism

The idea that people should be held responsible for their bad health decisions is part of a broader debate about whether (or how far) principles of distributive justice should be sensitive to considerations of personal responsibility (Davies 2022: 113). Proponents of responsibility-sensitivity often endorse "luck egalitarianism" (Albertsen and Knight 2015). In general, luck egalitarians claim that inequalities that are purely down to luck (i.e., circumstances over which people lack control) are unjust, but that inequalities that reflect choices for which people are responsible can be justifiable (Lippert-Rasmussen 2016). Luck egalitarians often appeal to the widespread intuition that it is fair for people to suffer or benefit from the foreseeable consequences of their own choices and unfair to ask others to bear the costs of these choices (Abad 2011), an intuition reminiscent of the saying "he's made his bed, so he should lie in it". This chapter discusses some difficulties faced by luck egalitarianism (and other responsibility-sensitive approaches) in the context of responsibility for health, focusing on the implications of free-will scepticism.

The idea that the person who is "responsible" for an action must "take responsibility" for the results rather than expecting others to do so may seem prima facie plausible, but it arguably equivocates between two senses of responsibility: "outcome responsibility" and "liability–responsibility". Drawing on the work of Hart (1968: 210–37), Vincent (2009: 46) explains that outcome responsibility is a backward-looking concept describing *past* actions as attributable to an agent. In contrast, a claim about liability responsibility makes a "prescription for the *future*" concerning what an agent must now do (or undergo) to "take responsibility" for what has happened.[4] Applying these concepts, a more precise restatement of the luck egalitarian claim is: "if you are outcome responsible for something then you (and not others) should take liability responsibility for it" (Vincent 2009: 46). However, Vincent argues that this restatement reveals that the consequent (referring to liability responsibility) does not follow automatically from the antecedent (referring to outcome responsibility). For the conclusion about liability to follow logically from premises about outcome responsibility, one would need to add "normative bridging premises". To bridge the gap, Vincent (2009: 47) argues, we would need "reactive norms", i.e., norms that "govern our reactions to outcome responsible parties". Reactive norms are familiar from the context of criminal liability: to bridge the gap between statements such as "X is *outcome responsible for crime Y*" and "X is justifiably *liable* to receive punishment Z", utilitarian norms might be invoked that appeal to the deterrent value of imposing punishment Z on those who are outcome responsible for crime Y. An alternative reactive norm might be the retributive claim that X *deserves* punishment Z. Vincent claims that utilitarian and retributive reactive norms are also likely to feature (implicitly or explicitly) in domains beyond criminal justice when attempting to link outcome responsibility with liability responsibility. This chapter discusses problems with trying to use desert-based arguments to bolster claims that people should "take responsibility" for their own health.[5]

Examples lend intuitive support to the argument that conclusions about liability responsibility do not follow *automatically* from claims about outcome responsibility. Consider Bob. On one occasion, Bob imprudently decides not to fasten his seatbelt; his car skids on an icy road and he hits a wall. Bob sustains severe brain injuries, which he would have avoided if he had been wearing a seatbelt. Bob will die unless he receives health care. Giving him access to public resources would seem to go against the luck egalitarian maxim, because giving him these resources would require *others*—taxpayers—to "take (liability) responsibility" for the consequences of an action for which *Bob* was (outcome) responsible.[6] However, this case raises questions about the logic of luck egalitarianism, because even if we agree that Bob is outcome responsible for his injuries, due to his decision not to wear a seatbelt, this does not entail that that he should bear liability responsibility for all the consequences and be left to die. Examples like this are used to illustrate the "harshness objection" against luck egalitarianism: the objection that it is sometimes too harsh to force people to suffer all the consequences of their responsible actions (see Gabriel De Marco's chapter in this volume). Such examples also highlight the logical gap between outcome responsibility and

---

[4] Emphasis added.
[5] Utilitarian arguments for responsibility-sensitive health policies are not addressed.
[6] Luck egalitarians might respond in different ways to this example depending on the version of luck egalitarianism that they endorse and how they respond to the harshness objection.

liability responsibility. Outcome responsibility does not seem capable of bearing the weight that the luck egalitarian wants to put on it by itself.

Luck egalitarians might try to fill in the missing premise(s) in a way that could both (a) bridge the gap between outcome responsibility and liability responsibility, and (b) enable luck egalitarianism to avoid excessively "harsh" outcomes, such as leaving Bob to die. It might be thought that the missing premise(s) should feature the concept of "desert". In fact, Segall (2015: 355) observes that "many luck egalitarians invoke desert, whether explicitly or implicitly". As I discuss below, desert is invoked by other responsibility-sensitive distributive principles besides luck egalitarianism. I focus on a concept of desert that I call "culpability-based desert".

3.4 Luck Egalitarianism, Culpability-Based Desert and Harming One's Health

Culpability-based desert functions in the following way. To say that someone *deserves* to bear burdensome consequences[7] in virtue of having performed a risky act/omission requires meeting at least three conditions: (1) that the person was morally responsible for performing the risky act/omission, (2) that the act/omission was morally wrong and (3) that the burdensome consequences are proportionate to person's moral responsibility and the gravity of the wrongdoing.[8] My description of culpability-based desert is similar to Eyal's account of the relationship between culpable risk-taking and disadvantages, where "'Culpable' choice is understood as a free and at least somewhat morally wrong choice" (Eyal 2006: 6).

I focus on the culpability-based conception of desert for two reasons.[9] First, examples involving burdens being placed on individuals because they meet the requirements of culpability-based desert are seen by philosophers and lay people as central rather than borderline cases of "desert" (Levy 2018; Cushman 2008). Competent language-users could reasonably disagree about whether borderline cases really are examples of "desert"; or else they could agree that such examples

---

[7] "Burdensome consequences", or "disadvantages" to use Eyal's (2006) terminology, could refer to the natural consequences of one's act (e.g., damage to one's health) or a burden imposed by society (e.g., being made to contribute to the costs of one's treatment or being made to wait longer for treatment) in lieu of having to suffer the full natural consequences of the act.

[8] This chapter focuses on deserving burdensome consequences, rather than deserving or failing to deserve benefits. One might argue that being denied (on desert grounds) access to resources above a certain level is a case of "failing to deserve a benefit" rather than deserving a burden and that one does not need to have committed a morally wrongful act, to "fail to deserve a benefit"; perhaps failing to perform a praiseworthy act would be enough. This chapter, assumes that health care is so fundamental that the disadvantages imposed by responsibility-sensitive health care policies should be considered burdens rather than failures to benefit.

[9] For an overview of different accounts of desert see: Feldman and Skow (2020).

only involve "desert" in a loose sense.[10] In central cases, the concept of desert is used in a way that conforms with a core meaning of the term.[11] Culpability-based desert is not considered *a loose sense* of "desert" even by its critics, such as moral responsibility sceptics. Moral responsibility sceptics accept that examples of culpability-based desert involve what we standardly mean by "desert". They just argue that, given certain facts about the world, human beings cannot meet the requirements for this type of desert. A second reason for focusing on culpability-based desert is that this concept is a relatively plausible candidate (compared to some other conceptions of desert) for a principle that could both (1) bridge the gap between outcome responsibility and liability responsibility and (2) exclude harsh cases. Ultimately, however, I cast doubt on whether responsibility-sensitive health care policies based on this kind of desert are justifiable.

Some theorists who discuss desert in the context of luck egalitarianism empha-sise that desert need not involve moral responsibility. Abad (2011) and Brouwer and Mulligan (2019) give the example of the most beautiful person *deserving* to win a beauty contest. This is only a borderline case of desert, involving desert in a loose sense; although it still shares some formal features with culpability-based desert. For example, there is a "fittingness" relationship between the desert-object (winning the contest) and the desert-basis (beauty); for culpability-based desert, there is a "fittingness" relationship between the desert-object (the burdensome consequences) and the desert-basis (being morally responsible for a wrongful act), which in the context of culpability-based desert could be explained in terms of proportionality.[12] The example of the beauty-contest winner only involves desert in a loose sense of the word because competent language-users disagree about whether it is a correct usage of the word "desert" to say that the most beautiful contestant deserves to win. While some might argue that the beauty-contest winner deserves to win because she satisfied the competition criteria of the competition, others might insist that, it is a misuse of the word to say that she "deserves" it, because, for example, she was just lucky to look that way. While Abad (2011) and Brouwer and Mulligan (2019) consider the case of the beauty-contest winner to be an example of desert, many writers would not, as, "by many accounts, desert seems to require agency, or the capacity to control one's own actions" (Ristroph 2006: 1301).[13] It becomes clearer that the beauty-contest

---

[10] There are different analyses of "borderline cases" Sorensen (2022). My description of borderline cases in terms of reasonable disagreements between language users is similar to Parikh (1994).

[11] For further discussion of the "core meaning" of terms, see Lasersohn (1999). Sometimes one word may have multiple senses, with each sense having a core meaning and borderline cases.

[12] For a discussion of such formal features of desert see Feinberg (1970: 221–50).

[13] It might be objected that such accounts are too narrow. I am not insisting that agency/control is a necessary condition for all senses of the word "desert". Rather, I maintain that generally, there is likely to be more agreement that someone "deserves" an outcome in examples where they exercise agency/control over the desert-base than in cases where they don't, and that the former are more central cases of desert. I also maintain that accounts of desert that require agency/control are of most relevance to desert-based rationales for health care policies that impose severe consequences on individuals.

example involves a loose sense of desert when the example is redescribed in a way that does not emphasise a label, such as "most beautiful", but instead emphasises the distribution of significant resources (which is the context most relevant to the current discussion). If we assume that competitions that award resources based on attractiveness are unobjectionable, it still seems to involve an odd word-choice to say that Zebedee *deserves* to receive a £1,000,000 prize because he had the genetic good fortune to have a conventionally attractive chin, whereas Xavier with his unconventional chin *deserves* to walk away penniless. I suggest that other supposedly morally neutral uses of the term "desert", such as an athlete deserving a medal, come closer than the beauty-contest example to a core meaning of desert because they have features that resemble being morally responsible (voluntarily choosing to train hard) for morally good actions (sport is assumed to be a valuable activity).

If "desert" is to provide a neat solution to the luck egalitarian's problems, the relevant sense of "desert" should help both (1) to rule out harsh cases and (2) bridge the gap between outcome responsibility and liability responsibility. Some senses of desert can do one job but not both. For example, "personhood-based desert" (Vilhauer 2013) can rule out harsh cases but cannot link outcome responsibility and liability responsibility. Personhood-based desert refers to what someone deserves in virtue of being a person, covering a diverse range of entitlements such as the right to fresh water or to receive a fair trial. We cannot forfeit through our actions things that we deserve based on our personhood. In contrast, action-based desert refers to what people deserve in virtue of their actions and omissions, not their status as persons. The concept of personhood-based desert might help explain why no one should be denied treatment for their life-threatening injuries: everyone deserves basic health care in virtue of being a person. Thus, personhood-based desert might rule out harsh cases. However, personhood-based desert cannot support the claim that people must take liability responsibility for the outcomes of their *actions*, because this claim invokes "action-based desert". Culpability-based desert is a type of action-based desert.

Other types of action-based desert (distinct from culpability-based desert) might link outcome responsibility and liability responsibility but cannot rule out harsh cases. According to one account, desert is defined in relation to social contribution (Brouwer and Mulligan 2019).[14] However, if this sense of desert is truly distinct from and does not implicitly rely on culpability-based desert, then it cannot rule out *all* cases which seem harsh (at least partly) for forcing people to suffer burdensome consequences when they were not morally responsible for running wrongful risks, or when running the risk was morally permissible or even praiseworthy, or when the burdensome consequences were disproportionate to their wrongdoing.

---

[14] Brouwer and Mulligan (2019: 1) argue that "luck egalitarians should consider supplementing their theory with desert considerations. Or, even better, consider desertism as a superior alternative to their theory."

It might be objected that desert does not need to serve both as a way of linking the two types of responsibility and as a way of ruling out harsh cases. Perhaps if desert just served the "linking" function, another principle could perform the function of ruling out harsh cases. Indeed, some luck egalitarians are pluralists who supplement their theory with a range of other values, such as beneficence, or concern for the worst off (e.g., Arneson 2000) which might be invoked to explain why people should not be forced to suffer severe consequences of their non-culpable choices. In reply, first, as Brouwer and Mulligan (2019: 2283) argue, "Occam's Razor tells us that we should not multiply entities without need". If we can find a concept of desert that performs both functions, that is an advantage. Second, Albertsen and Nielsen (2020) suggest that Anderson's (1999) claim that luck egalitarianism is internally inconsistent poses a problem for this pluralist solution. According to Anderson, egalitarianism assumes that each citizen should be treated with equal concern and respect, but a view that implies that someone who chose to run certain risks should be left to sink "to the depths" of severe misfortune fails to treat that person with equal concern and respect. Thus, if luck egalitarianism implies that people should suffer such harsh consequences, it cannot be rescued by invoking some other principle within a pluralist framework, because it has contradicted its initial claim to be an *egalitarian* theory. I would add that typically, in situations when one moral consideration is outweighed by another when making all-things-considered judgements, the initial moral consideration should have some intuitive force. We should feel the intuitions pulling in different directions but recognise that one ultimately has greater weight. However, I suggest there is no intuitive force to the initial claim that someone should be left to suffer severe misfortune because they non-culpably chose to run certain risks.

In contrast to these other senses of desert, culpability-based desert seems prima facie to have potential both to help justify linking outcome responsibility and liability responsibility, and to rule out harsh cases. The luck egalitarian idea that people should take liability responsibility for the consequences of actions for which they are outcome responsible seems to gain stronger intuitive support when we consider examples involving culpability-based desert, compared to cases where the prerequisites for this sense of desert are not met. Compare the car-crash victim, Bob (who merely failed to fasten his seatbelt and as a result sustained brain injuries and who does not meet the preconditions for culpability-based desert) with Celandine, who arguably meets those preconditions. Imagine that Celandine was fully morally responsible[15] for trying to run over an innocent

---

[15] There are different accounts of moral responsibility with different prerequisites, e.g., libertarian accounts which require indeterminism (e.g., Kane 1998) and compatibilist accounts which do not (e.g., Fischer and Ravizza 1998) and some theorists deny that anyone is ever morally responsible (e.g., Pereboom 2001; Caruso 2021a, 2021b). For the purposes of the example, I am relying on the idea that *if*

person; in her attempt to do this, Celandine crashed into a wall and sustained severe injuries. Although controversial, it seems more intuitively plausible to make Celandine bear (some of) the consequences of her actions than to make Bob do so (Schneiderman and Jecker 1996; West et al. 2003; Schneiderman 2011). Bob does not meet the prerequisites for culpability-based desert, because it is questionable whether failing to fasten one's seatbelt is morally wrong or proportionate to any consequence relating to the health care he receives. If we remove prerequisites for culpability-based desert (e.g., by also specifying that Bob was not morally responsible for failing to fasten his seatbelt, because he was acting under hypnosis) it becomes even less plausible that he should bear any burdensome consequences. Although many would consider it too harsh to allow Celandine to suffer all the consequences of her actions, and be left to die, theorists who invoke culpability-based desert might have the resources to explain why this is so, by referring to their proportionality scale. They might argue that no wrongful act is proportionate to the outcome of being completely denied medical treatment and being left to die by the roadside. They might argue that a proportionate response would be to deprioritise Celandine's treatment in a triage situation.[16] For example, some theorists (Gold and Strous 2017) and medical professionals (Gold et al. 2021) think that after a terrorist incident victims should be treated before the injured terrorist. Alternatively, Celandine might be required to pay some financial contribution toward her health care.

So far, I have argued that (prima facie) culpability-based desert stands a better chance of helping to justify responsibility-sensitive health policies than some alternative approaches. The luck egalitarian's attempt to infer liability responsibility from outcome responsibility fails to spell out certain premises. With the help of culpability-based desert these missing premises might be filled in:

P1:   X is morally responsible (to W extent) for course of conduct Y, which carries a risk of harm to X's health.

P2:   Course of conduct Y is morally wrong.

P3:   People who are morally responsible (to W extent) for morally wrongful course of conduct Y deserve to suffer proportionate burdensome consequence Z.

C: X is liable to undergo Z.

---

people can sometimes be fully morally responsible and *if* Celandine meets whatever conditions the reader believes would be needed for full moral responsibility, then it is more plausible that Celandine should take the consequences of her actions, than if she did not meet these conditions.

[16] Proportionality is functioning as what Olsaretti (2009: 167) calls "a principle of stakes", i.e., "an account of what consequences can justifiably be attached to features that are the appropriate grounds of responsibility".

Although it has more prima facie plausibility than other responsibility-sensitive approaches, ultimately this approach is problematic. The remainder of this chapter casts doubt on the idea that responsibility-sensitive health care policies based on culpability-based desert are justifiable. I challenge each of P1–P3, claiming respectively that there are reasons for doubting whether people are morally responsible for harming their own health, whether harming one's own health is morally wrong and that significant penalties for harming one's health would be a proportionate response.

3.5 Are People Morally Responsible for Harming Their Health?

A number of theorists have defended moral responsibility scepticism (e.g., Honderich 1993; Strawson 1994, 2015; Pereboom 2001; Callender 2010; Waller 2011; Vilhauer 2013; Shaw 2014; Alces 2018; Corrado 2018; Levy 2018; Focquaert 2019; Caruso 2021b). The sense of moral responsibility at issue in this literature is that which is relevant to culpability-based desert. Moral responsibility in this sense is backward-looking. Holding someone morally responsible in this sense implies that, because of a past action, they deserve a negative response, such as blame or punishment, solely because of the nature of the action and irrespective of consequentialist considerations. Moral responsibility sceptics typically endorse forward-looking senses of responsibility, such as Pereboom's (2001, 2021) account, which aims at improving an individual's future conduct by, for example, engaging in dialogue with the individual about a wrongful act that they have committed, drawing the individual's attention to what that action says about their character, providing reasons for acting differently, and so on. Moral responsibility sceptics also insist that wrongdoers (e.g., rapists or murderers) who pose a risk to others should not be allowed to roam free. However, the justification for restricting their liberty is based on forward-looking considerations, such as social protection, not on culpability-based desert. Moral responsibility sceptics would oppose penalties for self-harm based on the self-harmer's supposed culpability.[17] Some moral responsibility sceptics might endorse penalties based on utilitarian considerations, but these are beyond the scope of this chapter.

A debate has raged for millennia about whether humans possess the kind of free will that could provide a fair basis for holding people morally responsible in the backward-looking sense. One traditional challenge to the idea that humans possess this kind of free will arises from the view that all our decisions and actions

---

[17] "Self-harm" is used here as a shorthand for "damaging one's health", which might not constitute all-things-considered harm, e.g., if the individual values the health-damaging activity more than s/he values the aspect of her health that was damaged.

are determined by factors beyond our control, such as our biological constitution and environment. If determinism in this sense is true, factors beyond our control determine how our characters develop and which reasons we see fit to act upon. Given the influence of such factors, all our decisions (including each step in any rational deliberation leading up to them) must occur exactly as they do. These factors do not just make it more likely that we make particular decisions. They render it inevitable that we have the motivations and preferences that we have and that we reason and decide as we do. Most contemporary responsibility sceptics take a no-moral-responsibility-either-way position, according to which moral responsibility is incompatible with both determinism and indeterminism. One argument against indeterministic free will is that, rather than making us morally responsible, indeterminism would simply introduce randomness into our deliberative processes.

Another recent trend among moral responsibility sceptics is to focus on epistemic considerations (e.g., Pereboom 2001; Double 2002; Rosen 2004; Vilhauer 2009, 2013, 2023; Shaw 2014, 2021; Caruso 2018, 2021a, 2021b; Levy 2018; Corrado 2018). Although moral responsibility sceptics cannot provide 100 per cent proof that people are not morally responsible, it is hard to deny that there is significant room for doubting whether people are morally responsible. This doubt has important practical implications. It seems unfair to harm someone unless we are highly confident that harming them would be justified. This plausible notion underlies the principle—central to our criminal justice system—that people can only be punished if the case against them is proved beyond reasonable doubt. If a person's moral responsibility for performing an act is necessary to justify harming them, then according to the epistemic argument, it should be proved to a very high degree of certainty that this person was indeed morally responsible. However, the existence of the intractable debate surrounding moral responsibility suggests that there is not a sufficiently high degree of certainty that people are morally responsible, since respected experts in the field who have thought seriously about the matter fail to agree on whether people are morally responsible and on what the requirements for moral responsibility should be (Bourget and Chalmers 2014). There is a wide variety of challenges to the notion of moral responsibility (for an overview see Caruso 2018). Suffice it to say that with every hurdle new possibilities for error arise and the cumulative effect of this is to create considerable room for doubt. In Levy's (2018: 462) words, "it seems plausible that no one can confidently conclude that we know that every one of these challenges fails".

When deciding whether someone should suffer harm because they were responsible for making bad choices about their health, the severity of the harm seems relevant. The more severe the harm, the more certain we should be that the harm is justified. If the harm is very severe, we need a high degree of credence in the soundness of all elements of the argument for allowing the person to suffer it.

If the claim that they are responsible for their poor health is an element of the overall argument for harming them, we should be certain they were responsible. It is doubtful that we can be sufficiently certain, both because of the doubts about moral responsibility in general and because of the difficulties of establishing that the prerequisites for moral responsibility have been met in any particular case involving self-harm (Levy 2018; Brown et al. 2018). Harming one's health often involves going against one's strongest natural instinct (self-preservation), thwarting highly valued goals (one's own well being and flourishing), typically occurs under considerable stress and pressure, and often occurs under the influence of addiction which may have started below the age of responsibility (see, e.g. Wiss et al. 2020). Specific acts of self-harm often take place in the absence of witnesses, and damaging one's health is often associated with limited education about health issues and often coexists with or is a symptom of mental illness (see, e.g. Harwood et al. 2007). Each factor would make it difficult for medical professionals to establish with a high degree of certainty that any specific individual met the control and rationality prerequisites for moral responsibility at the specific time the individual engaged in health-harming behaviour. Given these doubts, imposing significant penalties for harming one's health seems unjustifiable, although, as I will discuss near the end of the article, the imposition of very minor burdens may be acceptable.

It might be objected that, despite these uncertainties, responsibility should be used as a "tie-breaker" in situations where one patient must lose out. For example, when there are two transplant candidates and only one suitable organ, perhaps the organ should go to the patient who seems less likely to be responsible for her illness. Indeed, the luck egalitarian might claim that failing to do so would be a "harsh" way of treating the patient who apparently was not (or was apparently less) responsible for her condition. However, responsibility would be a particularly problematic tie-breaker because, unlike other considerations, such as the likelihood that the treatment will be beneficial, attributing responsibility carries stigma. This stigma would affect patients and their loved ones when they are most vulnerable, which would be hard to justify given the uncertainties described above. Furthermore, such stigma has many counterproductive effects (Helweg-Larsen 2019)—including the impact on the doctor–patient relationship (Hansson 2018)—which could engender widespread mistrust in the health system felt by patients who fear they might be blamed, in addition to those who are actually blamed. Any value there might be in using as a tie-breaker the *possibility* that a patient *may* have been responsible for her illness could well be outweighed by the adverse consequences of doing so.[18]

---

[18] For discussion of other counterproductive effects of responsibility-sensitive policies besides stigma see (Levy 2018).

3.6 Is Harming One's Own Health Morally Wrong?

Three of the most influential justifications for holding people morally responsible for their own health ("pro-moral-responsibilisation arguments") have been identified and critiqued by Brown et al. (2019). They are: (1) *obligations of solidarity* to other health care system users, (2) *special relationships* (e.g., obligations to one's child or employer) and (3) *self-regarding obligations*. This section builds on Brown et al.'s critiques, addressing solidarity, special relationships and duties to oneself in turn, and provides additional reasons for doubting the soundness of these viewpoints.[19]

The *solidaristic obligations* argument appeals to the idea that public health care systems like the NHS are based on the relationship of solidarity between members of a community who share a common plight (e.g., all members face the possibility of becoming ill or injured in the future) and who decide to share the financial costs of their health risks. On this view, the creation of a publicly funded health system, based on solidarity, gives rise to reciprocal rights and duties. For example, the most vulnerable have a right not to be abandoned to their fate, but correspondingly, people who may benefit from the system have obligations to reduce costs to other health care system users by leading a healthy lifestyle. According to the pro-moral-responsibilisation version of this argument, people who fail to discharge their moral responsibility to look after themselves should be penalised. For example, smokers and the obese could be deprioritised on waiting lists for surgery.

When critiquing the solidaristic obligations argument, Brown et al. draw on two claims defended by Wilkinson (1999): first, there is little empirical evidence suggesting that those typically targeted by responsibilisation policies place a disproportionate burden on the health care system (e.g., smokers contribute to the economy via taxes on cigarettes and those who die young will not require expensive long-term care in old age); second, the logic behind penalising "lifestyle diseases" would result in an implausibly wide range of risky conduct being penalised. Furthermore, such policies may be tainted by class-based discrimination, since targeted behaviours (e.g., excessive drinking, smoking and obesity) are linked with socio-economic deprivation, while other equally risky behaviours (e.g., horse-riding, foreign travel and skiing) that are more often associated with wealthier classes are tolerated or celebrated.

I would add that, if any action can meet the control prerequisite for moral responsibility, it is likelier that this pre-condition is more frequently met when individuals engage in non-penalised activities associated with wealth than when stigmatised individuals engage in self-harming behaviours associated with poverty, since the latter are often performed under greater pressures and with the

---

[19] Like Levy (2018), Brown, Maslen and Savulescu claim that such arguments would only result in relatively few people being considered morally responsible for their health, as many of those who harm their health do not meet the control prerequisite for moral responsibility.

benefit of less information/education. Furthermore, if people can be held morally responsible, they are arguably more responsible or blameworthy for premeditated acts than for impulsive ones, and complex activities such as foreign travel and skiing are more likely to require careful conscious planning than purchasing cigarettes or ordering too many drinks in the pub.

My argument is not simply that responsibility-sensitive policies target the wrong behaviours. I am not proposing that health care systems should penalise skiing, horse-riding, and so on. Rather, I am raising a dilemma for responsibility-sensitive policies. If they penalise these kinds of health-damaging behaviour, these policies become impractical, intrusive and politically unviable. However, if they focus on the 'easier targets' the policies become arbitrary.

Furthermore, even if the control condition for responsibility were met when one engages in behaviour typically targeted by such policies, it is doubtful whether the solidarity argument can establish that self-harm is morally wrong as opposed to prudentially wrong, or, even if self-harm is morally wrong, it is doubtful whether it is sufficiently morally wrong to justify health-related penalties. One reason for saying that it is morally wrong to breach obligations of solidarity is because doing so is selfish: it involves according excessive weight to one's own interests. However, compared to more clear-cut cases of moral wrongdoing (e.g., intentionally harming another person for one's own financial gain), it is not straightforward to characterise people who harm themselves as selfish. The burden that self-harmers are willing to risk inflicting on themselves is typically much more severe than the burden they risk placing on any other individual to whom they owe solidary obligations. The smoker risks suffering and dying from lung cancer, whereas (even if smokers collectively cause a net loss to the NHS) the burden that any individual smoker places on any individual tax-payer will be comparatively minimal. It is plausible that it is more in accordance with the spirit of solidarity to abandon responsibilisation policies that risk stigmatising already vulnerable groups, than to impose penalties on self-harmers.

A second reason that harming one's health might be considered morally wrong is that self-harm could leave one unable to fulfil obligations stemming from *special relationships*. For example, some people who regularly drink excessively, or who damage their health through smoking or overeating, may be unable to discharge duties to their children, partner or employer. One might think that those who, through self-harm, fail to discharge such duties deserve to bear some health-related penalties. When criticising the special relationships argument, Brown et al. note that this argument is not "broadly applicable" as not everyone will be in a relevant relationship and obligations arising from such relationships would typically only require quite low levels of health to be discharged.

I would add that the class of cases in which it could be established that the patient was sufficiently morally responsible for a sufficiently wrongful act would be further narrowed due to the complex nature of both self-harm and special relationships. Loving parents who self-harm are often acting irrationally by thwarting the goals they value most: their own well being and that of their child. This raises questions about whether the rationality prerequisite for moral responsibility has been met (for a related argument, see Nir Eyal's chapter in this volume). In other cases, problems within the relationship, for example, domestic abuse or unfair and stressful work conditions, are the cause of the self-harm. In such cases, further penalising the abused spouse or employee to vindicate the supposed rights of the abusive spouse or employer would be unethical. The costs in time and resources it would take for the authorities to get clear about such issues could well be greater than any benefit the health system could derive from penalising the minority of self-harmers who could be proved to have been fully responsible for wrongfully breaching obligations within special relationships, even assuming that such proof is ever possible.

Cases where the breach of duty has severe consequences (e.g., allowing a child to come to harm) are already covered by criminal laws.[20] To impose additional health-related sanctions unfairly punishes someone twice. This consideration is recognised by the principle that prisoners should receive a level of health care equivalent to that provided to other members of the community (National Prison Healthcare Board 2018).[21] Moreover, imposing health-related sanctions for breaching obligations stemming from special relationships would likely penalise the very people whose rights the policy aims to protect, that is, those with whom the self-harmer is in a special relationship, such as family members. Family members are likely to be unpaid carers for the self-harming person. Regardless of whether the health system deems the patient less deserving of care, family members are unlikely to take the luck egalitarian line of "you made your bed so you must lie in it" and abandon their relative. So, for example, if a smoker were sanctioned by being deprioritised on a waiting list for surgery, this would likely result in a longer period in which the patient would face disability and discomfort which would intensify the caring burden placed on family members. This policy could also interfere with the patient's ability to care for their children whose care might then become an additional burden on other relatives. Thus, the policy would likely penalise the very groups whose interests supposedly justified it. Furthermore, such a policy might involve indirect discrimination[22] against women, as female relatives comprise the majority of unpaid carers (Brimblecombe et al. 2017) and such a policy could therefore have a disproportionate impact on them.[23]

The third argument for holding people morally responsible for their own health is based on the Kantian idea of *duties to oneself*. For example, on this view, a person might owe it to herself to nurture her talents. A person who smokes,

[20] *R v Gibbins and Proctor* (1918) 13 Cr App R 134.
[21] See also: United Nations General Assembly (1991: Principle 9).   [22] Equality Act 2010, s.19.
[23] For discussion of discrimination against unpaid carers see Tribe and Lane (2017).

overeats or drinks excessively might develop health conditions that interfere with her ability to nurture her talent for athletic, intellectual or artistic pursuits. In reply to this argument, Brown et al. note that Kantian claims about self-regarding duties are contentious and that even if individuals have self-regarding moral obligations, it is unclear whether these can be "legitimately enforced by the state" (Brown et al. 2019). I would add that penalising failures to fulfil duties to oneself would arguably be self-contradictory and incoherent. If X has self-regarding duties, it seems plausible that the existence of such duties must rely at least partly on the idea that X's flourishing is a good thing. It thus makes little sense to respond to X's failure to promote their own flourishing by taking measures that further interfere with their flourishing. One might reply that threatening to penalise someone for self-harm could be an effective deterrent. However, there is a lack of empirical evidence for that claim; in fact, stigma and threats seem to make self-harm more likely (Hansson 2018; Helweg-Larsen et al. 2019; Turner et al. 2020). Furthermore, arguments based on deterrence seem to appeal to consequentialist reasoning that go against the Kantian rationale behind self-regarding obligations.

3.7 Are Health-Related Sanctions a Proportionate Response to Harming One's Own Health?

The most highly-developed theoretical discussion of proportionate sanctions is in the literature on criminal punishment. A similar concept of proportionality is needed to justify penalising self-harm with reference to culpability-based desert. As argued above, without a proportionality constraint such policies would lead to health outcomes that intuitively seem too harsh in certain cases. Indeed, the current section argues that there are reasons for doubting that significant penalties for harming one's health could ever be a proportionate response (although very minor burdens might be less open to challenge).

In the context of imposing sanctions as a deserved response to morally culpable conduct, "proportionality" is a fittingness relationship between the sanction and the severity of the conduct, such that more serious misconduct attracts harsher penalties. Some think that the *kind* of penalty imposed should also be appropriate to the *kind* of wrongdoing, as reflected in the biblical idea of "an eye for an eye and a tooth for a tooth". In modern societies which reject physical punishments it is seldom deemed ethically acceptable to achieve a very close match in terms of *kinds* of misconduct and penalty.[24] An acceptable example might be requiring someone who had vandalised property to do community service that involved repairing

---

[24] Health-related sanctions are arguably a kind of physical punishment by omission. This raises concerns about degrading treatment (article 3 of the European Convention on Human Rights) which are beyond the scope of the current chapter. See Boruckiee (2019).

property. However, a penal system in which fines and imprisonment are typical punishments clearly prioritises proportionality of severity over proportionality of kind. Indeed, prioritising proportionality of severity seems intuitively plausible, as the injustice of disproportionate punishments such as life imprisonment for a minor assault seems to lie primarily in the mismatch of severity between crime and punishment, rather than a mismatch in "kind": assaulting the offender would not be an improvement.

Indeed, proportionality of kinds can sometimes be positively misleading, opening the door to disproportionality of severity, and may be a source of confusion in the responsibility for health debate. At first glance, it may seem fitting to respond to a person who harms their health by imposing a health-related sanction, such as deprioritising their treatment. However, such a response may be disproportionate in terms of severity and also may fail to fit with the underlying justification for holding the individual morally responsible for harming her health. Regarding severity: health-related sanctions such as delaying surgery or asking people to contribute to the costs of health care are significantly burdensome. Although they are less extreme than outright treatment denial, they come with substantial costs in terms of discomfort, reduced independence, stress and stigma at a time when patients are very vulnerable and when the policy's adverse impact on mental health might affect the physical outcomes. These sanctions seem disproportionate given the reasons discussed in the previous section for doubting that self-harm is (seriously) morally wrong and for doubting that any particular individual is (fully) morally responsible for self-harm.

Regarding the fit between the sanction and the underlying justification for holding the individual morally responsible, consider the solidarity justification, arguably the strongest of the three justifications for penalising self-harm discussed above (Davies and Savulescu 2019).[25] The solidarity argument is based on the idea that people who self-harm wrong their fellow citizens by placing an undue burden on the publicly funded health care system. However, it is puzzling why the scope of one's responsibility to one's fellow citizens should be delimited in this way (beyond a misleading *semantic* match, with "health" featuring as an element of the supposed wrongdoing and an as element of the sanction). There does not seem to be a strong *moral* reason for focusing narrowly on a putative solidary obligation to safeguard one's own health and to ignore how one conducts oneself as a citizen more generally. Imagine Carl, who helps to promote the health of others, for

[25] It is not entirely clear how (or whether) Brown et al.'s critique of this argument is compatible with Davies and Savulescu's (2019) "golden opportunities" argument, which defends imposing (albeit low-level) sanctions on those who breach solidary obligations in the context of health care. It is possible that these two articles can be reconciled with each other, on the basis that the former article opposes policies that in "a wide and unrestricted way" penalise breaches of solidary obligations, whereas the latter article defends restricted, low-level penalties for a narrowly defined set of patients who reject "golden opportunities" to preserve their health.

example, through cooking healthy meals for his children and educating them about healthy habits, doing voluntary work that promotes health in his community, relieving his spouse of other domestic duties so she has more time to exercise, and working as a conscientious doctor. However, Carl (perhaps due to the strain of his altruistic activities) has certain habits that damage his own health. Now, compare this individual to Damien, who does not have a social conscience, leads a selfish lifestyle, does as little as he can legally get away with to promote the health of his family or anyone else, and profits from selling or manufacturing unhealthy products, but takes perfect care of his own health. The narrow focus on damaging one's own health adopted by the solidarity approach discussed above would impose sanctions on Carl but not Damien. However, it is unclear why that is just. I am not suggesting that the state should do a wholesale investigation of each citizen's civic virtue and allocate health resources accordingly. But the impracticality and intrusiveness of doing so does not justify focusing narrowly on people who damage their own health. Such a narrow focus is unjustifiable because it would be arbitrary (as well as probably being almost as impractical and intrusive as investigating other aspects of civic virtue).[26] The arbitrariness of the narrow focus on damaging one's own health comes on top of the arbitrariness of policies that only penalise those who take risks with their health and are *unlucky enough to get sick*, while others who took the same risk have a stronger constitution and so face no penalty.

It might be objected that it would not be arbitrary for the health system to focus specifically on penalising people who damage their own health, because it is reasonable for there to be a division of labour in the way society administers sanctions, with different institutions/sectors (e.g., the civil courts and the criminal courts, etc.) dealing with different types of breaches of duty. In reply, first, it seems arbitrary for the health system to penalise damaging/neglecting one's own health but not to penalise damaging/neglecting other people's health, given that the health system seeks to promote health in general. Second, it seems arbitrary to endow a social institution with the power to impose sanctions for damaging one's own health given the doubts discussed in this chapter about whether such conduct is morally wrong, and given that there are many other types of conduct which seem morally much worse (e.g., Damien's conduct in the above example) which no institution has the power to penalise. Third, it seems arbitrary for one social institution (the health system) to allocate burdens as severe as illness or death based on a justification (the supposed immorality of damaging one's health) which

---

[26] Davies and Savulescu's (2019) and De Marco, Savulescu and Douglas's (2021) account of situations in which patients could be presented with a "golden opportunity" to change their lifestyles provide the most promising attempts at describing ways of taking responsibility into account while seeking to minimise intrusiveness, unfairness and practical difficulties. However, as they acknowledge, their approach still faces challenges. One difficulty is that in order to minimise unfairness, medical professionals must consider if the patient had a good justification or excuse for harming their health. However, these justifications and excuses seem both easy for patients who understand the system to fake and difficult for the most vulnerable patients to articulate.

is so debatable, whereas other social institutions (e.g., civil and criminal courts) impose sanctions that are much less severe (e.g., fines) for behaviour whose wrongfulness is much less disputable (rightly deeming more severe sanctions to be disproportionate).

This section now responds to some potential counterarguments against my claim that responsibility-sensitive health care policies are likely to create disproportionate outcomes. Albertsen and Nielsen (2020) have argued that it is incoherent to rely on the concept of "disproportionality" when criticising responsibility-sensitive health care policies, because one cannot invoke disproportionality without implicitly endorsing the very responsibility-sensitive approach that one was attempting to criticise. They also claim that a proportionality principle is not capable of ruling out harsh cases, since to say that it is disproportionate to attach certain burdens to certain instances of imprudence "must imply that responsibility-sensitivity in *some proportion* is appropriate for distributive justice. It also implies that we can imagine situations involving extremely bad consequences for the imprudent as a result of a proportionally bad exercise of responsibility which therefore does not disturb justice from the perspective of proportionality" (2020: 6).

However, neither of these supposed implications follows from the claim that it would be disproportionate to penalise people for poor health decisions.

In reply to their first claim (the incoherence claim): one can criticise a theory by appealing to a principle that is part of that theory without endorsing the principle or theory. The critic might simply be noting that the theory is internally inconsistent because its implications do not fit with its own principles. Thus, a critic of a theory that purports to attach proportionate consequences to responsible choices can coherently argue that certain consequences would be "disproportionate", without accepting the validity of this responsibility-sensitive theory and without accepting the proportionality principle. For example, if a particular retributive theory attempted to justify punishing shoplifting with death, critics could argue that this would be "disproportionate" by retributivism's own lights. Critics could coherently make this point without endorsing either retributivism or the retributive concept of proportionality. Similarly, critics might argue that imposing severe health consequences for imprudence would violate the proportionality principle that proponents of responsibility-sensitive health policies endorse without themselves endorsing responsibility-sensitivity or this proportionality principle. Alternatively, the critic might endorse proportionality, but interpret this concept in a way that does not depend on responsibility. For instance, some writers endorse the idea that a burden placed on an individual must be proportionate to the severity of an ongoing threat posed by that individual, regardless of the individual's responsibility status (e.g., Honderich 1993; Pereboom 2001; Caruso 2018; Shaw 2019). A carrier of an infectious disease might not be responsible for having that disease, but it might nevertheless be proportionate to quarantine that individual if the disease were sufficiently dangerous and infectious and if the quarantine were not too burdensome, and disproportionate to quarantine them if those conditions were not met. Finally, a critic might endorse responsibility-sensitive policies in certain contexts (e.g., criminal punishment), while coherently arguing that imposing burdens on people for damaging their own health is always disproportionate because such conduct is not wrongful enough to warrant state interference.

In reply to Albertsen and Nielsen's second claim, that proportionality principles cannot rule out harsh cases: it is unclear why they assume that a scale of proportionate burdens can have no upper limit. Albertsen and Nielsen (2020: 6) state that "we can imagine situations involving extremely bad consequences for the imprudent as a result of a proportionally bad exercise of responsibility which therefore does not disturb justice from the perspective of proportionality". It is puzzling why they do not consider the possibility that a proponent of proportionality might deem it unjust (because disproportionate) to impose "extremely bad consequences" on someone for mere imprudence no matter the extent of the imprudence. Analogously, a penal system might have a punishment scale that increased in severity in proportion to the severity of the crime until the maximum punishment allowable in that jurisdiction was reached (say life imprisonment). The penal system might presuppose that there is nothing a person can do in response to which a punishment more severe than life imprisonment would be proportionate. So too, the health care system could presuppose that there is no form of imprudence in response to which imposing/allowing "extremely bad consequences" would be proportionate. Of course, it has frequently been argued (often by critics of retributivism) that there is no principled way of devising a proportionality scale. Reasonable people can disagree about which kind of burden is a proportionate response to which kind of conduct and there is no obvious way to settle this disagreement. However, most people can agree about certain clear cases of disproportionality (e.g., the death penalty for theft, or being left to die because one didn't buckle one's seatbelt) and so the concept of proportionality (or gross disproportionality) might still be useful for ruling out these harsh cases. Furthermore, to return to the coherence point discussed in the previous paragraph, it would be coherent to object to responsibility-sensitive policies because (1) these policies rely on the notion of proportionality, (2) these policies imply that people should bear burdens that would be disproportionate *if* the concept of disproportionality were sound and (3) the concept of disproportionality is flawed because there is no principled way to devise a justifiable proportionality scale.

I have suggested that it would be disproportionate to impose significant penalties for harming one's health, such as deprioritising treatment for smokers and the obese or making such individuals contribute to the costs of their treatment. However, minor burdens might be less open to challenge. For example, taxes on cigarettes and minimum alcohol pricing may withstand the objections raised in this chapter. They are likely to be less stigmatic and emotionally damaging than other policies that impose burdens on those engaging in unhealthy behaviour, such as policies that cite a person's past irresponsible conduct as a reason for delaying surgery. Taxes and minimum pricing affect individuals at the point when they take the risk, rather than arbitrarily penalising only those risk-takers who are unlucky enough to get sick. They are not penalties that are only imposed when patients are at their most vulnerable. Taxing and minimum pricing also affect companies who sell unhealthy products, not just consumers, unlike policies which disproportionately impact socially disadvantaged groups and leave companies who profit from selling unhealthy goods untouched. Arguably, taxation and minimum pricing engage with a more forward-looking sense of responsibility, which is less vulnerable to the challenge from moral responsibility sceptics than the backward-looking moral responsibility that features in other potential policies. Because taxing and minimum pricing have their impact at the time when the consumer is deciding whether to take the risk, they have the potential to encourage the consumer to engage in more responsible future behaviour. In contrast, sanctions such as treatment deprioritisation that penalise the individual after the self-harm has already happened are more dependent on the idea of backward-looking responsibility (and any deterrent value they may have may be undermined by the counterproductive effects of stigma).

3.8 Conclusion

This chapter discussed some of the difficulties faced by proponents of responsibility-sensitive approaches in health care, highlighting the implications of moral responsibility scepticism. When discussing punishment, moral responsibility sceptics have often invoked an epistemic argument, maintaining that there are at least serious doubts about whether people are morally responsible (in the sense required for retributive punishment) and that, in view of this uncertainty, retributive punishment is unjust, given the serious harm it inflicts on offenders. This type of reasoning also implies that we should not take patients' responsibility for their poor health into account when deciding whether to give these patients treatments for serious illnesses and that the health system should not impose significant penalties on individuals for harming their own health. However, minor burdens, such as taxes on cigarettes and minimum alcohol pricing might be less open to challenge.

References

Abad, D. (2011), 'Desert, responsibility and luck egalitarianism', in N. Vincent, I. van de Poel and J. van den Hoven (eds), *Moral Responsibility* (Springer): 121–40.

Albertsen, A., and Knight, C. (2015), 'A framework for luck egalitarianism in health and healthcare', in *Journal of Medical Ethics* 41/2: 165–9.

Albertsen, A., and Nielsen, L. (2020), 'What is the point of the harshness objection?' in *Utilitas* 32/4: 427–43.

Alces, P. (2018), *The Moral Conflict of Law and Neuroscience* (University of Chicago Press).

Anderson, E. (1999), 'What is the point of equality?', in *Ethics* 109/2: 287–337.

Arneson, R. (2000), 'Luck egalitarianism and prioritarianism', in *Ethics* 110/2: 339–49.

Borucki, C. (2019), 'When healthcare goes up in tobacco smoke: A selective healthcare system from a (European) human rights perspective', in *Utrecht Law Review* 15/3: 6–26.

Bourget, D., and Chalmers, D. (2014), 'What do philosophers believe?', in *Philosophical Studies* 170/3: 465–500.

Brimblecombe, N., Pickard, L., King, D. and Knapp, M. (2017), 'Perceptions of unmet needs for community social care services in England: A comparison of working carers and the people they care for', in *Health and Social Care in the Community* 25/2: 435–46.

Brouwer, H., and Mulligan, T. (2019), 'Why not be a desertist?' in *Philosophical Studies* 176/9: 2271–88.

Brown, R., Maslen, H. and Savulescu J. (2018), 'Responsibility, prudence and health promotion', in *Journal of Public Health* 41/3: 561–5.

Brown, R., Maslen, H. and Savulescu, J. (2019), 'Against moral responsibilisation of health: Prudential responsbility and health promotion', in *Public Health Ethics* 12/2: 114–29.

Callender, J. (2010), *Free Will and Responsibility: A Guide for Practitioners* (Oxford University Press).

Caruso, G. (2018), 'Skepticism about moral responsibility', in E. Zalta (ed.), *The Stanford Encyclopaedia of Philosophy*. Available at: https://plato.stanford.edu/entries/skepticism-moral-responsibility/ [Accessed 18/08/22].

Caruso, G. (2021a), 'Retributivism, free will skepticism and the public health-quarantine model: Replies to Corrado, Kennedy, Sifferd, Walen, Pereboom and Shaw', in *Journal of Legal Philosophy* 46/2: 161–215.

Caruso, G. (2021b), *Rejecting Retributivism: Free Will, Punishment and Criminal Justice* (Cambridge University Press).

Corrado, M. (2018), 'Free will, punishment and the burden of proof', in *Criminal Justice Ethics* 37/1: 55–71.

Cushman, F. (2008), 'Crime and punishment: Differential reliance on causal and intentional information for different classes of moral judgment', in *Cognition* 108: 353–80.

Davies, B. (2022), 'Responsibility and the recursion problem', in *Ratio* 35/2: 112–22.

Davies, B., and Savulescu, J. (2019), 'Solidarity and responsibility in health care', in *Public Health Ethics* 12/2: 133–44.

De Marco, G., Douglas, T. and Savulescu, J. (2021), 'Healthcare, responsibility and golden opportunities', in *Ethical Theory and Moral Practice* 24: 817–31.

Double, R. (2002), 'The moral hardness of libertarianism', in *Philo* 5/2: 226–34.

Duff, R. (2001), *Punishment, Communication and Community* (Oxford University Press).

East and North Hertfordshire Clinical Commissioning Group (2017), 'East and North Hertfordshire NHS service changes: Decisions announced'. Available at: https://www.enhertsccg.nhs.uk/news/201710/east-and-north-hertfordshire-nhs-service-changes-%E2%80%93-decisions-announced [Accessed 7/04/22].

Equality Act (2010), (c.15). Available at https://www.legislation.gov.uk/ukpga/2010/15/contents [Accessed 30/08/22].

Eyal, N. (2006), 'Egalitarian justice and innocent choice', in *Journal of Ethics and Social Philosophy* 2/1: 1–18.

Feinberg, J. (1970), *Rights and Reason* (Springer).

Feldman, F. and Skow, B. (2020), 'Desert', in E. Zalta (ed.), *The Stanford Encyclopaedia of Philosophy*. Available at: https://plato.stanford.edu/entries/vagueness/ [Accessed 18/08/22].

Fischer, J. and Ravizza, M. (1998), *Responsibility and Control: A Theory of Moral Responsibility* (Cambridge University Press).

Focquaert, F. (2019), 'Free will skepticism and criminal punishment: A preliminary ethical analysis', in E. Shaw, D. Pereboom and G. Caruso (eds), *Free Will Skepticism in Law and Society: Challenging Retributive Justice* (Cambridge University Press): 207–36.

Gold, A., Greenberg, B., Strous, R. and Asman, O. (2021), 'When do caregivers ignore the veil of ignorance? An empirical study on medical triage decision–making', in *Medicine, Health Care and Philosophy* 24/2: 213–25.

Gold, A., and Strous, R. (2017), 'Second thoughts about who is first: The medical triage of violent perpetrators and their victims', in *Journal of Medical Ethics* 43: 293–300.

Hansson, S. (2018), 'The ethics of making patients responsible', in *Cambridge Quarterly of Healthcare Ethics* 27: 87–92.

Hart, H. (1968), *Punishment and Responsibility* (Oxford University Press).

Harwood, G., Salsberry, P., Ferketich, A. and Wewers, M. (2007), 'Cigarette smoking, socioeconomic status, and psychosocial factors: Examining a conceptual framework', in *Public Health Nursing* 24/4: 361–71.

Health and Social Care Act (2012), (c. 7). Available at https://www.legislation.gov.uk/ukpga/2012/7/contents/enacted [Accessed 30/08/22].

Health and Social Care Act 2022 (c. 31). Available at https://www.legislation.gov.uk/ukpga/2022/31/contents/enacted [Accessed 30/08/22].

Helweg-Larsen, M., Sorgen, L. and Pisinger, C. (2019), 'Does it help smokers if we stigmatize them? A test of the stigma-induced identity threat model among US and Danish smokers', in *Social Cognition* 37/3: 294–313.

Honderich, T. (1993), *How Free Are You?* (Oxford University Press).

Kane, R. (1998), *The Significance of Free Will* (Oxford University Press).

Lasersohn, P. (1999), 'Pragmatic halos', in *Language* 75/3: 522–51.

Levy, N. (2018), 'Responsibility as an obstacle to good policy: The case of lifestyle related disease', in *Journal of Bioethical Inquiry* 15/4: 459–68.

Lippert-Rasmussen, K. (2016), *Luck Egalitarianism* (Bloomsbury).

Moore, M. (1997), *Placing Blame* (Oxford University Press).

National Prison Healthcare Board. (2018), 'Principle of Equivalence of Care for Prison Healthcare in England'. Available at https://assets.publishing.service.gov.uk/govern ment/uploads/system/uploads/attachment_data/file/837882/NPHB_Equivalence_of_ Care_principle.pdf [Accessed 24 August 2022].

Olsaretti, S. (2009), 'Responsibility and the consequences of choice', in *Proceedings of the Aristotelian Society* 109/1: 165–88.

Parikh, R. (1994), 'Vagueness and utility: The semantics of common nouns', in *Linguistics and Philosophy* 17: 521–35.

Pereboom, D. (2001), *Living without Free Will* (Cambridge University Press).

Pereboom, D. (2021), 'Undivided forward-looking moral responsibility', in *The Monist* 104: 484–97.

Pillutla, V., Maslen, H. and Savulescu, J. (2018), 'Rationing elective surgery for smokers and obese patients: Responsibility or prognosis?', in *BMC Medical Ethics* 19/1: 1–10.

R v Gibbins and Proctor (1918), 13 Cr App R 134.

Ristroph, A. (2006), 'Desert, democracy, and sentencing reform', in *Journal of Criminal Law and Criminology* 96/4: 1293–352.

Rosen, G. (2004), 'Skepticism about moral responsibility', in *Philosophical Perspectives* 18: 295–313.

Royal College of Surgeons (2016), 'Smokers and overweight patients: Soft targets for NHS savings'. Available at https://www.rcseng.ac.uk/-/media/files/rcs/library-and-publications/non-journal-publications/smokers-and-overweight-patients–soft-targets-for-nhs-savings.pdf [Accessed 23/06/22].

Schneiderman, L. (2011), 'Response to open peer commentaries on "Rationing just medical care"', in *The American Journal of Bioethics* 11/10: 1–3.

Schneiderman, L., and Jecker, N. (1996), 'Should a criminal receive a heart transplant? Medical justice vs societal justice', in *Theoretical Medicine* 17: 33–44.

Segall, S. (2015), 'What's so egalitarian about luck egalitarianism?' in *Ratio*, 28/3: 349–68.

Shaw, E. (2014), *Free Will, Punishment and Criminal Responsibility* (PhD thesis, University of Edinburgh). Available at: https://era.ed.ac.uk/bitstream/handle/1842/ 9590/Shaw2014.pdf?sequence=2&isAllowed=y [Accessed 24/08/22].

Shaw, E. (2019), 'Justice without moral responsibility?' in *Journal of Information Ethics* 28/1: 95–114.

Shaw, E. (2021), 'The epistemic argument against retributivism', in *Journal of Legal Philosophy* 46/2: 155–60.

Sorensen, R. (2022), 'Vagueness', in E. Zalta (ed.), *The Stanford Encyclopaedia of Philosophy*. Available at: https://plato.stanford.edu/entries/vagueness/. [Accessed 18/08/22].

Strawson, G. (1994), 'The impossibility of moral responsibility', in *Philosophical Studies* 75/1: 5–24.

Strawson, G. (2015), 'The impossibility of ultimate responsibility?' in J. Dancy and C. Sandis (eds), *Philosophy of Action: An Anthology* (Wiley), 373–81.

Tribe, R., and Lane, P. (2017), 'Anti-discriminatory practice: Caring for carers of older adults with mental health dilemmas', in P. Lane and R. Tribe (eds) *Anti-discriminatory Practice in Mental Health Care for Older People* (Jessica Kingsley) 147–74.

Turner, M., Ford, L., Somerville, V., Javellana, D., Day, K. and Lapinski, M. (2020), 'The use of stigmatizing messaging in anti-obesity communications campaigns: Quantification of obesity stigmatization', in *Communication Reports* 33/3: 107–20.

United Nations General Assembly (1991), *Basic Principles for the Treatment of Prisoners*, UN Doc A/RES/45/111 (28 March 1991). Available at https://www.ohchr.org/sites/default/files/basicprinciples.pdf [Accessed 30/08/22].

Vilhauer, B. (2009), 'Free will and reasonable doubt', in *American Philosophical Quarterly* 46/2: 131–40.

Vilhauer, B. (2013), 'Persons, punishment, and free will skepticism', in *Philosophical Studies* 162/2: 143–63.

Vilhauer, B. (2023), 'Free will skepticism and criminals as ends in themselves', in M. Altman (ed.), *Palgrave Handbook on the Philosophy of Punishment* (Palgrave Macmillan).535-556.

Vincent, N. (2009), 'What do you mean I should take responsibility for my own ill health', in *Journal of Applied Ethics and Philosophy* 1: 39–51.

Walen, A. (2020), 'Retributive justice', in E. Zalta (ed.), *The Stanford Encyclopaedia of Philosophy*. Available at: https://plato.stanford.edu/entries/justice-retributive/ [Accessed 18/08/22].

Waller, B. (2011), *Against Moral Responsibility* (MIT).

West, J., Chao, S., Kelley, S., Schwartz, J., Bertsch, D. and Marsh, J. (2003), 'Organ allocation: A case for not transplanting the violent criminal', in *Seminars in Dialysis* 16/5: 362–4.

Wilkinson S. (1999), 'Smokers' rights to health care: Why the "restoration argument" is a moralising wolf in a liberal sheep's clothing', in *Journal of Applied Philosophy* 16: 255–69.

Wiss, D., Avena, N. and Gold, M. (2020), 'Food addiction and psychosocial adversity: Biological embedding, contextual factors, and public health implications', *Nutrients* 12/11: 3521. Available at https://doi.org/10.3390/nu12113521 [Accessed 24/08/22].

PART II

THE NATURE OF RESPONSIBILITY-SENSITIVE HEALTH CARE POLICIES

4

On Prevalence and Prudence

Nir Eyal

4.1 Introduction

Decades before social disparities in health, education and the economy became mainstream discussion points, progressive economist John E. Roemer proposed an elegant way for social planners to address unjust social disparities.[1] His proposal, which has now been honed and applied in multiple settings, is deeply sensitive to the machinations of real-world economics and public health. According to Roemer (1995: 4),

> An equal-opportunity policy must equalize outcomes in so far as they are the consequences of causes beyond a person's control, but allow differential outcomes in so far as they result from autonomous choice.
>
> (See also Roemer and Trannoy 2016: 1289ff)[2]

Note that Roemer's brand of luck egalitarianism does not hold persons responsible in justice for all their risky choices. It only does so inasmuch as these choices lie within these persons' control—to the extent that they are "autonomous." And Roemer puts forth a pragmatic interpretation of that basic ideal. Social planners who are luck egalitarians should hold persons responsible only inasmuch as these persons' choices break with common practices in their social "types." What I shall call a Roemerian type is a combination of personal characteristics, all conventionally deemed to lie outside the person's control. Roemer illustrates with smoking behavior, which increases risk for a host of adverse outcomes:

> Society must first decide what circumstances [*that is, in luck-egalitarian terms, what personal characteristics that lie or are eventually deemed to lie beyond the*

[1] For helpful comments, the author is grateful to audiences at the Institute for Future Studies and at the Rutgers School of Public Health's Department of Health Behavior and Social Policy, as well as the editors.

[2] Roemer later wrote that when people are *not* responsible for their disadvantage, he actually prefers a prioritarian pattern more than equalizing. That later clarification is largely tangential to the current discussion, whose focus is not the "pattern" but the "currency" of distributive justice. We are discussing here Roemer's suggested proxy for when people are responsible for their disadvantage, which would lose them some of their egalitarian *or* prioritarian entitlements.

Nir Eyal, *On Prevalence and Prudence* In: *Responsibility and Healthcare*. Edited by: Benjamin Davies, Gabriel De Marco, Neil Levy, Julian Savulescu, Oxford University Press. © Edited by Benjamin Davies, Gabriel De Marco, Neil Levy, Julian Savulescu 2024. DOI: 10.1093/oso/9780192872234.003.0005

person's control—NE] seem to be important in determining a person's smoking behavior: we might decide these are a person's occupation, her ethnicity and gender, whether her parents smoked, and her income level. For the purposes at hand, we would consider these five characteristics of a person to be factors beyond her control. The next step is to divide the relevant population into types, with each type consisting of the subset of the population having approximately the same values for all five characteristics.   (Roemer 1993: 150)

Since these characteristics and their combination are deemed to lie outside personal control,[3] so do the individual's choices, Roemer continues, so the individual is not responsible for them.[4] And *this* includes even risky personal choices, inasmuch as they coincide with these types (Roemer 1993, 1996, 2000; Roemer and Trannoy 2016). Therefore:

To take an extreme case, if all 60-year-old steelworkers smoked for 30 years, I would say that the choice of "not smoking" was not accessible to 60-year-old steelworkers: as a 60-year-old steel worker, one would have had effectively no opportunity except to smoke for 30 years. Given one's group, certain choices may be effectively, even if not physically, barred.   (Roemer 1995: 4)

Roemer also makes this case comparatively—where the relevant question is not whether there was sufficient/insufficient control for being held responsible, but whether there was more/less control than in other actors' cases. Imagine that a certain 60-year-old male steel worker has smoked for more years than a certain female college professor of that same age has, but less than most male steel workers of that age have. The professor smoked more than most female professors of that age have (Roemer 1993: 151ff; Roemer 1995). According to Roemer, if the

---

[3] So Roemer asks us to assume. Regarding some characteristics put forth by Roemer as "circumstances," and hence deemed to lie beyond the person's control, one might question whether they truly do, metaphysically or even as societies usually construe what we are responsible for. For example, is a person's occupation truly beyond personal control? Is it commonly considered to be beyond it? What about gender? Roemer provides interesting answers (e.g., Roemer (1993: 150, n.9); Roemer and Trannoy (2016: 1292–3)), and my own critique does not rest on questioning these mere illustrations. For the purposes of this chapter, one may further imagine that being a male steel worker (or a female college professor) is pretty much determined by one's parents' professions, one's elementary school, and so forth (or at least that a somewhat broader type like *working-class male*, which likewise correlates with smoking, is a circumstance outside personal control).

[4] On other conceptions of moral responsibility, personal freedom to do otherwise and, relatedly, control over the relevant matter, is not necessary for personal moral responsibility (Watson 1975; Frankfurt 1987; Bratman 2003; Hurley 2003). Indeed, one supporter of these conceptions criticized Roemer and other luck-egalitarians on precisely this matter (Hurley 2003). My own critique of Roemer is unrelated. Mine assumes with Roemer that we have luck-egalitarian responsibility for all and only for matters over which we have personal control, and to the extent that we do. (Note, however, that at one point Roemer (1995: 6) endorses Ronald Dworkin's view that, in Roemer's words, "the distinction between what a person is and is not responsible for is not the same as the distinction between what she has and has no control over.")

combination of being a 60-year-old, either male or female, and either a steel worker or a college professor is deemed to lie outside a person's control, that has an important implication. In that case, the person to be held more responsible for their own risk of smoking-related lung cancer, per luck egalitarianism, with potential impact on his or her entitlement to redress, is the college professor not the steel worker. The professor takes greater smoking risks than people in circumstances like hers; the steel worker takes fewer such risks than people in circumstances like his. At least as a proxy rule for luck-egalitarian planners, Roemer maintains, some individuals' unhealthful choices should be assessed relative to their social types or classes, not directly or relative to everyone. In short, "two people, of different types, have exercised a comparable degree of responsibility if each is at the pth percentile of his type-distribution, for some p" (Roemer 1993: 152).

This important proposal, which Roemer calls "normalizing by type," works nicely under many conditions. But this chapter exposes its limitations. For luck-egalitarian purposes, I shall argue, it can be irrelevant whether a risky choice is prevalent in one's Roemerian type.

Section 4.2 expounds the grounds for Roemer's pragmatic proposal. Section 4.3 begins my critique by illustrating that a risky choice that is "typical" (my short for "prevalent in the relevant person's Roemerian type") can nevertheless retain personal control and responsibility and reduce the need for social indemnity in luck-egalitarian justice. Section 4.4 illustrates that a risky choice that is "atypical" (that is, "rare in the person's Roemerian type") can remain outside the person's control, and therefore stay a matter of luck-egalitarian redress. Section 4.5 proposes an alternative proxy to assist luck-egalitarian planners: the relevant risky choice's degree of what I call "clear imprudence." Section 4.6 charts potential practical and theoretical applications.

4.2 Roemer's Grounds for Associating Atypicality with Personal Control

Roemer, recall, believes that for practical purposes, risky personal choices should be assumed to be low on personal control (and hence, on luck-egalitarian responsibility) inasmuch as they correlate with Roemerian types. Roemer defends that belief with the case of the two smokers:

> Here is the motivation. Within a type, the variation in smoking behavior is taken, by society, to reflect the different degrees of responsibility its members have exercised. For, by definition, society has already accounted for, in the definition of type, all the circumstances affecting smoking behavior that it considers to be beyond a person's control. Now let us compare two people of different types,

both of whom are at the median of their respective type-distributions for years of smoking...My egalitarian ethic says that both should receive the same degree of social indemnity: if all the medical expenses of the professor are covered by society, then the same should be the case for the steelworker.

(Roemer 1993: 150–1)

Put succinctly, "That a 60-year-old Black male steelworker is more likely to have smoked for 30 years than a 60-year-old White female college professor is a statistical fact not due to the autonomous choices of individuals, but to group" (Roemer 1995: 4).[5]

Roemer's point here is not that a social grouping is never also an individual characteristic. The fact that Vlad is part of a group of individuals who are war criminals does not mean that Vlad had no control over his crimes. The point is, rather, that a Roemerian type is grouped together precisely by imputed lack of control. To abstract from a Roemerian type is to remove from consideration only factors over which the person has no control. This is why "in comparing the degrees of effort of individuals across types, the rank measure in effect sterilizes the distribution of raw effort of the influence of circumstances upon it" (Roemer and Trannoy 2016: 1294). Again, "circumstances" are by definition outside the person's control or so deemed by society.

Thus, in the pure case in which all 60-year-old male steel workers smoke, Roemer seems to argue:

Premise 1:   For everyone who is a 60-year-old male steel worker, being a 60-year-old male steel worker lies outside one's personal control.

Premise 2:   All 60-year-old male steel workers are smokers.

Premise 3:   If 1 and 2 are true then, *at least as a practical rule*, for everyone who is a 60-year-old male steel worker and a smoker, being a smoker lies outside one's personal control.

Conclusion: At least as a practical rule, for everyone who is a 60-year-old male steel worker and a smoker, being a smoker lies outside one's personal control.

Thus, if you grant (as I propose you do) premises 1 and 2, namely, that being a 60-year-old male and a steel worker lies outside the person's control and that 100 percent of 60-year-old male steel workers smoke, then premise 3 offers egalitarian planners a proxy rule: for practical purposes, a choice to smoke lies outside the control of each male steel worker. He has no responsibility for it.

---

[5] See also "society has already factored out everything that it considers to be beyond [black male steelworker] Fernando's control, in so far as his smoking behavior is concerned, by assigning him to a group" (Roemer 1995: 6).

A word on the unusual "At least as a practical rule" clause. When can we say that, as a practical rule, proposition *p* obtains? One answer is *epistemological*. That answer is, roughly, "When *p* is always or nearly always true." Another answer is *thoroughly pragmatic*. That answer is, roughly, "When acting as though *p* is always or nearly always bringing about the desired results."

The epistemological answer helps us avoid false beliefs most of the time. The thoroughly pragmatic answer helps us avoid undesired outcomes most of the time. Many factors beyond the typical truth value of the beliefs we act on affect the desirability of the typical outcomes of our actions. Such factors include how these beliefs will typically be received (will the truth offend and agitate?) They also include how counterproductive it would typically be to act on their upshots (will the truth induce stress and impulsive risk taking?) It also matters how undesirable a false positive would be, and how undesirable a false negative would be.

In our setting, one might ask whether Roemer's so-called practical approach or proxy has an epistemological function or is thoroughly pragmatic. While Roemer does not address this question, my strong sense is that for him, the pragmatic strengths of his proposal come largely from its alleged epistemological strength. His approach is meant to identify accurately most of the time who has control and responsibility, and how much. Roemer nowhere alludes to factors like the reception of certain beliefs and the potential counterproductivity of trying to act on them, or the undesirable effects of a false positive or a false negative. So I believe that even if Roemer would recommend his approach as a practical rule in a thoroughly pragmatic sense *too*, that's primarily because it gets at the truth most of the time (in addition to the convenience that demographic data are often available to planners). I shall therefore interpret "at least as a practical rule..." in what I called the "epistemological" sense, as "it is true at least most of the time that..."

In more general abstract form, Roemer's implicit argument in relation to the pure case of male steel workers seems to be, for certain properties *A* and *B*, that premises 1' and 2' would lead to a certain conclusion through an implicit hedge assumption 3':

Premise 1':   For every individual who is an *A*, being an *A* lies outside that individual's personal control.

Premise 2':   All *A*s are *B*.

Premise 3':   For every individual who is an *A*, if being an *A* lies outside that individual's personal control and all *A*s are *B*, then *at least as a practical rule*, being a *B* lies outside that individual's personal control.

Conclusion':   For every individual who is both an *A* and a *B*, *at least as a practical rule*, being a *B* lies outside that individual's personal control.[6]

---

[6] Compare: "If we are to indemnify individuals against their circumstances, we cannot hold them responsible for being members of a type with a poor distribution of effort" (Roemer and Trannoy 2016:

And the abstract thought behind a particularly pure case like that of the female professor, in which only one such professor smokes, seems to be, for properties C and B:

Premise 1":   For every individual who is a C, being a C lies outside that individual's personal control.

Premise 2":   Only a single C is a B.

Premise 3":   For every individual who is a C, even if being a C lies outside that individual's personal control, if only a single C is a B then *at least as a practical rule,* being a B lies within that individual's personal control.[7]

Conclusion":   For every individual who is both a C and a B, *as a practical rule,* being a B lies within that individual's personal control.

Extensions are then needed in order to move from these pure case syllogisms in which *all* type members are B or *only a single* type member is a B, to more realistic mixed cases in which only a half of type members, or a little more than a half, or a little less, are B. But the two pure cases are a good start, and my argument below does not exploit any opening about the difference between these pure cases and more realistic mixed cases.

In addition to this abstract reasoning, Roemer's recurring emphasis on going beyond "formal" equality of opportunity (Roemer 1995: 3) and mere "physical" barriers to free choice (Roemer 1995: 4), and instead heeding "real" opportunity and "effective" barriers, displays thorough realism and sensitivity to concrete detail. As public health scholars know, when in theory a person has full opportunity to avoid a certain health risk, manifest and covert bottlenecks can still hamper its avoidance. A smoker eligible for smoking cessation training might be unable to afford it. One with two jobs and a family might lack the time to attend even free cessation training. If there is no transportation to the free clinic, if publicity about the training is not in a language that the smoker reads, if there is strong physical or psychological addiction, barely-resistible advertising for cigarettes, formidable peer pressure to smoke, or a combination of many such barriers, quitting is often (near-) impossible. For most practical purposes, the smoker will count as lacking control and responsibility over whether he or she quits. Roemer's proposal may be to compare the smoker to a group of peers in order to sift out the consequences of that smoker's autonomous choices to take a risk from environmental and other pressures that may trigger risky choices that lie outside his or her

1293–4). In the extreme case in which all type members make the risky choice, "all circumstances which are not within the individual's jurisdiction of responsibility have been accounted for" (Roemer 1996: 243).

    [7] Compare: "the differences... *within a class* are entirely the responsibility of individuals: these differences are due solely to the voluntary choices of persons" (Roemer 1996: 243).

autonomous choice. For example, if everyone in the smoker's workplace smokes then, Roemer might explain, there is strong evidence that something is going on in that workplace that ramps up the pressure to smoke, even if social planners have not put their fingers on what precisely it is. It may turn out to be the unaffordable costs of cessation training, language barriers, the omnipresence of irresistible cigarette ads in that workplace, formidable peer pressure distinctive to it, or still other powerful risk factors for continuous smoking. But even without the ability to tell what background pressures apply to the smoker's social type, planners can already say with confidence that there exist powerful background pressures which the smoker did not create and for which he or she is not responsible. Depending on the percentage of other longtime smokers in that smoker's Roemerian type, this proof of *some* pressures may suffice to establish that over this matter, he or she lacks "autonomous choice" (Roemer 1995: 4).

This, I believe, captures Roemer's rationale for his pragmatic advice for luck-egalitarian planners.[8] There is a large grain of truth in Roemer's advice and general approach. But the following critique also exposes limitations that seem to have evaded previous analyses.[9]

4.3 My First Objection: Risky Choices That Are Typical but Within Personal Control

My first objection to Roemer's approach is that a risky choice that undermines personal control and hence also luck-egalitarian grounds for redress can be common or even universal in the chooser's type. Imagine:

Very special factory: In a certain co-ed steel factory, no worker smokes and all are well paid. Management is paid sensationally, but promotion to management is rare. A new CEO changes much of that. He creates a new club in the factory,

---

[8] Roemer at one place (1996: 242–4) adds that his proposed distribution of resources "will give no individual a cause for reasonable regret." This may seem like an independent rationale, perhaps along Scanlonian lines. However, both the term "reasonable" and the term "regret" require sharpening, and here they seem to come down in combination to something like the absence of Scanlonian complaints. Yet Roemer's reasoning for such a Scanlonian determination would presumably adduce considerations like the ones that I ascribed to Roemer above (no other considerations are mentioned). Since the Scanlonian determination seems to stand or fall on the considerations above, we might as well focus directly on evaluating the considerations above—as I go on to do.

[9] Another type of novel criticism of Roemer was suggested to me by Dan Hausman and by Mark Budolfson, respectively. Their critiques confer around very small or arbitrarily defined social types. Imagine getting the female professor off of the personal responsibility hook by pointing out simply that in the small type that is *her own household*, "everybody" (i.e., she and her partner) smokes; or doing so through a large but arbitrary or even "gerrymandered" type designed to include only longtime smokers, say, *residents of floor 2 of 152 Washington Street or floor 5 of Walker Street*. My own objections do not rely on this type of criticism, which might be resolved through specifying a certain criterion for a valid social type, say, that it comprises only of familiar broad sociological categories.

SmokerMan, that is open only to male workers. And it quickly becomes clear that schmoozing over cigarettes in the club nearly guarantees a worker's promotion to management and the related sensational pay. As a result, all male workers take up cigarette smoking, around the same time as each other. They realize full well that smoking is addictive and detrimental to their health. But smoking enables them to easily secure lucrative promotion that would be rare otherwise. The positive effects of the resulting sensational pay on their overall welfare prospects slightly exceed even the very negative direct effects from the unhealthfulness of smoking. This is not a calculation error on club members' parts, or something they choose without understanding or the ability to deliberate rationally over the matter. While non-smoking would have dramatically protected their welfare if other things were equal, in this very special case, taking up smoking slightly improves net welfare prospects for all male workers.

Intuitively, in *very special factory*, the resulting high prevalence of smoking among male steel workers notwithstanding, smoking remains within the control of many (including even those male workers who now smoke, albeit a little less than their peers). Male workers' universal choice to smoke could be and in many cases is autonomous and genuine. That choice is not forced: no one is threatening these male workers with penalties should they refuse to smoke. Remaining only on a good salary not a sensational one is not so bad as to leave a worker no acceptable choice except to smoke for promotion. And these workers can and do choose with full information and make no calculation errors. Their choice to pick up smoking is not a result of miscalculation, hypnosis, unfounded fear of being fired unless they smoke or the like. Tellingly, prior to the association of smoking with lucrative promotion, none smoked. They pick up smoking because they figure correctly that smoking is likely to give them a net prospective improvement. For these reasons, my intuition is that these particular smokers' risky choices are not excused. If luck egalitarians support higher insurance premiums for willing smokers, they ought to do so for these smokers. While on some approaches to free will and responsibility no person is ever in control of anything, there is no special reason to deny control and responsibility to these particular smokers.

An opponent might point out, quite correctly, that the situation created by the new CEO violates equality of initial opportunity (Vallentyne 2002) between male and female workers. Women are unfairly denied the opportunity to make the choice that secures lucrative promotion. That is correct, but neither here nor there for my point—that in this unjust situation towards women it is fair to hold the men responsible for their choices to smoke.

Very special factory questions the claim that when a member of a Roemerian type takes a risk, but the level of that risk remains below the average or the median for that type, that member is never especially in control of their actions (and therefore, for luck egalitarians, never especially responsible for them).[10]

Admittedly, refuting that claim is not enough in order to bring down Roemer's proxy rule. A proxy rule may remain successful so long as refutations remain rare or unimportant, and the proxy rule gives the right answers on most cases or on the most important cases. And indeed, *very special factory* is a very special case.

Consider, however:

Not-so-special rescue: All boat owners like sailing to a certain remote atoll in the middle of the ocean. The atoll is exquisite, unforgettable. The sail is enjoyable and nearly always goes well. Occasionally, however, there is a storm on the way back, and an incredibly expensive social mission to save boaters' souls becomes necessary. As is widely known, that rescue mission is readily provided by the boaters' solidaric nation, and nearly always succeeds. So boaters' risks to life and limb are widely known to be very small and heading to the atoll, to be highly beneficial to them in prospect. Heading there is not strictly irresistible, however—no one is physically coerced, threatened, or forced by bad alternatives to head to the atoll, lacking in knowledge or in the ability to calculate risks, or the like. Even if whether one owns a boat is determined by one's background, heading to the atoll remains optional, and a morally responsible choice.

From a luck-egalitarian standpoint, would it be fair for the boaters' nation to hold boaters financially responsible for their choice to take this remote risk of needing rescue, in sailing to the atoll? That could take the form of charging those in need of rescue for some of the dire financial costs of any rescue, or exacting a significant premium for a license to sail to the atoll, which finances rescues.

In this not-so-special case, boaters who choose to sail to the atoll intuitively have control over the matter, or at least they can have as much control as people have over most matters. The reason seems to be that boaters' "gambles" in sailing to the atoll are rational and prudent. While there is risk in sailing there, the risk is very small, and large aesthetic benefits are guaranteed. Intuitively, therefore, sailing there can remain within a boater's control, as well as in his or her responsibility. It can remain so even for boaters who take somewhat fewer risks than is the average or the median in their Roemerian type, *boaters*. Most members of that type will have sailed to the atoll many times.

We are now in a position to see the falsity of Roemer's implicit hedge assumption for the first pure case in which all type members take the relevant risk. Recall that in its abstract formulation, that hedge assumption stated:

---

[10] The *very special factory* case was inspired by Thomas Scanlon's (1989) *Wisconsin voters* case.

Premise 3′:   For every individual who is an *A*, if being an *A* lies outside that individual's personal control and all *A*s are *B*, then *at least as a practical rule*, being a *B* lies outside that individual's personal control.

Set aside for a moment the "as a practical rule" clause. Without that clause, the assumption is clearly false, our two examples illustrate. Even when all *A*s are *B*, being a *B* does not always lie outside these *A*s' control—there may be other reasons why all *A*s are *B*. While it is remarkable that strictly all *A*s are *B*, and one possible explanation for that would have been that *A*s lacked control over the matter, sometimes the explanation is different. Sometimes it is that the *A*s willingly chose to be *B*s when they had the real opportunity to be either *B*s or *non-B*s because they believed correctly that to be *B* would prospectively benefit them. That is the most plausible explanation for the risky choices both in *very special factory* and in *not-so-special rescue*, hence the assignment of maximal control and responsibility in these cases.

It is true that the "as a practical rule" clause in premise 3′ makes evaluating that premise much harder. But in Section 4.1 above, we saw that that clause should be interpreted here as "it is true most of the time that . . . " If so, inasmuch as Roemer's approach fails to get at the truth most of the time then, by its own lights, it fails. While my *very special factory* example is a mere proof of concept, uninstructive about what happens most of the time, the *not-so-special rescue* example is meant to be more representative of normal events. Surely other prudent gambles exist. Whenever risky gambles are prudent, and there is no independent evidence of lack of control (e.g., no independent evidence of duress or manipulation), a choice typical of one's Roemerian type could easily remain within the person's control, not outside it. Thus, Roemer's proxy fails the epistemological test of capturing the truth most of the time.

4.4 My Second Objection: Risky Choices That Are Atypical but Without Personal Control

An imprudent choice can also be rare in the chooser's Roemerian type yet command luck-egalitarian priority for compensation, because it possibly or probably lies outside the person's control. Imagine:

> *A mundane reality*: Allocators and planners know that 1 percent of the population, randomly distributed across the Roemerian types they know, carries gene *X*, which, they also know, makes smoking simply irresistible, compromising the person's autonomy to avoid smoking. They cannot test who has that gene. But they note that some members of Roemerian types where smoking is rare have smoked for many years. In particular, although 60-year-old female professors

hardly ever smoke, some smoke for many years—way above the median or the average for that Roemerian type. The allocators and planners wonder whether these members might have that genetic conditioning. If they do, that would undermine their autonomous control over the matter.

In *a mundane reality*, the choice to smoke for many years is rare in the relevant Roemerian type. But that case also shows that when smoking for many years is rare, that does not even begin to settle whether the individual chose to smoke for many years autonomously. He or she may be smoking for many years, atypically for their social class, because of a personal factor that compromises their autonomy to avoid smoking, albeit one that does not coincide with their Roemerian type (in this case, a genetic impulsion to smoke).

Projecting onto Roemer's own classic example, it would be premature to deem a female college professor who, peculiarly for her type, has smoked for many years, in control and hence personally responsible for her smoking. Such a female professor's peculiar choice may just mean that she is dealing with pressures from which other female professors are free. Unbeknownst to allocators, she may be genetically inclined to addiction. Or she could be a victim of subliminal advertising, addicted to smoking since before the age of maturity and responsibility,[11] depressed, preschizophrenic, friends only with smokers or subject to other overwhelming influences from which most female professors are free, and for which allocators lack a test.

Such situations are quite mundane. Allocators rarely or never know each individual's level of exposure to every strong risk factor for smoking. Depression or preschizophrenia could be either undiagnosed or protected medical information. The professor's friends network will not normally figure on any public record. And Roemer is painfully aware that datasets are always very partial, especially in developing countries (Roemer and Trannoy 2016: 1317–18, 1327–8). And even when some allocators know a candidate recipient's case in detail, advance social planners clearly lack that knowledge. In short, an advance decision to penalize any female professor for having smoked a lot well may penalize a person who was especially unable to choose otherwise, in the name of equality of opportunity.

Roemer might answer that if, objectively, the professor has gene X *and* having gene X compels smoking, then, objectively, that professor belongs to a Roemerian type where smoking is prevalent, after all—she belongs to the type *female college professor who has gene X*. While her genetic combination might be unknown to allocators or to anyone else, her risky choice to smoke is not above average for her type, and, as per our intuitions about such genetic conditions, objectively she

---

[11] This example is Pedro Rosa Dias's (2010: 253).

should not be held responsible for it. Thus, Roemer may continue, the current complication is merely epistemological or practical, not real.

This answer is not open to Roemer. For the most part, he is discussing, not the circumstances that are objectively important in determining behaviors like smoking but the ones that, as he puts it, "seem to be" (Roemer 1993: 150) important in determining such behaviors. Like many other pragmatic rules, Roemer's seeks to serve agents with incomplete knowledge, here and now. That's precisely why he can compare current-day Denmark and current-day Hungary (Roemer and Trannoy 2016: 1295), or assess whether opportunities were particularly unequal in Egypt on the eve of the Arab Spring (Assaad et al. 2018)—presumably, only given what the data we now have tell us. Roemer (1993: 158) is also specifically emphatic that there are plenty of things that social planners do not know.[12] Some passages readily admit that current type definitions spring from our ever-limited knowledge, and may change as our knowledge improves:

> Suppose our medical technology were sufficiently advanced to ascertain the degree of this predilection for all persons. This, indeed, should also be a component of type, as it is clearly beyond the person's control... the development of medical technology, in many cases, will cause society to add new components to the list constituting the definition of type. As this happens, some actions that formerly appeared to be matters of personal responsibility come to be seen as due to circumstances beyond the person's control.   (Roemer 1995: 6)

This critique of Roemer will arise whenever planners' knowledge of the relevant area is limited. During the (infinitely?) long transition to complete knowledge, social planners will have only partial knowledge of the relevant scientific laws and of individuals' relevant characteristics. Thus, in Roemer's own example, contrary to Roemer's own account, a social planner who lacks that knowledge could not assume that the female professor should be held fully responsible for her smoking. There will always remain the possibility that the professor's divergence from her type is only a reflection of planners' incomplete understanding of how much

---

[12] It is true that at one point, Roemer (1996: 242; my italics) presents the "circumstances" that define Roemerian types saying, "Let us assume that we *know* (as a matter of biological and social *science*) what circumstances are influential in the formation of a person's conceptions of... a successful life." Since knowledge that p assumes p, one could argue that, risk factors per planners' *knowledge* are all also objective risk factors. But at many points, Roemer reocmmends the application of his approach with no delay, and hence, well before scientists and social planners may develop complete knowledge of (a) the circumstances that objectively influence the relevant risky choice (e.g., of all the genes that influence it) and (b) of which individuals are in these circumstances (e.g., of who carries the genes known to be influential). Roemer's position may be that we only rarely know the answers to these questions, and that his own appeal to concrete circumstances is merely for illustrative purposes, not because we ever know them enough to impute responsibility: "My task in this article is to articulate that ethic in a concrete form, so that it could, in principle, be applied in actual social situations by an egalitarian planner" (Roemer 1993: 147).

various circumstance types compel smoking and/or of that particular professor's exposure to such circumstances.

Planners' limited knowledge of applicants' particular circumstances is not only a matter of general human nonomniscience. It is also in the nature of advance planning. Planners set rationing rules without familiarity with the determinate circumstances that enable first-order rationers to enact these rules. Thus, Roemer's own targeting of social planners is partly why his proxy rule misfires.

Roemer may now respond that I am expecting too much of his proxy rule. Generalizations about people always admit of some exceptions, and generalizing works only part of the time, yet to refuse to generalize is an overreaction if Roemer's method gets most cases right. Allocators are better off using any low-resolution information at their disposal (say, about large demographic categories), he could insist, even when higher-resolution information (say, about individuals' genetic inclinations) would have altered some judgments.

But given the current stage of scientific knowledge, the proxy that Roemer proposes might mislead *typically*. First, Roemer seems to imagine social types that are very low-resolution indeed. The number of risk factors for smoking that he asks readers to imagine is three (workers' age, profession and gender) or, at one point, five (1993: 150). And Roemer's applied analyses of real-world equality of opportunity often compare a single determinant to the relevant choice. An example is the level of education of one of a person's parents as a predictor of his or her high income (Roemer and Trannoy 2016: 1306f.). Surely there can be more autonomy-limiting influences on a person than one to five influences! Roemer may explain that, to keep the calculations manageable, planners must use only a small subset of relevant risk factors; he sometimes suggests that the considered characteristics are only the "main" risk factors for the choices he examines. But then he should not describe (as he does: Roemer and Trannoy 2016: 1306) situations in which the particular circumstance he checked does not predict an advantage as ones in which all differences in that advantage between individuals are a matter of individual "effort" and not "circumstance."[13] In fact, such false attributions of responsibility should be common; at least in public health, contributors to many poor outcomes and to many risk behaviors are each small in absolute terms. The combination of a few such small influences would leave large gaps in prediction, leaving plenty of room for unknown risk factors for the relevant risky choice that lie outside the person's control. Roemer

---

[13] Compare: "the frequency distribution of years smoked for a group is a characteristic of the type, not of any person. People are not responsible, by hypothesis, for the group they are in; hence they cannot be held responsible for the frequency distribution of years smoked that is characteristic of their group. Exactly at what point in the distribution Fernando sits, however, is, by definition, a consequence of his autonomous choice: for society has already factored out *everything* that it considers to be beyond Fernando's control, in so far as his smoking behavior is concerned, by assigning him to a group" (Roemer 1995: 6; my italics).

and a coauthor seem to admit as much.[14] Planners who rely only on a few risk factors, each of which is small in absolute terms, would be unable to tell reliably whether what looks like an atypical risky choice reflects the crudeness of their own type definitions or a voluntary choice by the candidate.

This critique may initially seem to misfire because Roemer's real-world ana-lyses' main focus is equality of opportunity between Roemerian types, not between individuals.[15] A fairer distribution between Roemerian types is a part of what Roemer hopes for his pragmatic approach to achieve—(e.g., Roemer 1993: 156; Roemer and Trannoy 2016). Importantly, however, fairer distribution between individuals, in which their types only inform the discussion, is another major goal (Roemer and Trannoy 2016: 1306), and emphatically the theoretical foundation of how we distribute between groups.

Besides, Roemer (1995: 4) states, in relation to individual smokers: "I propose that we seek a distribution of socially financed medical care which is equal across groups, for all those who exercised a comparable degree of responsibility in regard to smoking." In other words, it is true that in the first instance, his proposal pertains to distribution between groups. But then, individual group members receive equally as counterparts who exerted the same degree of effort compared to their own types. That ultimate distribution is to individuals, and necessarily based on individual-specific information.[16]

Another possible response to my critique is that on some approaches to equality of opportunity, not every circumstance beyond an individual's control, which constrains his or her options in life, curtails his or her equal opportunity; only ones comprising the basic structure do so. A society in which futures are blighted by many members' sheer genetics (and not by discrimination, stigma, or the like) can remain perfectly just. But that response is not available to Roemer, whose conception of equality of opportunity is luck egalitarianism. Roemer's declared goal is "to find that policy which nullifies, to the greatest extent possible, the effect of circumstances on outcomes, but still allows outcomes to be sensitive to effort" (Roemer and Trannoy 2016: 1294). As this quote illustrates, no

[14] "The data sets that enable one to measure inequality of opportunity usually contain information on only a small set of circumstances (such as the education or income of the parents). Consequently, if one measures effort as the residual determinant of outcomes, once these few circumstances have been accounted for, it appears as if differential effort is massively responsible for outcomes. In fact, luck, meaning the effect of unobservable circumstances, plays a large role" (Roemer and Trannoy 2016: 1305, see also 1317).

[15] As Peter Vallentyne and Marc Fleurbaey suggested to me.

[16] The dearth of individualized information also threatens some of Roemer's determinations that one system has more luck-egalitarian equal opportunities than another. Even if in country A, smoking turns out to correlate with parental income (say) somewhat less than in other countries, country A-related opportunities may remain less equal overall. Assume that smoking correlates in country A very closely with an unobserved (or unobservable) individual circumstance. Imagine for example that depression goes almost completely untreated and unregistered in country A because of strong stigma. Then, a country-A smoker with high parental income should not be assumed to have made a fully autonomous choice to smoke. His or her decision to smoke may well have been driven by depression.

distinction is made between circumstances that count for this purpose and ones that do not.

Finally, note that some scholars, including Fleurbaey and Schokkaert (2009: 84–5) and Rosa Dias (2010: 262–3), raise the somewhat related challenge that the precise causal mechanism from a person's compelling circumstances to his or her risky choices is often "unobservable." For example, we might observe correlation between poor mental health or low educational achievement and smoking, strongly suggesting a causal connection, without knowing the causal pathway through which these circumstances increase a person's tendency to smoke (or, one might add, the ways in which both these circumstances and the person's smoking may emanate from a shared causal source). But there is also a difference between these scholars' challenge and my objection. For such scholars:

> If people are not to be held responsible for their socioeconomic background, can they then be held responsible for their smoking behavior, if this behavior turns out to be very highly correlated with (and therefore perhaps caused by) their social background? In our opinion, the only adequate way to tackle this difficult issue is the construction of good structural models of behavior, which in prin-ciple should create the possibility to disentangle the social class and the pure life-style effect.   (Fleurbaey and Schokkaert 2009: 84)

My proposed approach differs in several ways. First, in scope: my approach adduces not so much unobservable mechanisms leading from the known circum-stance to the risky choice, e.g., pathways through which educational achievement influences choices on smoking, but unobservable and unobserved circumstances that may be independent from the known circumstance, e.g., gene X and its impact on these choices. That impact can remain independent of any impact of gene X on educational achievements. Second, according to my approach, even the confirmation of a causal connection between a risky choice and one's circum-stances would not confirm one's lack of control over that choice. Finally, my approach does not suggest that what must happen is a psychological modeling of behavior—below I propose an alternate proxy, namely, the choice's level of prudence: for me, a choice that is clearly imprudent suggests inadequate control even if the precise psychological mechanism that led to it remains unknown.

To summarize my points thus far, when a person makes riskier choices than most members of his or her Roemerian type that is admittedly a red flag. But that flag can indicate either reckless autonomous choice or barriers to autonomous choice, and not the former only. Therefore, riskier choices than is common in one's type do not prove or even strongly suggest personal control and luck-egalitarian responsibility for resulting disadvantage. Making riskier choices than one's peers can just as well constitute preliminary evidence of limitations on autonomy that remain opaque to planners.

Return now to the abstract form of Roemer's hedge assumption in relation to the female professor:

Premise 3":   For every individual who is a *C*, even if being a *C* lies outside that individual's personal control, if only a single *C* is a *B* then *at least as a practical rule*, being a *B* lies within that individual's personal control.

I have now argued that it is premature in principle and in practice alike to ascribe control to a person just because we are unable to correlate his or her actions with some property known to block personal control. He or she may have other properties that block personal control. In *a mundane reality*, that could be a certain genetic disposition. In Roemer's analysis of the relation between the education level of one's least educated parent and one's earning over a certain income, it could be the education level of one's other parent, or parental income, or any of many other unanalyzed parameters. So Premise 3" is false. Roemer's practical rule will mislead in a significant portion of cases.

I will ultimately embolden this claim. In the case of smoking, for example, not only do we not know that smokers like the female professor in Roemer's case have chosen to smoke autonomously. Chances are they did not choose autonomously: as I shall argue, the *highly detrimental* nature of smoking for most smokers chips at the plausibility of thinking that a smoker's choice to smoke was autonomous. It makes it more plausible to view choices to smoke, whether typical or atypical, as instead an indication of barriers to free choice that are real but opaque to allocators and planners. The next section develops this thought into an alternative proxy for inadequately controlled choices, which do encounter luck-egalitarian responsibility.

4.5 From Prevalence in One's Type to Clear Imprudence

How is it possible that, in *very special factory*, universal smoking among male steel workers remains an autonomous choice? The reason seems to me to be as follows. In that very unusual factory, *SmokerMan* makes the choice to smoke prudentially defensible for each worker. When, for agents of a certain Roemerian type, a certain choice is prospectively beneficial on balance, the fact that all or most of them make that choice raises no eyebrows. We do not need special explanations, such as that the choice was compelled. The more general lesson is that in order to determine an individual's level of autonomy in making a choice (and relatedly, his or her luck-egalitarian personal responsibility for that choice), it helps to know how good or bad that choice is in prospect for that individual. Thus, as a proxy for the use of luck-egalitarian social planners, the best "cut" might not be between risky choices that are prevalent in one's Roemerian type and ones that are rare. Elsewhere I have argued at length that the right "cut" could be between risky choices that are clearly imprudent and ones that are on balance prudent or otherwise rationally defensible (Eyal, under consideration).

In particular, when someone makes a risky choice that on balance is prospectively bad for them both self-regardingly and also for their other, altruistic or personal, goals, then I call such choices "clearly imprudent." Clearly imprudent choices are unjustifiable rationally. They call for a special explanation, and those include duress, incomplete information, limited rationality and the like. The less rational the gamble, the stronger the indication for (severely) compromised autonomy in the process that led to the choice.

Take regular smoking (as distinct from smoking in the highly unusual situation of lucrative promotion guaranteed by smoking). Smoking kills a quarter of smokers prematurely, often after years of serious discomfort and severe physical or mental disability. Yes, smoking can be pleasurable and carry various social and personal meanings. It does not always result in significant harm to health. But let us grant, for the sake of argument, that regular smoking is on balance prospectively *very bad* for the smoker. It is on balance not just a clearly-imprudent choice, but a highly imprudent one. If so, I want to argue, that nearly suffices to establish that regular smoking is driven by something different than an autonomous choice. We might not know what is driving someone's smoking—whether it is ignorance, manipulation, addiction, genetic proclivity, peer pressure, hyperbolic discounting or a still different influence. But we pretty much know that whatever is driving it is a factor that curtails autonomy. Hence, even for individuals in whose Roemerian types the prevalence of smoking is unknown, or known to be low, and even when planners' knowledge of the risk factors for smoking (and their respective strengths) or of an individual's ranking on them remains very limited, we can usually conclude safely that an individual's choice to smoke regularly was inadequately autonomous—simply because regular smoking is so bad for smokers. The peculiarity of the fact that people smoke although smoking is bad for them already points in the direction of nonautonomous influences as explanations for why they do. There may or may not exist instances of genuine weak will in which prospectively bad gambles are made with full autonomy. But genuine weak will on matters affecting one's own interests is arguably the exception, especially when choices are prospectively very bad indeed. As a proxy rule, planners and other allocators can therefore assume that if a person's choice is very bad for that person—arguably a common reality when it comes to both regular smoking and most other risk behaviors that worry public health officials—then that choice is probably not genuinely in that person's adequate control, but primarily attributable to constraining influences. Thus, while we have found the degree of prevalence of risky choices in one's type wanting as a guide for luck-egalitarian planners, a proxy that they should explore more is risky choices' degree of clear imprudence.

Admittedly, Roemer's proposed proxy has some pragmatic advantages over mine. In particular, belonging to this or to that social demographic is relatively descriptive and data confirming or excluding it are available and relatively uncontroversial. By contrast, judging which choices are on balance prudent is controversial and perhaps not the job of social planners. The correct judgments may also vary for different individuals depending on individual-level situations of which planners are unapprised. However, as I argue elsewhere (Eyal, under consideration), doctors and public health officials already need to judge what tends to be prudent or rationally condonable for their patient constituencies, and what tends to be clearly imprudent. This is how they decide, for example, what is the proper ambit of preventive medicine and what would be overreach: while for people who consume enough calories, any amount of cake is unhealthful, to recommend and start costly campaigns to stop people from ever tasting any cake would be overreach, because tasting cake once in one's life has a tremendous marginal welfare benefit in nonhealth terms, well worth the unhealthfulness of that lone crumb. To determine correctly that such a campaign would be overreach, doctors and public health officials must make judgment calls on the correct balance of health and other interests. These existing judgments are precisely about what choices would be clearly prudent for most of us on balance. These judgments can be applied to our area as well.

4.6 Applications and Implications

The proposed approach has both practical applications and theoretical implications. Let me mention briefly three that merit further exploration.

4.6.1 Initially-Difficult Cases

In many places around the world, certain clinics have come to possess a public image as "feminized" social spaces, which unfortunately can put off male patients.[17] HIV clinics in southern Africa, for example, are often socially construed as spaces unsuitable for men—perhaps because many are former maternal health clinics, an image that stuck (Dovel et al. 2020; Radunsky et al. draft). Such "gendering" of spaces and workplaces (West and Zimmerman 1987; Acker 1990; Spain 1993; Dovel et al. 2015) recalls how certain Wall Street companies' lap-dance parties may have put off potential female employees for years. It may also recall how female-unfriendly conversations in the American Philosophical

---

[17] I am grateful to Alex Radunsky for related conversations and for the cites included here.

Associations' so-called "smokers" may have contributed to putting off potential female philosophers for decades.

When an HIV clinic's feminized social image wards off male patients, there are straightforward public health reasons to try to change that social image or to entice these male patients to the clinics otherwise. But a separate question also arises, about distributive justice. Few if any luck egalitarians would harshly recommend denying any HIV patient of treatment or its financing. But reduced access to health services would by luck-egalitarian lights warrant especially-full social indemnity, more readily than a history of autonomous risky choices. So does HIV clinics' feminized social image hamper the fair access of male patients to health advantage? If it does, that would somewhat reduce men's luck-egalitarian responsibility for any bad outcomes from their neglect to get HIV-tested and potentially start treatment promptly. Or does that social image reduce only men's desire and autonomous choices to utilize what remains perfectly accessible HIV care, preserving their luck-egalitarian responsibility for the resulting health disadvantage?

Initially, the answer as to how autonomous are most of these men's choices to avoid feminized HIV clinics is not obvious. For one thing, the precise psycho-logical mechanisms driving these men's choices are opaque. Is the feminized image of these clinics a serious threat to these men's social standing in social settings we hardly understand, a merely perceived threat, a well-founded per-ceived threat, sheer machismo or even outright misogyny?

I believe, however, that luck egalitarians can already answer the question whether these men's choices can be understood as autonomous enough for the assignment of control and luck-egalitarian responsibility, with some confidence. The choice to avoid HIV clinics and thereby refrain from HIV testing despite signs that one might have gotten infected and need treatment tends to be calamitous for the individual. It delays life-saving treatment at least until the virus has done some damage to one's body. And it typically gains little or nothing. Putting one's head in the sand about a potential infection will not make it go away. Ultimately, visiting the clinic will become necessary, along with any HIV stigma encountered from its use, and along with any shame and any social risks from having frequented a feminized space. If so, the choice to keep away from HIV clinics would seem, for most everyone who makes it, clearly imprudent. My proxy rule moves from that judgment quite directly to the conclusion that a man who chooses to avoid the HIV clinic, whether a victim of informal social sanctions against "effeminate" men, of his own false and unfounded prediction of such sanctions, of his intern-alized machismo, of an uncontrollable sense of mortification or of still other drivers, is rarely in sufficient control of this risky decision for personal responsi-bility over the outcomes. Luck egalitarians must conclude that in all justice he cannot be adequately responsible for his woeful choice. Avoiding badly needed HIV care must be driven by barriers to autonomous decision making of one sort or another. So men who avoid HIV clinics with a feminized social image deserve full indemnity at the bar of luck-egalitarian justice.

My proposed proxy, to assess whether risky choices are clearly imprudent, thereby helps us make headway on an otherwise difficult question. It could likewise help with other health policy questions that initially seem opaque for luck egalitarians who lack direct evidence on the circumstances in which a choice was made. So long as the risky choice is on balance clearly imprudent, luck egalitarians should usually deem that choice sufficiently uncontrollable and hence, in all justice, an excusable one.

4.6.2 Free Will

A major mystery of Jewish medieval philosophy was how it is possible that "All is expected but the authority has been given" ("הכל צפוי והרשות נתונה", *Ancestral Chapters* [פרקי אבות] III.19). That is to say, how is it possible that an omniscient God can predict everything, including people's future choices, yet these choices remain free?

The framework I am developing points to a possible answer. In *very special factory*, an omniscient being could predict that all male workers will take up smoking. So could an omniscient being with regard to all boat owners in *not-so-special rescue*. But these male workers and boaters could retain freedom of choice over these matters. The predictable choices in these examples can be free. The male workers could if they so wished avoid smoking (and workplace promotion) and the boaters could avoid sailing to the atoll (and enjoying its beauty).

We may therefore respond to medieval colleagues that a perfectly free choice can be perfectly predictable.

4.6.3 Nudging

Libertarian-paternalists claim that choices that are predictably prevalent in the presence of effective "nudges" can remain perfectly free (Thaler and Sunstein 2021). How much can my line of reasoning warrant that claim? Take a concrete case. A behavioral psychologist predicts on good grounds that altering cafeteria organization and placing salad and not pizza first in line and at eyesight would make a majority, or even 100 percent, of clients, all of whom used to choose pizza over salad, choose salad over pizza. She turns out to have been right. Does that show that, contrary libertarian paternalism, the clients' choice was unfree?

My first observation is that this alone would not clarify that the cafeteria reorganization makes clients choose (healthfully but) unfreely. Whether they choose unfreely would depend in complex ways on the precise psychological mechanism that prompts the change in their purchasing patterns. If, for example, the mechanism is that clients slavishly pick whatever item they see first, and salad is displayed first, then clients will have had no real choice except to pick salad. Suppose however that clients are only prompted to give *serious consideration* to the first item they see. Then, given that cafeteria clients, too, want to live long and that they know (let us assume) that salad is likelier to help them achieve that, it is possible that 100 percent would choose salad, yet that their choice remains fully free and autonomous.

Importantly, the latter account, which understands the customers' choice as free, would not have been available had choosing salad been a prospectively bad gamble for cafeteria clients—say, if the salad were (secretly) poisoned. Then we would have to conclude that, while the exact psychological mechanism needs to be characterized, it simply cannot involve fully autonomous choice. A full explanation would need to invoke clients' ignorance about the poison, forced feeding or some other barrier to autonomous choice. Likewise, when a cafeteria is organized to induce buying pizza not salad (say, because owners' profit margins are higher for pizza), then, assuming that salad is better for most clients overall, we can already tell that for most clients, a harmful choice architecture defeats both autonomy and the luck-egalitarian ground for holding these clients accountable for their choice.

Thus, when changing cafeteria choice architecture pretty much guarantees clients' choices, their choices can remain free, but normally not if these choices are clearly imprudent. It may seem strange that the same cafeteria choice architecture that tends to result in a certain choice might be deemed free or unfree based on whether that choice is clearly imprudent, or not. But this is a sheer reflection of the use of proxies in the absence of independent evidence on that choice's degree of freedom.

4.7 Conclusion

The prevalence of a choice in one's Roemerian type is sometimes an inadequate proxy for its luck-egalitarian excusability. On such occasions, Roemer's proposed proxy can be usefully complemented or replaced by a different proxy. In particular, when a risky choice is clearly imprudent, that suggests lack of control and hence, per luck egalitarianism, an excusable choice. This is not to deny the important kernel of truth in Roemer's brilliant analysis, and the usefulness that his proposed proxy rules retains in many situations. Future work should characterize when Roemer's proxy remains helpful and when substitutes are needed. That work could characterize further the best ways to combine Roemer's emphasis on the relevant risky choice's prevalence in one's social type and my proposed emphasis on the clear imprudence of that choice.

References

Acker, Joan. 1990. "Hierarchies, Jobs, Bodies: A Theory of Gendered Organizations." *Gender & Society* 4 (2): 139–58. doi: 10.1177/089124390004002002.

Assaad, Ragui, Caroline Krafft, John Roemer, and Djavad Salehi-Isfahani. 2018. "Inequality of Opportunity in Wages and Consumption in Egypt." *Review of Income and Wealth* 64 (s1): S26–54. doi: https://doi.org/10.1111/roiw.12289.

Bratman, Michael E. 2003. "A Desire of One's Own." *Journal of Philosophy* 100 (5): 221–42.

Dovel, K., S. L. Dworkin, M. Cornell, T. J. Coates, and S. Yeatman. 2020. "Gendered Health Institutions: Examining the Organization of Health Services and Men's Use of HIV Testing in Malawi." *Journal of the International AIDS Society* 23 Suppl 2: e25517. doi: 10.1002/jia2.25517.

Dovel, K., S. Yeatman, S. Watkins, and M. Poulin. 2015. "Men's Heightened Risk of AIDS-Related Death: The Legacy of Gendered HIV Testing and Treatment Strategies." *AIDS* 29 (10): 1123–5. doi: 10.1097/QAD.0000000000000655.

Eyal, Nir. under consideration. "Luck-egalitarian priority for the imprudent."

Fleurbaey, Marc, and Erik Schokkaert. 2009. "Unfair Inequalities in Health and Health Care." *Journal of Health Economics* 28 (1): 73–90.

Frankfurt, Harry. 1987. "Identification and Wholeheartedness." In *Responsibility, Character, and the Emotions: New Essays in Moral Psychology*, edited by Ferdinand David Schoeman. Cambridge University Press.

Hurley, Susan L. 2003. *Justice, Luck and Knowledge*. Oxford: Oxford University Press.

Radunsky, Alexander P., John R. Weinstein, Rifat Atun et al. draft. "Male HIV Testing in Rural Malawi: Qualitative Insights into HIV Under-Testing and Testing by Partner Proxy."

Roemer, John E. 1993. "A Pragmatic Theory of Responsibility for the Egalitarian Planner." *Philosophy & Public Affairs* 22 (2): 146–66.

Roemer, John. 1995. "Equality and Responsibility." *Boston Review* 20 (2): 3–7.

Roemer, John E. 1996. *Theories of Distributive Justice*. Cambridge, MA: Harvard University Press.

Roemer, John E. 2000. *Equality of Opportunity*. Cambridge, MA: Harvard University Press.

Roemer, John E., and Alain Trannoy. 2016. "Equality of Opportunity: Theory and Measurement." *Journal of Economic Literature* 54 (4): 1288–332. doi: 10.1257/jel.20151206.

Rosa Dias, Pedro. 2010. "Modelling Opportunity in Health Under Partial Observability of Circumstances." *Health Economics* 19 (3): 252–64.

Scanlon, Thomas. 1989. "A Good Start: Reply to Roemer." *Boston Review* 20 (2): 8–9.

Spain, Daphne. 1993. "Gendered Spaces and Women's status." *Sociological Theory* 11 (2): 137–51.

Thaler, Richard H, and Cass R Sunstein. 2021. *Nudge (The final edition)*: Yale University Press.

Vallentyne, Peter. 2002. "Brute Luck, Option Luck, and Equality of Initial Opportunities." *Ethics* 112: 529–57.

Watson, Gary. 1975. "Free Agency." *Journal of Philosophy* 72: 205–20.

West, C., and D. H. Zimmerman. 1987. "Doing Gender." *Gender & Society* 1 (2): 125–51.

5

Responsibility, Health Care, and Harshness[1]

Gabriel De Marco

5.1 Introduction

Suppose that two patients are in need of a complicated and expensive heart surgery. Further suppose that they are identical in various relevant respects: for example, state of the heart, age, likelihood of successful surgery, and so on. However, they differ on one feature: for one of these patients, Blair, the need for the heart surgery is largely, if not fully, due to her lifestyle—she is a life-long smoker—whereas the other, Ingrid, has not had this lifestyle, nor any other that is associated with this illness.

Many people think that:

1. We are sometimes responsible and blameworthy for our actions and their consequences.
   Some of those people also think that:
2. We should take this into account when making decisions in the context of health care.[2]

For the purposes of this paper, I assume that 1 is true, and further stipulate that Blair is blameworthy for her illness.[3] For the purposes of this paper, when I discuss an agent who is responsible for something, the reader can also assume that she is blameworthy. This fact about blameworthy Blair, combined with the fact that Ingrid is innocent with regard to her illness—she is not blameworthy for it—suggests that at least in some contexts, we could, and maybe should, treat them differently. If 2 is true, we have reason to implement a policy that adopts, and reflects, this claim in the context of health care in particular. Call such a policy, specific to the context of health care, a *Responsibility-Sensitive Policy*, or *RSP* for short.[4]

---

[1] I am grateful to Ben Davies and Neil Levy for comments on an earlier draft of this paper.
[2] For some empirical studies gauging people's views on this, see Chan et al., Chapter 1, this volume.
[3] A further assumption I make is a common one: S is blameworthy for A only if S is responsible for A.
[4] In this debate, people sometimes discuss different notions of responsibility, and it is not always clear which one is intended. For instance, sometimes people speak of responsibility without any sort of qualification, whereas other times, people speak of moral responsibility, prudential responsibility, culpability, etc. In this paper, I am concerned with sketching possible relationships between

Gabriel De Marco, *Responsibility, Health Care, and Harshness* In: *Responsibility and Healthcare*. Edited by: Benjamin Davies, Gabriel De Marco, Neil Levy, Julian Savulescu, Oxford University Press. © Edited by Benjamin Davies, Gabriel De Marco, Neil Levy, Julian Savulescu 2024. DOI: 10.1093/oso/9780192872234.003.0006

It is commonly thought that, assuming there is an appropriate response to blameworthiness for something or other (e.g., an action, a habit, or an outcome), it will be negative in some way or other. In the context of health care, this might involve, for example, responses in terms of a reduction of claims to certain goods, or of a difference in the treatment one receives. Proponents of RSPs might disagree on what differences in treatment are required or permissible, and this might be because of disagreements concerning a variety of other positions, some of which I discuss below.

In this paper, I aim to clarify the possible relationships that might hold between claims of responsibility and aspects of an RSP; enough to allow me to make some points about what is sometimes called the harshness objection. On this objection, one argues against a particular RSP, or RSPs in general, by pointing out that it would be too harsh. In Section 5.2, I distinguish two possible implications of the claim that an agent is blameworthy, including different sorts of negative responses. In Section 5.3, I elaborate on one of these implications, identify three dimensions along which negative responses can be more or less severe, and discuss the relationship between these and RSPs. In Sections 5.4–5.6, I present three main sources of harshness: harsh reductions of claims to goods, harsh treatment, and an RSP's procedure for determining whether, and the degree to which, a patient is blameworthy. Clarifying these different sources of harshness will help to evaluate when the harshness objection applies to an RSP. In Section 5.7, I wrap things up.

5.2 Responsibility and RSPs

Some think that when we say that someone is blameworthy for something, this implies that it would be noninstrumentally good—it would be good regardless of its effects—for the blameworthy agent to suffer or be harmed in some way as a response to her blameworthiness. This is a thought that we encounter often, particularly when we talk about retributive justifications for punishment.[5] Call this the *Strong Implication* of responsibility.[6]

Although many think that punishment should reflect this implication, one might be hard-pressed to find proponents of RSPs that think an RSP should reflect it. A policy implementing the sorts of responses suggested by the Strong

---

responsibility, the nature of RSPs, and a particular objection to them. Most, if not all, of what I say will apply to different notions. Thus, I use the general "responsibility."

However, sometimes "responsibility" is intended to refer to something like an obligation. For instance, if we say that "a lawyer has a responsibility to her clients," "responsibility" is intended to refer to some sort of duty or obligation that the lawyer has to her clients. This is not the sense of "responsibility" I intend to use.

[5] For some discussion, including various other understandings, see (Clarke et al. 2015, pp. 9–12).

[6] With regards to *moral* responsibility, it has been suggested that minimally, it may be in some sense good that the blameworthy agent feel guilt (see, for instance (Clarke 2013, 2016)).

Implication would be quite difficult to justify. When it comes to distribution of health care resources, such an RSP might hold that we are justified[7] in ensuring that someone like blameworthy Blair suffers or is harmed, even if doing so does not benefit anyone else. For example, we might be justified in letting blameworthy Blair suffer a bit before we operate, or in making sure that the operation would not be as good as the one we do on innocent Ingrid, even if we could easily avoid doing these things.

Another implication we might be thinking about when we discuss responsibility is that, when we say that someone is blameworthy for something, this implies some sort of *reduction thesis*: perhaps that, at least in some contexts, the agent's claims to certain goods have been reduced, that we should not weigh her interests equally, that the strength of her rights has been reduced to some extent, or that the strength of our prima facie or pro tanto obligations to her has weakened. Call this the *Weak Implication*; when we say that someone is blameworthy for some item, this implies some sort of reduction.[8] For simplicity's sake, the discussion will focus on reductions of claims to certain goods; however, most of what I say could be understood in terms of the other sorts of reductions as well.

The Strong and Weak Implications are consistent; one could think that blameworthiness for some item both implies that the agent now has some reduced claim *and* that it would be noninstrumentally good for the agent to suffer or be harmed in response to her blameworthiness. Yet, one might think that an RSP should only reflect the Weak Implication, and discussion of RSPs tends to focus on this Weak Implication, even though theorists are not always explicit about this.

An RSP that merely reflects the Weak, but not the Strong, Implication will justify treating blameworthy agents differently than nonblameworthy ones, at least in some cases. Yet, in contrast to an RSP that reflects the Strong Implication, one that merely reflects the Weak Implication will not justify going out of our way to cause harm and/or suffering to a blameworthy agent.[9]

Even when limiting discussion to RSPs that only reflect the Weak Implication, there can still be substantial differences between different RSPs. They can differ, for instance, on how severe the responses are, or on the nature of the reductions justified by blameworthiness. These differences may result in differences in the relative merits of different policies, and how vulnerable they are to versions of the

---

[7] When I say that some action is justified, I mean that the action is either permissible or required. If an action is unjustified, then it is impermissible.

[8] There is an even *weaker* implication that I will not consider here. Blameworthiness for some item could merely imply that it is fitting to judge the agent as blameworthy, or to hold certain reactive attitudes such as resentment towards her, while stopping short of making any more substantial negative response appropriate. The sorts of negative responses to blameworthiness involved in RSPs tend to be more substantial.

[9] For examples of some related distinctions, see (Duus-Otterström 2012; Olsaretti 2003, pp. 14–16; Segall 2009, pp. 16–18; Vallentyne 2003; Zimmerman 2015, p. 54).

harshness objection. It is thus worthwhile to consider some of the ways that reduction theses can differ, with a focus on claims to health care.

5.3 Reductions and RSPs

First, notice that claims to some good or other can be reduced to different levels. In the extreme case, one's claim can be reduced to zero; that is, one can lose one's claim to certain goods. According to one type of reduction thesis, then, blameworthiness for some unhealthy behavior, and/or its outcome, can result in one's completely losing one's claim to some health care resource. Call this an *unlimited reduction thesis*. Alternatively, one might think that although blameworthiness can reduce one's claim to some health care resources, one cannot lose the claim in virtue of blameworthiness; there is some nonzero limit on the levels to which one's claim can be reduced in virtue of blameworthiness. Call this a *limited reduction thesis*.

An RSP on which blameworthy Blair's claim to surgery can be lost in virtue of her blameworthiness—one which accepts an unlimited reduction thesis—will be more severe than an RSP according to which her blameworthiness results in less of a claim than Ingrid, yet she retains a claim nevertheless, and could not lose it in virtue of her blameworthiness—one which adopts a limited reduction thesis. Different reduction theses can differ on the limits they place on the reduction; other things equal, the more limited the reductions made on the basis of blameworthiness, the less severe the reduction thesis, and the less severe an RSP reflecting this thesis.

Second, there are various claims to goods that we might have in mind when formulating a reduction thesis. Suppose that blameworthy Blair's claim(s) have been reduced in virtue of her blameworthiness. One might just have in mind the claim to heart surgery, or perhaps, more generally, claims to the different resources that may be required for treating negative health outcomes associated with her lifestyle. A reduction thesis that only allows for reductions to claims to these sorts of goods may be called a *discriminate reduction thesis*.[10] A different reduction thesis might hold that in virtue of her blameworthiness, Blair's claims to *any* health care resources can be reduced. Such a reduction thesis would be relatively indiscriminate, and we might call it an *indiscriminate reduction thesis*.

Suppose, for instance, that in an automobile accident that was not her fault, blameworthy Blair ends up breaking a bone. According to an indiscriminate

---

[10] We can imagine all sorts of different ways to carve out the relevant set of claims that are reduced. For instance, one could adopt a reduction thesis that applies to Blair's claim to heart surgery, but not to her claim for medication that can help to mitigate some of her heart problems. This would be an even more discriminate reduction thesis.

reduction thesis of the form just mentioned, her claim to goods required to treat the broken bone may be reduced. According to a discriminate thesis along the lines mentioned above, on the other hand, she has just as much of a claim to these goods as anyone else. The distinction between discriminate and indiscriminate is a matter of degree; other things being equal, the fewer the claims that are reduced by an agent's blameworthiness, the more discriminate (and less *in*discriminate) the reduction thesis. Further, I suggest that other things being equal, the more discriminate the reduction thesis, the less severe it is, and the less severe an RSP reflecting the thesis.

Third, reduction theses can differ on the contexts in which the reduction occurs. The discussion up to this point has treated reduction theses as *contextually general*; the agent's claim to a relevant good will be reduced, and this can justify different treatment in a variety of health care contexts. A different version of the thesis might be *contextually specific*, such that it only applies in a limited set of contexts.

To illustrate, consider a variation of the original case of Ingrid and Blair. First, suppose that the situation is such that we can operate on either Blair or Ingrid, but not both, and suppose there is a further difference between Blair and Ingrid: Ingrid's chances of a successful surgery are somewhat lower than Blair's. Now consider a very general reduction thesis, on which Blair's claims are reduced in all contexts, including this one. Depending on how we should weigh the strength of claims against the chances of success, and depending on how much the claim was reduced, an RSP adhering to this general reduction thesis might suggest that we operate on Ingrid rather than Blair in this modified case.

Suppose, instead, that we are considering a reduction thesis according to which a patient's reduced claim to some health care resources would only be used as a tie-breaker—in cases where all other relevant features are equal, and we cannot treat both patients equally. In this modified version of Blair and Ingrid, this RSP would recommend that we operate on blameworthy Blair, given that the non-responsibility features are not equal (so we do not have a tie). Reduction theses can differ on how contextually general or specific they are; the more contexts in which the reduction applies, the more contextually general it will be, and other things being equal, the more severe the reduction thesis, the more severe an RSP reflecting the thesis.

We have, at least, three dimensions along which reduction theses can differ. Yet it is important to note that a single RSP may reflect reduction theses with different features applying in different contexts, or for different blameworthy items. Consequently, there are many ways that RSPs can differ, and things can get complicated very quickly. For instance, one might hold that for those lifestyles that involve addiction, and thus a reduction or perhaps absence of control, unlimited reduction theses would be false while at the same time holding that unlimited theses apply to blameworthiness for lifestyles that do not involve addiction. Or proponents of RSPs can disagree on the existence of resultant luck; they might disagree on whether bad luck in the outcomes of our actions can affect our blameworthiness. If one rejects the claim that such luck can affect our blameworthiness, then one might be inclined to accept the view that reduction theses apply to blameworthiness for unhealthy lifestyles, but not to their outcomes which are due to luck.[11] Proponents may instead (or also) disagree about the nature of rights or claims. Some proponents of RSPs may hold, for example, that a person's right or claim to life-saving health care treatment is inalienable. Regardless of their commitments about the nature of responsibility, such proponents will likely adopt, at most, a limited reduction thesis concerning one's claim to life-saving treatment; if the claim is inalienable, then one cannot eliminate it on the basis of blameworthiness.

Thus, proponents of RSPs can agree on some reduction theses, while disagreeing on others. If one takes an RSP to be partly constituted by the reduction theses it reflects, we have a large variety of potential RSPs. RSPs can differ in virtue of reflecting different commitments with regard to these reduction theses.

It is also important to note that holding the view that health care policies should be sensitive to patients' responsibility does not commit one to the claim that health care policies should *only* be sensitive to patients' responsibility. Policies already take a variety of factors into account, and suggesting that we should implement an RSP is merely to suggest that patients' responsibility for their lifestyles, and/or their health outcomes, should be one of those factors.[12] My aim is to say something general about RSPs, and proposing an RSP is consistent with different theoretical commitments elsewhere. Thus, I do not discuss specific views of justice, fairness, or equality, even though such views will often have an impact on what sort of RSP one thinks is appropriate.[13] Instead, I focus on types of reduction theses.

However, there is one sort of RSP which may not clearly fit this picture. Sometimes, people suggest a policy on which those who choose to engage in an unhealthy lifestyle are either taxed—for example, by imposing a tax on

---

[11] One can, after all, have an unhealthy lifestyle that significantly increases the likelihood of a negative health outcome and still not have that outcome. Such a person might be blameworthy for the lifestyle, and/or for taking on associated risks, but not the negative outcome (since she did not have the negative outcome).

[12] For an extended discussion of this, and an example of how a Luck Egalitarian view of distributive justice can be combined with other normative claims to avoid some versions of the harshness objection, see (Segall 2009, chapters 4–5).

[13] One common way of arguing for an RSP is to appeal to Luck Egalitarianism (LE), and show how this view implies that we should adopt an RSP. Yet different versions of LE support different sorts of RSPs (Albertsen and Knight 2015, p. 167; Duus-Otterström 2012). Further, avoiding the harshness objection might lead proponents of LE to violate principles of LE (Voigt 2007), or to appeal to nonegalitarian considerations (Segall 2009, chapters 4–5). Finally, although one way of arguing for an RSP is via LE, one need not adopt LE to be in favor of an RSP. For some discussion, see (Herlitz 2019).

cigarettes—or pay a higher premium on health care. These sorts of suggestions are not generally accompanied by the idea that those who are blameworthy have reduced claims.[14] Such a policy is consistent with the reduction of claims. Perhaps blameworthiness for an unhealthy lifestyle reduces some claims, and the purpose of the tax, or the increase on premiums, is to bring the strength of the claims back to normal levels. However, it is not clear that one needs to adopt this picture when proposing such a policy.[15] For the purposes of this paper, I will not be considering these sorts of RSP.

With this framework in place, we can now consider the harshness objection to RSPs. As I argue below, the strength of this objection will depend on what sort of RSP is being objected to. To be clear, I do not intend to rebut this objection; it seems to quite clearly apply to at least some views. My main goal is rather to delineate some of the ways that this objection can work, and how the strength of the objection will relate to the nature of the RSP under consideration, by investigating possible sources of harshness.

5.4 Abandonment and Harsh Reductions

According to the gloss on the harshness objection above, it is an objection either to RSPs in general, or to some specific (forms of) RSPs, on the basis that implementing this RSP would be harsh. Of course, pointing out that a person will be responded to negatively, as a result of their blameworthiness, will not be enough to show a problem with an RSP. Responding negatively in virtue of one's blameworthiness, after all, is simply a part of responsibility practices.[16] In order for the harshness objection to have force, the negative response under consideration must be *too* negative, or *too* severe. As I will be understanding the objection, it states that the negative response under consideration is negatively disproportionate; it is more severe than what would be proportional.[17]

---

[14] At least, the blameworthy do not have reduced claims *once they have paid.*

[15] If this *is* the picture behind such policies, then perhaps some of what I say in the discussion of the harsh-reduction objection will be of relevance.

[16] For a related point, concerning Luck Egalitarianism in particular, see Albertsen and Nielsen's response to the "bad consequences" version of the harshness objection (Albertsen and Nielsen 2020, pp. 432–3).

[17] Albertsen and Nielsen give a detailed discussion of four versions of the harshness objection to Luck Egalitarianism (Albertsen and Nielsen 2020). This disproportionality interpretation is their third version. Their first version, on which the harshness is counterintuitive, is consistent with the others; what distinguishes it might be more a matter of how we justify the claim that something is harsh, not the content of that claim. The second version interprets "harsh" as negative consequences which, as I just suggested, is a nonstarter; and they offer a similar response. Their fourth version appeals to some form of inconsistency in Luck Egalitarianism, which, as they point out, would need to be developed significantly in order to be convincing.

Sometimes, this objection is stated in terms of claims or rights; the idea is that the RSP would reflect a reduction of claims or rights that is too harsh. Call this a *harsh-reduction objection*. Other times, it is stated in terms of treatment; the RSP would result in treatment that is too harsh. Call this a *harsh-treatment objection*. These are obviously related. If an RSP reflects a reduction in a patient's claim that is too harsh, then acting perfectly in accordance with this reduction will often result in treatment that is too harsh. Yet, there may be some circumstances in which these come apart. For example, if there are not enough resources to go around, then a reduced claim for a scarce good may result in one not getting that good at all. We will return to this latter sort of case in the next section.

The most common version of the harshness objection, which gives us the starkest way of showing that an RSP is overly harsh, is what is sometimes called the *abandonment objection*. According to this objection, RSPs in general, or some RSPs in particular, would result in the complete loss of blameworthy patients' claims to health care, and thus justify not treating these patients at all.

Often, the objection is presented as a critique of a broader view of distributive justice. Consider, for instance, Elizabeth Anderson's famous critique of Luck Egalitarianism, where she considers the case of a helmetless, uninsured motorcyclist who is at fault for an accident that resulted in head injuries. While objecting to a specific RSP based on a version of Luck Egalitarianism, she suggests that, according to this RSP:

> the faulty driver has no claim of justice to continued medical care.
>
> (Anderson 1999, p. 296)

Marc Fleurbaey uses a similar case in his discussion of the view on which we should aim to equalize opportunities across individuals in society, except for those inequalities due to agents' responsible behavior. Fleurbaey argues that:

> [i]f you freely and deliberately make the slightest mistake that can put you in a very hazardous situation, a society complying with equal opportunity will quietly let you die.   (Fleurbaey 1995, p. 40)[18]

In such societies, a patient's claim to health care could be lost, and thus lack of treatment could be justified, even in cases where we can help. Sometimes, instead, the abandonment objection is aimed at RSPs in particular. For example:

> [Implementing an RSP] can have harsh consequences, especially in a health setting. Ascribing responsibility for past choices means that people whose poor

---

[18] For other versions, see (Daniels 2007, pp. 76–7; Duus-Otterström 2012; Segall 2009).

health follows from their past choices have no right to be helped—this leads to proposals which give these people lower priority to certain treatments.

(Vansteenkiste et al. 2014, p. 68)

Or, for instance:

[a]n often cited criticism against responsibility ascription in a healthcare context is that it is too harsh to deny people public healthcare simply because their disease is caused by voluntary risk exposure.

(Bærøe and Cappelen 2015, p. 838)

Finally, of a smoker who has developed coronary heart disease:

[m]any think it would be a harsh judgement to deny him the procedure because the disease could be said to be self-inflicted.

(Cappelen and Norheim 2006, p. 314)

Let us suppose, along with these authors, that losing one's claim or right to medical care is too harsh of a result, at least for some blameworthy items (e.g., actions, habits, outcomes, etc.). Thus, RSPs with unlimited reduction theses relating to these blameworthy items—on which one can fully lose one's claim— would be too harsh in their reduction.[19] Even on this supposition, the objection will not count against RSPs in general, since a policy can be sensitive to responsibility without adhering to an unlimited reduction thesis. An RSP on which a patient's claim or right can be reduced, but not lost, in virtue of blameworthiness—one reflecting a limited reduction thesis—avoids this objection.[20] Insofar as one's claim to some good or other can be reduced in virtue of one's blameworthiness, the policy will still be sensitive to responsibility. Some have responded to the abandonment objection precisely by pointing out how a Luck Egalitarian view, or at least an RSP, can avoid justifying the abandonment of individuals in such cases.[21]

Yet, although the harshness objection tends to focus on some form of abandonment, usually in terms of a lost claim on goods, there are, I suggest, other ways for an RSP to be too harsh. Consider a variation of the case of the helmetless

---

[19] Though there may be some disagreement concerning claims of disproportionality. For example, Segall, citing (Arneson 2004), suggests that abandoning the motorcyclist would not "fit the crime" (Segall 2009, p. 70), yet Fleurbaey seems to suggest that it is proportional, at least to the degree of risk he has chosen (Fleurbaey 1995, p. 40).

[20] This point is made in broad strokes, but notice that, if it turns out that abandonment is too harsh for only *some* lifestyles or outcomes, then an RSP can avoid this objection insofar as it reflects limited reduction theses for just those outcomes.

[21] See, for example (Albertsen 2020; Albertsen and Nielsen 2020; Bærøe and Cappelen 2015; Segall 2009; Voigt 2007, 2013).

motorcyclist. Suppose that, not only does the accident result in the motorcyclist suffering head injuries, it also results in a pedestrian's breaking his left pinkie-finger. When a paramedic arrives, she has to determine whom to treat first. She can treat the pedestrian's finger well enough, while still having time to help the motorcyclist afterwards. Should she? I suspect that most of us will say that the paramedic should direct her attention to the motorcyclist first.

However, an RSP which reflects a limited reduction thesis with regard to the medical care for the motorcyclist's injuries might still reflect a reduction thesis on which the claim has been reduced so much that he should not get priority in this case. Although this would not amount to abandonment of the motorcyclist, it might still be too harsh.

This case helps to illustrate how we might get other forms of the harshness objection, ones which do not rely on abandonment. In the previous section, I laid out three features of reduction theses that might be reflected by an RSP. I further suggested that these features can be a matter of degree, ranging from more to less severe. What I wish to suggest now is that the more severe the reduction theses reflected by a particular RSP, the more likely it is to face problems with the harshness objection. If the reduction, on a given RSP, is too severe, then a nonabandonment version of the harshness objection can still apply. Other things being equal, the greater the difference between the reduction suggested by the RSP and the reduction that would be appropriate, the stronger the harshness objection to that RSP is.

Thinking of the harshness objection in this way helps to show how an RSP can be overly harsh in terms of the other features of reduction theses mentioned above. Consider how discriminate a reduction thesis is. The less discriminate the reduction thesis, the higher the amount of claims to specific goods that are reduced; conversely, the more discriminate a reduction thesis is, the fewer the amount of claims that are reduced. An RSP can be too harsh not only in the amount that it reduces a specific claim by, but also in the amount of claims it reduces.

Similarly, the strength of the harshness objection will vary depending on the contextual generality or specificity of the reduction theses an RSP adheres to. Suppose an RSP adheres to a fully contextually general reduction thesis, on which the reduction applies in all contexts. The response to blameworthiness on this RSP will be fairly severe, since being blameworthy for the relevant item will result in reduced claims throughout every context. The modified case of the motorcyclist above shows how both the contextual generality and the limit of a reduction thesis can combine to result in a harsh-reduction objection. Changing the limit, without changing the contextual generality, might result in the motorcyclist getting priority over the pedestrian; but so might changing the contextual generality, such that it does not apply in this context. Other things being equal, an RSP that is committed to reduction theses that are more contextually specific—specific to, say, contexts in which there are not enough resources to help everyone, or to tie-breaking scenarios—will not have as severe of a response. There will still be many contexts in which the blameworthy patient has an equal claim to the relevant goods. Other things being equal, the more contextually general the reduction thesis, the more severe the response; and the more severe the response, the more likely that it is unduly harsh.

Before moving on, it may be useful to make a point about the dialectic concerning these other forms of the harshness objection that do not rely on abandonment. Cases of abandonment that involve significant needs, like the original case of the motorcyclist, are, I suspect, the ones that are most intuitively harsh; most, I presume, would have the intuition that we should not leave the helmetless motorcycle driver to die by the side of the road (assuming that we could help). In particular, it seems implausible that the driver has fully lost his claim to care in this circumstance, and thus, it seems implausible that an unlimited reduction thesis, with respect to this blameworthy item, would be true.

To suggest that a limited reduction is too harsh, however, will likely require more argument, since we may not have very clear intuitions about limited reductions. The modified case of the motorcyclist above, though perhaps intuitively harsh, is also an extreme form of harshness. However, we may not get such clear intuitions once we change features of the case. For instance, the more significant that we make the pedestrian's injuries, or the less significant that we make the motorcyclist's, the more difficult it might be to get clear intuitions. Intuitions in such cases will be complicated by the fact that we will often, as in the modified case, not be comparing identical needs, or claims to the same goods.

5.5 Limited Resources and Harsh Treatment

Harshness in the reduction theses an RSP reflects, in any of the ways mentioned above, can straightforwardly lead to harsh treatment. Yet, there is a potential harshness objection that may still apply *even if* the reduction is proportionate and otherwise permissible. This will be because a reduced claim, even if reduced by a proportionate amount, can lead to more severe consequences in some circumstances. This can happen when the reduction has an outsized effect in treatment, and most obviously seen in contexts of limited resources.[22]

Suppose that, instead of heart surgery, Blair and Ingrid both need lung transplants, yet there is only one pair of lungs available. Supposing that Blair's blameworthiness has resulted in a reduced claim, in this context, then Ingrid is

---

[22] There can be other ways for reduction theses to have outsized effects, though I will not develop this here. Roughly put, this can happen when the differences in possible treatments do not map on perfectly to the differences in possible degrees of blameworthiness.

more likely to get the lung transplant than Blair is. Even if an RSP gets the reduction thesis just right here, the result will be greater: a refusal of lungs.[23] Or consider a reckless driver who gets into an accident in which he, and his innocent passenger, receive similar injuries. Emergency personnel can only save one. As Segall puts it, a policy which would give priority to the innocent person in cases of limited resources would imply that "a small measure of recklessness may lead to what amounts to a death sentence" (Segall 2009, p. 70). In this case, the harsh-treatment objection would seem to apply, even if the harsh-reduction objection does not. Although the driver might retain his proportionally reduced claim to treatment, the nature of the situation can translate a slight reduction into lack of treatment.

What is a proponent of an RSP to say about such scenarios? One thing she might say is that in such scenarios, *someone* is going to go without life-saving treatment; harshness, and abandonment, is unavoidable. Given this, cases of limited resources are a problem for *any* policy, regardless of whether it is sensitive to responsibility or not.

In taking a similar line, Albertsen asks us to consider what the source of this unfortunate situation is:

> It is the organ scarcity rather than the luck egalitarian policy which causes harshness. Any distributive principle would result in not treating someone.
> (Albertsen 2020, p. 590)

> The harshness here stems from the shortage rather than luck egalitarian policies.
> (Albertsen 2016, p. 334)[24]

The fact that *someone* will receive harsh treatment is due to the fact that there are not enough resources to go around. Yet one might object that this does not fully resolve the issue, since the fact that *Blair*, or *the reckless driver*, will receive harsh treatment is also largely due to the RSP itself. The proponent of the RSP might insist, as Albertsen does, that it is not implausible to allow responsibility to count as a factor here (Albertsen 2016, p. 334, 2020, p. 590). Duus-Otterström seems to agree:

> When we cannot assist all, it makes sense to let responsibility determine how we prioritise—more sense, in fact, than alternative methods such as deciding between potential recipients via a lottery or (still worse) giving the imprudent priority.   (Duus-Otterström 2012, p. 154)

---

[23] This is not to say that this amounts to a complete refusal of treatment, since she may still be able to get other forms of treatment that are available in the absence of a lung transplant. However, these forms of treatment may not be as good as a lung transplant would have been.

[24] Albertsen further seems to deny that such a policy would be overly harsh (Albertsen 2016, p. 334, 2020, p. 590).

The proponent of an RSP can surely agree that it would make sense to take responsibility into account in some cases, and probably most can agree that we should not, as a matter of policy, give the imprudent priority in these particular cases.[25] Perhaps one thought here is that although someone is going to get negatively disproportionate treatment, not giving the lungs to Blair is *closer* to what would be proportional for her than for Ingrid, assuming that they have similar needs.

After making similar suggestions, Segall suggests that someone might still feel uneasy about such a policy. If one has this sort of worry, Segall suggests, one might prefer the use of a weighted lottery (Segall 2009, p. 71). The weights here could be modified to track degrees of responsibility.[26] On this route, an RSP need not allow some blameworthiness to *always* translate into disproportionate treatment in cases of limited resources. Nor need it allow a reduction in one's claims to result in one's not having a chance at treatment in such cases.

Thus, although there is a possibility of negatively disproportionate treatment even for RSPs which do not involve disproportionate reductions, the proponent of an RSP has some means available for responding. And if one thinks that always giving priority to the blameless individual in such cases is too harsh, then the use of a weighted lottery may help.

5.6 Harshness and Determination Procedures

Finally, even if we develop an RSP with reduction theses that are appropriately proportional to blameworthiness for the relevant items, and even if we have an adequate solution to the problem of harsh treatment in cases of limited resources, there is still the potential for another source of harshness. As others have pointed out, determining whether a particular patient is blameworthy for some particular item is either something that is not feasible, or would produce significant tensions with other moral concerns—for example, by being objectionably intrusive, or by undermining the doctor–patient relationship.[27]

When introducing the case of Blair and Ingrid, I stipulated that Blair is blameworthy and Ingrid is innocent. Now suppose, instead, that Blair is innocent as well, and we have mistakenly judged her to be blameworthy. In this case, even if we have a true reduction thesis, we would still get harshness with respect to Blair.

---

[25] Though see Eyal, Chapter 4, this volume.

[26] This may introduce another source of harshness: if the weights given in the lottery are not proportional to the reductions in claims, or are not proportional to the degree to which the agents are blameworthy, then we might get negative disproportionality.

[27] For discussion, see Levy, Chapter 7, this volume, and (Bærøe and Cappelen 2015; Brown 2013; Brown et al. 2019a, 2019b; Buyx 2008; Cappelen and Norheim 2005; Cohen and Benjamin 1991; Daniels 2007; Davies and Savulescu 2019; De Marco et al. 2021; Feiring 2008; Friesen 2018; Ho 2008; Levy 2019; Schmidt 2009; Sharkey and Gillam 2010; Vathorst and Alvarez-Dardet 2000; Véliz 2020).

Reducing her claim *at all* would be an overly harsh reduction, and modifying our treatment on the basis of judgments of blameworthiness, and corresponding reductions on the basis of it, would result in overly harsh treatment.[28]

An RSP will have an implicit or explicit commitment to a procedure, or perhaps procedures, for determining whether a patient is blameworthy for the relevant item; call such a procedure a *determination procedure*. As this modified case of Blair suggests, such a procedure, if not perfectly accurate, can result in an overly harsh reduction and treatment. One might, of course, point out that any realistic policy will result in some mistakes, and again, harshness resulting from epistemic limitations is, to some extent, inevitable. Yet, some harshness may also be avoidable, insofar as we may be able to improve accuracy. I suggest that, other things being equal, the more likely the determination procedure is to result in a false judgment of blameworthiness, the more likely the RSP is to result in overly harsh reductions, treatment, or both (and the more often that it will).

This last point concerns judgments of *whether* a patient is blameworthy for the relevant item. However, it is also plausible that we can be blameworthy to varying degrees, and which responses are proportional to an agent's blameworthiness will not just track *whether* the agent is blameworthy, but also the *degree* to which she is.[29] This leads to a slightly different potential source of harshness. Suppose, again, that Blair is blameworthy, but she is blameworthy to degree *d*. Further suppose that the determination procedure got it right in determining that Blair is blameworthy, yet resulted in the judgment that Blair is blameworthy to degree *d+1*. In this case, again, we may end up being overly harsh. Thus, the less well-calibrated the determination procedure is to degrees of blameworthiness, the more likely it is to result in incorrect judgments concerning degrees of responsibility. And other things being equal, the more likely a procedure is to exaggerate the degree to which an agent is blameworthy, the more likely the RSP is to result in overly harsh reductions, treatment, or both (and the more often it will).[30]

Finally, another potential source of harshness arising from the determination procedure comes not from falsely determining that a patient is blameworthy for some item, nor in exaggerating the degree to which a blameworthy patient is blameworthy; rather it comes from inaccuracy in the other direction.[31] Suppose that a procedure wrongly determines that a blameworthy patient is not

---

[28] Notice that, even if one accepts the claim that we will need to have overly harsh treatment in (at least some) cases of scarce resources, the solutions to this problem, presented above, would not seem to help in this case.

[29] How we are to best understand claims of degrees of blameworthiness is a complicated matter. For some discussion, see (Coates 2019).

[30] It may be worth noting that, insofar as we can have harshness in cases that do not involve limited resources (by reducing claims too much, say), harshness may be more likely to arise from (and may arise more often) from inaccuracy that exaggerates the degree to which a patient is blameworthy.

[31] I am grateful to Ben Davies for pushing me to consider and develop this point.

blameworthy. This would result in the lack of a negative response to the blame-worthy patient. In isolation, this would not be a case of harshness.[32] However, consider again cases of limited resources. If, for instance, we should give the lungs to Ingrid over Blair, or perhaps weigh the lottery slightly in Ingrid's favor, and we do not do this because the determination procedure falsely determined Blair to be innocent, then we may end up being too harsh with Ingrid. Thus, when it comes to avoiding harshness via the calibration of the determination procedure, there is the potential for harshness, to other patients, in virtue of wrongly determining that a patient is blameless, and perhaps from underestimating the degree to which they are blameworthy.

5.7 Concluding Thoughts

The purpose of this paper has not been to respond to the harshness objection, but rather to explore some of the relationships between the different notions relevant to the harshness objection to RSPs, and to identify sources of harshness. As I have argued, an RSP risks being overly harsh in reducing a particular claim by too much, by reducing too many claims, or by reducing claims in too many contexts. Further, these forms of harshness need not involve any form of abandonment, leaving an opening for many more versions of the harshness objection than those typically found in the literature. Yet given the many ways that RSPs can differ, it will be difficult to find a particular version of the harshness objection that applies to RSPs in general.[33]

Further, even if the reduction theses an RSP reflects are just right, one might still get overly harsh treatment in cases of limited resources, and overly harsh reductions and treatment in virtue of a determination procedure that is not properly calibrated to degrees of blameworthiness. These last two potential sources arise from trying to develop an RSP that is realistic. Given the nature of real-life situations, we will sometimes be limited in terms of the resources we have available, or our abilities to use them; and given the potential difficulty in determining an individual's blameworthiness, we will be limited in the extent to which we can calibrate our determination procedure to degrees of blameworthi-ness. Realistically, that is, any RSP will be condemned to *some* harshness. Which sorts, and amounts, of harshness one is willing to accept will depend, of course, on a variety of other commitments.

---

[32] One might, however, also be concerned with being *lenient*, which we might understand as a positively disproportionate response. Yet accepting the claim that being overly harsh is problematic, I think, does not commit one to the claim that leniency is similarly problematic.

[33] At least so long as we maintain that agents can be blameworthy for their health-related actions, lifestyles, and/or their health outcomes.

References

Albertsen, A. (2016), "Drinking in the Last Chance Saloon: Luck Egalitarianism, Alcohol Consumption, and the Organ Transplant Waiting List," in *Medicine, Health Care and Philosophy* 19/2: 325–38.

Albertsen, A. (2020), "Personal Responsibility in Health and Health Care: Luck Egalitarianism as a Plausible and Flexible Approach to Health," *Political Research Quarterly* 73/3: 583–95.

Albertsen, A., and Knight, C. (2015), "A Framework for Luck Egalitarianism in Health and Health Care," in *Journal of Medical Ethics* 41/2: 165–9.

Albertsen, A., and Nielsen, L. (2020), "What Is the Point of the Harshness Objection?" in *Utilitas* 32: 427–43.

Anderson, E. (1999), "What Is the Point of Equality?" in *Ethics* 109/2: 287–337.

Arneson, R. (2004), "Luck Egalitarianism Interpretated and Defended," in *Philosophical Topics* 32/1/2: 1–20.

Bærøe, K., and Cappelen, C. (2015), "Phase-Dependent Justification: The Role of Personal Responsibility in Fair Health Care," in *Journal of Medical Ethics* 41/10: 836–40.

Brown, R. (2013), "Moral Responsibility for (Un)healthy Behaviour," in *Journal of Medical Ethics* 39/11, 695–8.

Brown, R., Maslen, H., and Savulescu, J. (2019a), "Against Moral Responsibilisation of Health: Prudential Responsibility and Health Promotion," in *Public Health Ethics* 12/2: 114–29.

Brown, R., Maslen, H., and Savulescu, J. (2019b), "Responsibility, Prudence and Health Promotion," in *Journal of Public Health* 41/3: 561–5.

Buyx, A. (2008), "Personal Responsibility for Health as a Rationing Criterion: Why We Don't Like It and Why Maybe We Should," in *Journal of Medical Ethics* 34/12: 871–4.

Cappelen, A., and Norheim, O. (2005), "Responsibility in Health Care: A Liberal Egalitarian Approach," in *Journal of Medical Ethics* 31/8: 476–80.

Cappelen, A., and Norheim, O. (2006), "Responsibility, Fairness and Rationing in Health Care," in *Health Policy* 76/3: 312–9.

Clarke, R. (2013), "Some Theses On Desert," in *Philosophical Explorations* 16/2: 153–64.

Clarke, R. (2016), "Moral Responsibility, Guilt, and Retributivism," in *Journal of Ethics* 20/1: 121–37.

Clarke, R., McKenna, M., and Smith, A. (eds.) (2015), *The Nature of Moral Responsibility: New Essays* (Oxford University Press).

Coates, D. (2019), "Being More (or Less) Blameworthy," in *American Philosophical Quarterly* 56/3: 233–46.

Cohen, C., and Benjamin, M. (1991), "Alcoholics and Liver Transplantation," in *JAMA*, 265/10: 1299–301.

Daniels, N. (2007), *Just Health: Meeting Health Needs Fairly (1st ed.)*, (Cambridge University Press).

Davies, B., and Savulescu, J. (2019), "Solidarity and Responsibility in Health Care," in *Public Health Ethics* 12/2: 133–44.

De Marco, G., Douglas, T., and Savulescu, J. (2021), "Health Care, Responsibility and Golden Opportunities," *Ethical Theory and Moral Practice* 24/3: 817–31.

Duus-Otterström, G. (2012), "Weak and Strong Luck Egalitarianism," in *Contemporary Political Theory* 11/2: 153–71.

Feiring, E. (2008), "Lifestyle, Responsibility and Justice," in *Journal of Medical Ethics* 34/1: 33–6.

Fleurbaey, M. (1995), "Equal Opportunity or Equal Social Outcome?" in *Economics & Philosophy* 11/1: 25–55.

Friesen, P. (2018), "Personal Responsibility Within Health Policy: Unethical and Ineffective," in *Journal of Medical Ethics* 44/1: 53–8.

Herlitz, A. (2019), "The Indispensability of Sufficientarianism," in *Critical Review of International Social and Political Philosophy* 22/7: 929–42.

Ho, D. (2008), "When Good Organs Go to Bad People," in *Bioethics* 22/2: 77–83.

Levy, N. (2019), "Taking Responsibility for Responsibility," *Public Health Ethics* 12/2: 103–13.

Olsaretti, S. (2003), *Desert and Justice* (Oxford University Press).

Schmidt, H. (2009), "Just Health Responsibility," in *Journal of Medical Ethics* 35/1: 21–6.

Segall, S. (2009), *Health, Luck, and Justice* (Princeton University Press).

Sharkey, K., and Gillam, L. (2010), "Should Patients with Self-Inflicted Illness Receive Lower Priority in Access to Health Care Resources? Mapping Out the Debate," in *Journal of Medical Ethics* 36/11: 661–5.

Vallentyne, P. (2003), "Brute Luck Equality and Desert," In *Desert and Justice* (Oxford University Press), 169–86.

Vansteenkiste, S., Devooght, K., and Schokkaert, E. (2014), "Beyond Individual Responsibility for Lifestyle: Granting a Fresh and Fair Start to the Regretful," *Public Health Ethics* 7/1: 67–77.

Vathorst, S., and Alvarez-Dardet, C. (2000), "Doctors as Judges: The Verdict on Responsibility for Health," in *Journal of Epidemiology & Community Health* 54/3: 162–4.

Véliz, C. (2020), "Not the Doctor's Business: Privacy, Personal Responsibility and Data Rights in Medical Settings," in *Bioethics* 34/7: 712–8.

Voigt, K. (2007), "The Harshness Objection: Is Luck Egalitarianism Too Harsh on the Victims of Option Luck?" in *Ethical Theory and Moral Practice* 10/4: 389–407.

Voigt, K. (2013), "Appeals to Individual Responsibility for Health: Reconsidering the Luck Egalitarian Perspective," in *Cambridge Quarterly of Health Care Ethics* 22/2: 146–58.

Zimmerman, M. (2015), "Varieties of Moral Responsibility," In R. Clarke, M. McKenna, and A. Smith (eds.), *The Nature of Moral Responsibility* (Oxford University Press), 45–64.

PART III
RESPONSIBILITY FOR HEALTH

6

Informed Consent and Morally Responsible Agency

Dana Kay Nelkin

6.1 Introduction

There is an intriguing overlap to be found in recent accounts of the conditions of informed consent in medical contexts on the one hand, and of the conditions of morally responsible agency on the other hand.[1] In particular, notable accounts in both areas prominently feature reasons-responsive capacities. And debates over the details in each area also exhibit important parallels, such as disagreements about what *kinds* of reasons one must be responsive to. These observations lead naturally to the question of whether there is a common framework that applies in the case of morally responsible action and in the case of informed consent. Making this question even more salient are recent and insightful discussions that connect each of these domains indirectly through the foundation of autonomy.[2]

In this paper, I begin by drawing parallels between debates in both areas and defend the idea that we should focus on a particular family of substantive views that appeal to high-enough quality opportunities for decision-making which are determined by both agents' reasons-responsive capacities and situational factors. Notably, even if reasons-responsiveness conditions take center stage in both accounts of responsible agency and of informed consent, it remains an open question just how similar the conditions are in each area, and what sorts of situational factors are relevant to each. There are a number of subtly different options in each area, including whether to require awareness or only the capacity to be aware of relevant

[1] This paper was first presented at the Health Care and Responsibility Workshop, Oxford Uehiro Centre, and then to the Centre for Moral, Social and Political Theory Seminar at the Australian National University. The paper is much improved as a result, and I am very grateful to Christian Barry, Claire Benn, Garrett Cullity, Nicky Drake, Brian Hedden, Katie Steele, Nir Eyal, Julian Savulescu, Neil Levy, Walter Sinnott-Armstrong, Lok Chan, Richard Holton, and Zoë Fritz. The paper grew out of my experiences serving on an IRB and teaching a course on Bioethics, Capacity and Consent at UC San Diego over several years, and I am grateful to Michael Kalichman and the students in those courses for making these experiences possible. Finally, special thanks to Gabriel De Marco, Garrett Cullity, and Sam Rickless for very helpful written comments.

[2] For example, see Knutzen (2020), especially section 2, for a subtle argument that autonomous agents are responsible ones, and Eyal (2018) and (2019) for a rich assessment of the widespread idea that autonomy is the fundamental justification of informed consent policies.

Dana Kay Nelkin, *Informed Consent and Morally Responsible Agency* In: *Responsibility and Healthcare*. Edited by: Benjamin Davies, Gabriel De Marco, Neil Levy, Julian Savulescu, Oxford University Press. © Edited by Benjamin Davies, Gabriel De Marco, Neil Levy, Julian Savulescu 2024. DOI: 10.1093/oso/9780192872234.003.0007

facts, and whether those facts include both empirical facts and facts about value or only the former. Here I focus largely on identifying key choice points, as well as homing in on whether and what kind of requirement there should be in either domain when it comes to the capacity to track value. I will argue that there are independent reasons to treat responsible agency and eligibility for informed consent similarly, all else equal, so we need principled reasons for divergence.

In Section 6.2, I present some cases at the heart of the two recent debates, about informed consent and responsible agency respectively. Exploration of these cases will illuminate the appeal of an overarching framework, and, at the same time, some key choice points to address in filling it out. In Section 6.3, I set out and defend the framework as it applies to responsible agency, and in Section 6.4, I explain how it can also illuminate our understanding of informed consent. I also consider some ways in which considering the two areas of inquiry together can help illuminate our understanding of responsible agency. In Section 6.5, I consider objections, both theoretical and practical, as well as further questions.

Before going further, however, it is important to spell out how I understand both informed consent and responsible agency and to make some preliminary points about each. When it comes to the former, I will begin by adopting the suggestion of Nir Eyal (2018) and (2019) that informed consent is shorthand for informed, voluntary and decisionally-capacitated consent. To further fix the concept, as Eyal also notes, without informed consent medical intervention is impermissible when someone is capable of giving it (at least under most circumstances). Finally, I take it that it is further fixed by its being an attempt to instantiate a principle concerning respect for persons, conceived of as autonomous agents.[3] As the Belmont Report (1978) expresses the point in its first "basic principle" for research involving human subjects, the Principle of Respect for Persons, "individuals should be treated as autonomous agents." In turn, in the section in which the principles are applied, the Report concludes that:

> respect for persons requires that subjects, to the degree that they are capable, be given the opportunity to choose what shall or shall not happen to them. This opportunity is provided when adequate standards for informed consent are satisfied.
>
> (National Commission for the Protection
> of Human Subjects of Biomedical and Behavioral Research 1979)

Thus, respect for persons underwrites the importance of obtaining informed consent, and does so via the requirement that people be treated as autonomous agents.

---

[3] In this last way of fixing the concept, I depart somewhat from Eyal who offers reasons for skepticism that autonomy is what really underwrites informed consent policies. See also O'Neill (2002) and Manson and O'Neill (2007) for other examples of such skepticism. I return to this point of disagreement with Eyal briefly in Section 6.5.

One further point regarding informed consent is that discussions of what exactly is required for informed consent closely track discussions of what is sometimes called "decision-making capacity." Being "decisionally-capacitated" is often treated as *one element* of informed consent, as we have just seen. Further, once the details of decision-making capacity are spelled out, it can begin to look even more similar to informed consent itself. One reason for this is that as Buchanan and Brock (1989) have pointed out, it is most fruitful to think of decision-making capacity as indexed to a particular decision rather than as a global capacity. After all, a person might have such capacity relative to some decisions and not others. But once relativized to a particular decision, questions about whether a person comprehends and appreciates the salient facts are often treated as relevant to whether someone does have the capacity, and indeed understanding and appreciation are treated as elements of the capacity. For example, as Hawkins and Charland (2020, section 4.2) spell it out: "obviously, in order to be capable of consenting to or refusing a given intervention, a subject must have some basic understanding of the facts involved in that decision."[4] This suggests that a condition of being "informed" is already absorbed into decision-making capacity.

It is also important to emphasize that the absence of informed consent does not necessarily make intervention impermissible; if someone is not in a position to provide it, intervention may be permissible, as in the case of children or those who are not conscious or are otherwise relevantly impaired. Moving in the other direction, the presence of informed consent does not necessarily make medical intervention permissible either. We can see this by noting that there are other conditions that are relevant. The Belmont Report itself mentions both principles of beneficence, which directs us to promote well-being in particular ways, and of justice, or "fairness in distribution." Someone might present us with informed consent, such as a case of willing self-sacrifice, where it would nevertheless be wrong for us to proceed on the grounds that doing so would cause an unjustified harm and failure to promote well-being. Thus, meeting an informed consent condition is neither necessary nor sufficient for overall permissibility of acting in the way consented to.

Finally, as for responsible agency, I understand it to be a kind of agency that makes it appropriate to hold agents to account and to blame and praise them in a robust sense when they act badly and well respectively. While moral responsibility is most often at the center of debates, there is renewed interest in seeing the moral

---

[4] At times there is apparent ambiguity about whether it is a capacity to be informed or actually being informed that is required for the relevant capacity, and as far as I can tell, both claims are made. Interestingly, though, a similar question attaches to the question of whether the informed condition in informed consent requires actual uptake or just availability or something in between. I will return to this issue in Sections 6.3 and 6.5.

as one domain among others when it comes to responsible agency.[5] Exactly what conditions one must meet to be accountable is a matter of intense debate, and I turn briefly to it in Section 6.3.

6.2 Some Cases and Parallels

Anorexia Nervosa is a challenging condition to treat, and it also raises particularly difficult questions related to informed consent and decisional capacity. According to the *Diagnostic and Statistical Manual V* (DSM-V), a person must meet all of the following conditions in order for the diagnosis to be apt:

(1) Restriction of energy intake relative to requirements leading to a significantly low body weight in the context of age, sex, developmental trajectory, and physical health. Significantly low weight is defined as a weight that is less than minimally normal or, for children and adolescents, less than that minimally expected.

(2) Intense fear of gaining weight or becoming fat, or persistent behavior that interferes with weight gain, even though at a significantly low weight.

(3) Disturbance in the way in which one's body weight or shape is experienced, undue influence of body weight or shape on self-evaluation, or persistent lack of recognition of the seriousness of the current low body weight (American Psychiatric Association 2013).

One of the most striking aspects of at least some of those with the condition is the co-existence of their tendency to perform very well on cognitive comprehension and reasoning tests and their apparent lack of appropriately weighty value for their own valuable lives.[6] As one participant in a study put it to an interviewer, "[a]lthough I didn't mind dying, I really didn't want to, it's just I wanted to lose weight, that was the main thing" (Tan et al. 2003: 702).

The participant's failure to appropriately weigh prudential value here raises the question of whether she in fact fulfills the conditions for decision-making capacity, despite high scores on cognitive tests and apparent understanding of the treatment options. When such a patient refuses what she understands to be life-saving treatment, it is natural to ask whether this decision should be treated as fully autonomous and whether it would not in fact be a violation of respect for persons if others were to intervene and provide it. In other words, is such a person

---

[5] This understanding of moral responsibility corresponds to accountability. For more detailed discussion of the concept, see Watson (2004) and Nelkin (2016).

[6] See Tan et al. (2003) and (2006) and Holroyd (2012) whose discussion of anorexia includes assessment of whether people with serious conditions meet the conditions for mental capacity as defined by the Mental Capacity Act of 2005 in the UK.

in a position to give informed consent, such that withholding it necessarily prohibits treatment? In turn, this raises the question whether in addition to comprehension of the relevant facts about the condition, prognosis, and treatment, it is also essential to have understanding of the relevant values and their relative weights. In other words, are the true decision-making capacity conditions "value laden"?[7] And if so, in what way?

For various reasons that we will investigate in more detail in Section 6.4, the value-neutral approach has been in the ascendant when it comes to the literature on medical decision-making. But as the patients with anorexia illustrate, this position has serious costs in intuitive plausibility. And the problem does not arise from a mere conflict between two values, autonomy and well-being; there is a tension insofar as the autonomous decision-making capacity of participants in the study by Tan and colleagues seems itself to be compromised. Despite the fact that for some participants, the evaluative orientation associated with the condition seems to be part of their very identity and sense of self, the fact that their evaluation of their options appears to be "distorted" undermines confidence that they have a robust capacity to exercise autonomous agency in the circumstances.

I return to this debate in Section 6.4, but for now, I turn to a debate within the moral responsibility literature that bears a striking parallel.

Although it is put in different terms, questions about whether agents must be able to track value or correctly make value judgements take center stage in debates about what it takes to be morally responsible and blameworthy in particular, as well.[8] An extreme test case is provided by the psychopath, as stipulated to be a person who lacks any sort of moral understanding and cannot grasp that others' interests provide reasons for action or omission.[9] On one influential view, psychopaths are nevertheless morally responsible and blameworthy when they act badly and are perhaps the paradigm of such a category. On this sort of view, they are blameworthy because they act out of malice or in some way express a judgment that others' interests *do not* matter. Their inability to judge otherwise is simply irrelevant to *our* judgment that they are blameworthy and the appropriateness of responding to them with blame and the reactive attitudes such as resentment and indignation.[10] On a competing view that I favor, the capacity to understand moral facts is required for moral responsibility and blameworthiness. If they simply lack the capacity to understand a moral fact, this is not relevantly

---

[7] See Freyenhagen (2009), Owen et al. (2009), Holroyd (2012), and Hawkins and Charland (2020) for discussion of this question.

[8] In Nelkin (2015), I also consider whether one must have an understanding of moral value to be praiseworthy in different senses, but most of the literature focuses on blameworthiness as a way of being morally responsible.

[9] There is some good evidence that this is not a mere stipulation, but is an accurate description of psychopaths as diagnosed by the main diagnostic tool, the Hare Checklist Revised. See Blair (2008), for example.

[10] See Scanlon (1998); Talbert (2008); and Smith (2015). See Watson (2011) for discussion.

different from a lack of capacity to understand an empirical fact such as what the implications of an action would be. To put the debate in terms that parallel that of decision-making capacity, it is a debate between a value-neutral and a value-laden approach to moral responsibility and blameworthiness. On one side are those who deny that responsible agents must have the capacity to track moral value; on the other are those who recognize a requirement of such a value-tracking capacity.

It is also important to note that there are less extreme cases that raise similar questions and feature agents who suffer from more localized moral impairments. For example, "incorrigible racists" might be thought to have moral understanding in general, but have so-called "blind spots" and be unable to apply it in certain domains. Importantly, there may be no such actual people, but the theoretical case is an interesting one on its own and historical cases have often been used as illustrations.[11] Here I believe that such people, as stipulated, are eminently criticizable insofar as they are clearly morally deficient, but it remains much less plausible that they are *accountable* for their actions. Finally, there are cases of moral ignorance, in which people are ignorant of the moral facts in the way that they might be ignorant of empirical facts. This ignorance can range from the simple (e.g., not understanding that a person's life is more important than another person's property right to an intact cabin window) to the very subtle (e.g., not knowing how much weight to give to a customer's request to leave out an ingredient for religious reasons, compared to the weight of other customers' interests in getting their food in a hurry).[12] Here, too, there is a lively debate about whether those who are morally ignorant through no fault of their own can be morally responsible and blameworthy when they engage in wrongdoing.

In both the debate about informed consent and the debate about moral responsibility, we face a choice about whether the incapacity to understand or accept either empirical or value facts compromises the kind of agency in question. In the first case, the relevant value judgments are primarily prudential, and in the second, the relevant value judgments are moral. In each case, there is a choice point about whether to go value neutral or value laden.

In addition to the important parallel in key choice points in the two debates, there is an additional reason for thinking that it can be instructive to consider them together. It is that both debates are connected to a third, namely, the debate about the nature of autonomy itself. As we have seen, from its inception the idea of "informed consent" has been associated with autonomy, and moral responsibility and autonomy have also been closely linked, even if perhaps more at some times than others.[13] While "autonomy" itself is a contested and arguably

---

[11] See, for example Zimmerman (1997); Rosen (2004); Guerrero (2017); and Rudy-Hiller (manuscript).
[12] See Albuquerque (manuscript) for a similar case.
[13] See Buss and Westlund (2018) for a very useful overview of the literature on personal autonomy that notes many places in which there is intersection with issues of responsibility and accountability.

ambiguous term,[14] we consistently find important parallels in debates about its nature that perhaps underlie those in both of the debates under discussion here.

Thus, in the informed consent and the moral responsibility debates, we find important parallels, and a connection of each to what is at least sometimes thought to be the more foundational phenomenon of autonomous agency. Finally, there is yet one more reason for thinking that the two debates are fruitfully considered together, namely, that when it comes to the kind of medical decision-making in the informed consent cases, it seems intuitively correct that we care about whether the agents in question are in an important sense *responsible* and accountable for their decisions. Is the decision up to you in such a way that you are accountable for it? This description seems apt in both domains. For all of these reasons, then, it is worth exploring whether a shared framework applies in both domains. In the next section, I set out a framework for moral responsibility, and in Section 6.4, explain how it can illuminate one difficult issue in the informed consent debate.

6.3 The Quality of Opportunity Account

When it comes to moral responsibility, there is a large divide between two camps that corresponds to the divide between value-laden and value-neutral accounts of autonomy.[15] On the one hand, there are views that fall under the umbrella of "quality of will" accounts. And on the other, there are views that fall under the umbrella of "control" accounts. Quality of will accounts are a varied group, but they share the idea, mentioned above in connection with psychopaths, that what matters is the quality of will one expresses in one's actions.[16] If one is able to express malice, say, or one is able to express a judgment about one's ends, then one can be responsible and, depending on the action and the ends in question, blameworthy. Such views require sophisticated cognitive capacities, as even to make judgments about what is and is not valuable is a cognitive achievement. But one need not have *moral* understanding, and, in fact, one might be impaired with respect to recognizing moral reasons for action such as that otherwise someone's interests would be undermined, or they would be caused pain. Thus, the psychopath or, even more clearly, others with more localized moral impairments are responsible and blameworthy for their actions on this kind of account.[17]

---

[14] See Dworkin (1988) and Arpaly (2004).

[15] It is important to note that some accounts of moral responsibility do not fit neatly into either of these groups, such as some communicative accounts. I leave for another time the interesting question of how such views might best be thought of in relation to debates about autonomous agency.

[16] See for example Talbert (2008) and (2022); Smith (2015); Shoemaker (2015); and Sripada (2017).

[17] Despite proponents of quality of will views' claims, it is not actually clear that psychopath s are responsible and blameworthy even on this account, for it is not clear that they are capable of malice or

In contrast to this sort of view are views that take control of a sort to be necessary for moral responsibility either instead of or in addition to the expression of a certain quality of will. Typically, these views take it that one must have had certain capacities underwriting the ability to have avoided the wrongdoing when one is blameworthy, or the ability to act well for good reasons. While some interpretations of the control required take the lack of determinism to be essential, others are compatibilist and take the control, and hence responsibility itself, to be compatible with determinism. For present purposes, I set aside the question of compatibilism, and focus on the debate amongst compatibilists who are quality of will theorists and those who are control theorists.

In the rest of this section, I will set out a version of a control view that I believe best accommodates cases of responsible and blameworthy action, and that also best captures the conceptual connections that have often been made to other notions such as desert, freedom, and autonomy. The "quality of opportunity" account is centered on the idea that whether one is responsible at all depends on one's having a good enough opportunity to act for good reasons, and when one is morally blameworthy, a good enough opportunity to act well morally for good reasons. The account of blameworthiness is scalar, in that one's blameworthiness can be mitigated to the extent that one's opportunity is compromised or of a lower quality. For example, if it is really difficult to do the right thing in the circumstances, one might still be blameworthy, but less so than if it had been easier. What can affect the quality of an opportunity? I take it that there are two main factors that might be thought of as internal and external, though they can also interact. The "internal" factor comprises our skills and capacities for recognizing and weighing reasons, and translating that recognition into the adoption of reasons and action is an essential factor. (This package of capacities is often known as "normative competence.") If I am skillful at identifying reasons, it should be easier in at least one sense for me to do the right thing, and that can make my opportunities better. On the other hand, even if I have incredible capacities, the world is not always so cooperative, and situational factors might still cause me to face a low-quality opportunity for doing the right thing. For example, the existence of duress or socioeconomic factors can lower the quality of an opportunity from what it would otherwise be. This two-factor model has the resources to account for an otherwise heterogeneous set of possible mitigating factors, or excuses. One might have impaired capacities that make it harder to act for good reasons, or one might face difficult situational factors, or both. What unites all of these kinds of excuses is that

---

related instantiations of quality of will. For an argument against this application of the account to psychopaths, see Nelkin (2016). But for now what is important is the account itself and that it does count those with more localized impairments as blameworthy.

they in some way compromise the quality of one's opportunity to act well and for good reasons.[18]

In addition to accounting well for cases in an explanatory and unifying way, the account also makes sense of the connection of responsibility to other phenomena such as desert, free will, and autonomy. As I've argued elsewhere, it seems that having such an opportunity is necessary for one's being deserving of a negative response. And this sort of view—along with other control views—explains why historically discussions of moral responsibility and free will have been linked. Having the control that allows one to be able to avoid wrongdoing or act well just is a particular kind of freedom. I have thus far been deliberately noncommittal about what autonomy requires, but insofar as it is connected to respect for persons on the one hand, and the ability to deliberate, the opportunity to make choices (per the Belmont Report) and determine their own lives, the connection between the quality of opportunity account of responsible agency and autonomy emerges.[19]

With the outlines of the account in place, at least two important choice points remain. The first concerns the role of awareness of reasons. So far, recognition of reasons appears in a capacity or competence condition. One need only be able—in the situation—to recognize reasons; there is no explicit requirement that one must actually be aware. It is here that we face a significant choice. On some versions of this sort of view, what is essential and sufficient for having a high enough quality opportunity to act well is that one could—and should—have been aware of the relevant reasons, even if one was not culpable oneself for one's current state of ignorance.[20] But it is difficult to say how one has a true opportunity, or real control, if one is unaware of the reasons in question and is not responsible for one's own ignorance. This leads to the idea that awareness of a kind *must* be implicated in any instance of responsible agency. But that awareness need not be awareness of the reasons at the time of action (or omission), and this thought leads to what is often called a "tracing" condition.

On most well-known such theories, one's moral responsibility for a later action, omission, or consequence traces back to a prior decision or act of agency such that one could foresee at that time the (risk of the) later action, omission, or consequence (see, e.g., Fischer and Tognazzini 2009). In work with Sam Rickless, we develop a more minimalist account (Nelkin and Rickless 2017). On the view we

---

[18] See, for example, Wolf (1990); Nelkin (2011) and (2016); Brink and Nelkin (2013); Vargas (2013); and Brink (2021). For a view that focuses more on the internal conditions of reasons-responsiveness, see Fischer and Ravizza (1998).

[19] Quality of will views of various sorts can also make a claim to a connection with autonomy, where autonomy is understood as a kind of self-expression or authenticity of choice. So, the ability to make this connection is not unique to the view at hand. But I believe that the quality of opportunity view, or control views more generally, make better sense of the connection to respect for persons. See Knutzen (2020).

[20] See Vargas (2013); Clarke (2014); and Brink (2021).

favor, whether an agent is morally responsible and blameworthy for X at time T2 depends entirely and solely on whether there was a prior time, T1, at which the agent had the *opportunity* to do something that, as she reasonably believed, would significantly raise the likelihood of avoiding X. Note that what is key here is the having of a prior opportunity to avoid the consequence; having such an opportunity does not require any act of decision or agency. We call this the "opportunity tracing view," and it makes explicit a condition on the quality of opportunity account.[21]

As we will see, exactly which path we take in response to the question of the role of actual awareness will have implications when we transfer the framework to the debate about informed consent. But it is also important to keep in mind that different answers, while disagreeing on some verdicts, will also be in agreement that many cases of acting in ignorance of reasons are exempt from responsibility. As long as it is not the case that one *should* have known, one cannot be responsible and blameworthy when acting badly in ignorance.[22]

A second key choice point is where exactly to draw the threshold for responsible agency. Given that it is a scalar notion, we can ask whether the threshold is to be drawn at bare opportunity, no matter how low quality or not. As I have presented the account, the threshold for responsible agency is only when an opportunity is high enough in quality. But where exactly? Here, too, I do not have a specific algorithm to offer, although a test, or indicator, is that it is where it would not be unreasonable to expect a person to act well. It might be thought that in the moral case, it is not a bad result that there is no bright line available for use at the threshold while we need one in the legal responsibility context, as well as in the informed consent and decision-making capacity contexts. I return to this question in Section 6.5.

Before turning to the question of how to understand informed consent, let us note some key features of the account spelled out so far. The quality of opportunity view is one that is value laden as opposed to value neutral (or substantive as opposed to procedural), insofar as it requires normative competence. In other words, it requires one to be able to recognize the reasons that there are, and to be able to adopt and act on the reasons one has identified. There is no restriction on the nature of reasons here. To be morally responsible, one must have the ability to recognize and respond to moral reasons. It requires that one have opportunities to act well even if they are not taken. In this way, we cannot necessarily read back

---

[21] There are accounts that are skeptical of tracing, and on these one is only responsible for the earlier act (or omission) at the time that one meets the basic conditions of responsible agency. See King (2009) and Agule (2016). See also Vargas (2005) for a challenge to tracing accounts. A full defense of tracing must await a further occasion.

[22] For an idea in the domain of autonomy that might also be thought of under the tracing rubric, see Savulescu (1994: 198): "It is true that a person can autonomously choose to lead an impulsive life... Autonomy of choice can be inherited if the parent choice was autonomous."

from what one actually does and with what quality of will to whether one is responsible or not. What matters is what one did given the quality of one's opportunity, which is in turn a function of one's reasons-responsive capacities and one's situation.

6.4 Bringing the Framework to the Informed Consent Debate (and Back Again)

As Hawkins and Charland (2020: section 5) set it out, the assumption of "value neutrality" is one of five widely accepted assumptions that "underlie virtually all contemporary work on decisional capacity," itself a key element in informed consent. And it is not difficult to see why. For "[i]t is usually assumed that the only way to satisfy the requirement of value neutrality is to draw a sharp distinction between process and outcome" (2020: section 5.4). According to the value-neutral picture, whether someone satisfies the conditions for decision-making capacity in making a particular choice depends on whether it is the result of a process that meets a threshold for general cognitive capacity, one that is sometimes called "internal" rationality. To meet this requirement, one must have consistency among one's beliefs and values, be able to prioritize and update one's priorities in light of new information, and be able to recognize which options serve one's ends. In contrast, according to the value-laden picture, as typically under-stood, whether someone satisfies the conditions depends on whether the outcome matches the right values (or what enough reasonable people would choose).[23] But it is often taken as axiomatic that we should not simply decide that someone lacks the relevant capacity when they choose contrary to medical advice (Ganzini 2005; Hawkins and Charland 2020), and for good reason. If we were to simply read off of the decision whether the decisionally capacitated condition was satisfied, there would be no value added from knowing the agent's choice. We could just decide ourselves what the best decision is, and act accordingly. And yet, as we saw earlier, the Belmont Report rightly represents the principle of Respect for Persons as independent from a Principle of Beneficence. There would be no point in taking Respect for Persons to indicate a separate factor to be taken into account if we were to decide that respect for persons is only in play when the agent makes the decision that is in their best interest.

Thus, if our only options are, on the one hand, process in the form of internal rationality and, on the other hand, reading back from outcome, and we value respect for persons as an independent factor in guiding our actions, then we should select process in the form of internal rationality. However, this leaves

---

[23] See Culver and Gert (2004) for the latter.

us with challenging cases, including that of the serious cases of anorexia described earlier.[24]

But there is a third option that preserves an independent role for autonomous agency underlying respect for persons and that provides flexibility in the challenging cases, and the framework provided by the quality of opportunity account shows the way. The basic idea is that to meet the relevant condition, one must have the opportunity and, a fortiori, the *capacity* to track the reasons there are, without necessarily exercising that capacity well or seizing one's opportunity. In this way, it does not allow for reading back from outcomes alone to whether the requisite agency was in place, but it is at the same time value laden in that the *ability* and opportunity to track actual value is in place. Just as someone who is responsible for their actions can be blameworthy precisely by acting badly while retaining the capacity and opportunity to avoid doing so, so one can have the relevant capacity to make a particular medical decision and not exercise it well. Nevertheless, the capacity itself in both cases is a value-laden one. The key is to prize apart value ladenness from outcome.

A natural first question at this point is how we determine capacity or competence, if it cannot simply be read off of performance. In principle, this should not necessarily be different from other ways we attribute capacity. Missing one problem on a math test does not show incapacity; but missing all of them repeatedly is some evidence of incapacity. Similarly, if we learn that someone has a short-term memory impairment that could *explain* the sub-par performance in such a way that points to incapacity, we would have even more evidence of incapacity. At the same time, we must be very careful here. As Hawkins and Charland also point out, diagnosis itself is not an automatic determiner of incompetence. And this is surely right. Once again, we can find analogies in the responsibility literature. One might act in a way that includes the exercise of an impaired capacity in doing wrong. But this is not an automatic excuse; if one could have worked around it in some way, appealing to other mechanisms, then one can

---

[24] See Freyenhagen (2009) for this kind of dilemma as applied to the case of anorexia in particular. He offers two kinds of options for the relevant capacity to make decisions: value neutral (either "reflective endorsement" or "procedural rationality") and value laden (a version of reasons-responsiveness). The latter appears to require that the choice be the correct one (or within a limited correct range) to have the relevant capacity. It is modeled on a Kantian view according to which autonomous choice is a matter of one's conduct needing "to be within the bounds of the morally permissible to count as (genuinely) autonomous" (2009: 466, citing Meyers (1989) and O'Neill (2003)). Also see Holroyd for a similar line of reasoning. As she puts it, if the participants in the anorexia study are making an evaluative mistake, then they are unable to weigh information in a way relevant to decision-making capacity (2013: 158). In concluding, she writes that if the line of reasoning she sets out is correct, then "it is a mistake to suppose that whether or not one meets the conditions for capacity can be determined purely formally, without reference to the content of an individual's choice or commitments" (2013: 167). Now "determine" is ambiguous as between a metaphysical and an epistemic reading. On the latter, it seems that the content of the choice is relevant only as evidence of incapacity, not insofar as it entails it. At points, Holroyd seems to adopt only the weaker epistemic reading.

still be responsible and blameworthy.[25] And on the positive side, people often can and do find workarounds for all sorts of problematic mechanisms such as implicit bias, confabulation, and self-deception in ensuring that they act well, despite internal obstacles. We find a clear parallel in the domain of decision-making related to health. In many cases, people appear to make bad decisions, even ones that are based on under- or over-valuation of relevant considerations, and yet appear to meet conditions for autonomous agency. For example, there are people who continue to smoke cigarettes despite being advised by their medical providers that their risk is high if they continue, receiving an offer of a treatment to help them quit, and despite the high value they place on continued life of high quality. Clearly, we do not normally read back from a decision that strikes us as the wrong one to the idea that such agents are incapacitated relative to them. Perhaps we are wrong about particular cases and plausibly there are many factors operating in our thinking about such cases, but the fact that this is such a ubiquitous kind of case suggests that we distinguish between the content of decisions and capacity quite frequently, even when the decisions are bad.

What should we conclude about the most challenging cases of anorexia with which we began? The first step is to apply the capacity account that includes the capacity to track value. This avoids an approach that directs us to read back from the outcome but also avoids an approach that takes cognitive competence to be sufficient for informed consent. That does not mean it is obvious just *how* it applies, and there is much work yet to be done. Here I think we need to know much more, but there is some reason to doubt that the impairment in recognizing reasons is such that the participants in the study have the ability to work around it. Remaining uncertainty means that we can err in either direction, and it might be that ultimately, policy-level reasons should guide us when we are in this situation. Erring on the side of assuming autonomous agency might be best under many circumstances, for example, if we risk erring in both directions. Importantly, however, we are now engaged in a different conversation, namely, one about which kind of error to guard against most strongly.[26] But simply because we face

---

[25] Yet another interesting parallel between the debates concerning informed consent and responsible agency arises here. Among reasons-responsiveness theorists about responsibility, there is a divide between those who advocate for a "mechanism-based" approach and those who advocate for an "agent-based" approach. On a mechanism-based approach, what matters is whether agents act on a reasons-responsive mechanism; on an agent-based approach, what matters is whether agents are themselves reasons-responsive. Precisely because it seems that we can sometimes work around mechanisms that are not reasons-responsive and thereby are not excused, agent-based views seem to have an intuitive advantage. In the informed consent case, it seems equally counterintuitive that simply because one acts because of a medical condition, it is not automatically disqualifying if one could work around it,. This parallel reinforces the point in support of agent-based accounts of responsibility. For the most well-known mechanism-based view, see Fischer and Ravizza (1998).

[26] It might be that a large part of the explanation of why theorists have narrowed down the options to the two mentioned earlier—process in the form of "internal" rationality, and outcome—is that more meta-level considerations about where to err have infiltrated the debate about how to understand the

uncertainty about whether someone has, but fails to exercise, a capacity to choose well in some cases, or even where the right values lead, we should not be led automatically to conclude that we are always in a state of uncertainty.

Finally, there might be reasons in addition to respect for persons and their autonomous agency not to intervene against a person's wishes.[27] As we saw at the outset, lack of informed consent is not sufficient for intervention, and this can be true even in a case in which it appears that intervention would promote the agent's well-being.

Before turning to objections, it is worth noting an important way that the illumination goes in both directions. Cases involving people who are impaired in tracking *moral* value are often ones from whom we recoil. This makes it under-standable that our first reactions are often blame in its most robust forms. But on control views, such as the quality of opportunity view, many of these reactions are understandable but inapt. While many negative reactions on our part are no doubt warranted, the ones that presuppose that they are accountable for their actions are not, given their inability to act well. Thus, defenders of such views, including myself, must grapple with a kind of resistance that comes from the natural desire to express one's shock and dismay at confronting someone who truly cannot track the value of others' interests. But thinking about cases in which there is impair-ment in the ability to track the value of one's own life, or to give it its due weight, elicits a much different sort of reaction. The fact that we worry that such a person is not really capable of making an autonomous decision, precisely because of that inability to track value, should give us pause. There are two possible reasons for this. On the one hand, it could be that the lack of capacity to track moral value is itself an impairment to autonomous agency and thus that insofar as that underlies morally responsible agency, responsible agency is itself undermined. On the other hand, it could be that though the lack of capacity to track moral value does not impair autonomous agency in general, it undermines automony in the specific domain relevant to moral responsibility. If so, we have a parallel to the way that the lack of capacity to track prudential value undermines autonomy relevant to the domain of informed consent. Whether the lack of capacity to track moral value undermines autonomous agency in general or simply its exercise in a particular domain, this lends support to the idea that control in the form of

very conditions for autonomous agency in place in medical decision-making. While completely understandable, there is value in separating them out, even if ultimately we will need to confront both kinds of questions.

[27] See Eyal (2018) and (2019) for a list of possible justifications for informed consent that are presented as alternatives to autonomy. Here, I take it that respect for autonomous agency is the primary justification for informed consent policies, but I agree that there can be many other reasons for accommodating others' wishes even when they are not autonomous agents. Children are one kind of case in which we often respect their wishes despite their not being fully autonomous agents, in order to help them learn and cultivate the skills they will need to become fully autonomous. There might be other sorts of reasons operative in the case of those with anorexia. See, for example, cases discussed by Jaworska (1999) and Fistein (2012).

opportunity, and in turn, of the ability to recognize moral value, is also required for moral responsibility and blameworthiness.

6.5 Objections and Further Questions

While I have argued that the quality of opportunity framework offers an important insight when it comes to the informed consent debate, there are important objections to consider. First, one might argue that if this is meant to be the entire story about informed consent, it cannot do the job. For one thing, as a practical matter, informed consent is supposed to provide guidance to physicians and other health care providers as to what they are obligated to do. This naturally leads to the thought that informed consent should encompass whatever is needed for a provider or investigator to have fulfilled their obligations.[28] We might think here that providing opportunity is insufficient, particularly when the stakes are high. Even on the more demanding understanding of opportunity I offered above, it only requires awareness at an earlier point that there are reasons for seeking salient specific reasons; it does not require actual uptake of those specific reasons. This is an important challenge and there are several points to make in reply.[29]

The first is simply to concede that more is required, but to say that quality of opportunity is an essential necessary condition that is along one dimension *stronger* than many particular proposals. That is the dimension of value ladenness. It allows for value to enter the picture at the level of *capacity* without necessarily making value entirely a matter of outcome.

A second reply is slightly less concessive, at least at the outset, and it is to point out that informed consent is not sufficient on any account for permissible intervention of a kind consented to. A paradigm illustration is a case in which someone provides informed consent for, say, participation in research, for which they know all the relevant facts, but where the procedure or intervention is not in their best interest. (Imagine a case in which the experiments are not well-designed, for example, and harm would come to the participants; in such a case, it would be impermissible for the researchers to proceed even with informed consent in hand.) But there are other sorts

---

[28] For example, in raising a challenge to the idea that autonomy is what grounds informed consent, Eyal (2019: section 2.2) wonders if a physician or investigator can "really be blamed" for an agent who had an excellent opportunity to comprehend the relevant facts and choose on the basis of that apprehension, but as a result of her own neglect, does not take it. If, as on the quality of opportunity framework, having had the opportunity is sufficient for the relevant agency, *and* if the physician or investigator still has a duty to try again or recruit other candidates, *and* if informed consent must capture all the duties in question, then it is clear that ensuring the conditions for informed consent requires more than just autonomous choice on the framework just described.

[29] Thanks to Gabriel De Marco for very helpful discussion of this point.

of cases where autonomous choice is not sufficient for permissibility. For example, consider cases in which friends make sub-par decisions because they don't take enough time to think through the options. They might even make decisions that benefit their friends. It is not that the decisions are not autonomous; they have all the decision-making capacities one could want. But as a friend who can see what is happening, one might be obligated to gently suggest that they take a moment and go over their options, rather than simply accepting a benefit autonomously offered. This can be a tricky area to navigate, precisely because one wants to respect, and show respect for, their autonomy. Returning to the medical context, there is a large and important literature on just what providers' and researchers' obligations are to ensure uptake (and not just availability) about the relevant empirical facts, and I cannot begin to do justice to it here.[30] The point I wish to make here, however, is simply that it is possible that providers would not violate agents' rights by failing to ensure uptake if they have high enough quality opportunities and easily accessible available information, while at the same time would fail in providers' own *duties* not to do more to ensure actual uptake. Our duties can extend beyond what others have a right to claim from us.[31] One might then see informed consent as fundamentally about the protection of autonomy-related rights, while other principles support additional duties on the parts of providers.[32]

On the other hand, while this reply might be right in theory, and it captures importantly distinct factors in generating our bottom-line obligations to those who are candidates for medical interventions, it is simply too difficult to make these distinctions in practice. At the end of the day, providers need to know what their obligations are, period. And given that informed consent is described as *informed* after all, it is of course natural to look here for all of one's duties regarding the conveyance of information. This is eminently understandable, but it does not undermine the importance of distinguishing the separate factors. That work still helps us see that respect for persons and for autonomous agency is a driving justification for policies requiring informed consent, but that this justification is

---

[30] See, for example, Manson and O'Neill (2007).

[31] For example, duties of beneficence and rescue are, arguably, of this kind. One can have a duty to rescue someone who is drowning at low cost to oneself even if the person drowning has no right to assistance. Or, to take another kind of example, it might be that one ought to forgive someone who has done all she can to make up for the wrong, even though she has no right to one's forgiveness.

[32] It might also be that patients or research participants have other rights—in addition to autonomy-based ones—when it comes to treatment from providers. Just as friends have special obligations to friends, and corresponding rights to be treated in certain ways, it is also highly plausible that the provider–patient and researcher–subject relationships entail special rights and obligations. In both cases, it is essential to separate these from autonomy-based rights and duties. Further, it is important to note that ensuring actual uptake of information is one practical way to ensure that the agent's opportunity is sufficiently good. Thus, while uptake is important in its own right, it also contributes to our evidence that the opportunity itself is a good one.

perhaps exhausted once quality of opportunity is in place. At this point, other justifications come into play to fill out additional aspects of informed consent.[33]

A second challenge is a more general theoretical one. Why try to bring a shared framework to these different areas of inquiry? After all, there are competing views in each area, and perhaps we should simply take each on its own terms. Here my reply has two parts.

First, as we have seen, there is mutual support available if we are able to employ a shared framework in each context insofar as there is already reason to see the two domains as connected. In addition to the intuitive connection discussed earlier, it is now also possible to draw yet another connection. From the very introduction of autonomy into the rationale for informed consent policies, we see an essential link to respect for persons. It is because persons are autonomous agents that they deserve respect, and we would violate their rights by failing to do so. It is also the case that responsible agency is tied to respect. While it might initially seem like a good thing when people withhold blame from others, a strong case can be made that doing so, when one acts badly, can in fact be patronizing or disrespectful. Interestingly, this same language of feeling patronized is used in both contexts.

For example, Tan and her colleagues write of the participants in the study on anorexia nervosa and decision-making:

> The participants performed on the MacCAT-T [a standardized test of compe-
> tence] to a high standard, which was comparable to the healthy population
> control group in a previous study using the MacCAT-T (Grisso et al., 1997). In
> fact, as all the participants were already highly conversant with the facts of their
> disorder, the exercise of going through information about anorexia nervosa and
> its treatment, with the systematic checking prescribed by the MacCAT-T, was
> experienced as onerous and patronizing to the participants and awkward and
> painful to carry out for the interviewer.   (Tan et al. 2003: 704)

The idea that the participants might feel patronized in the situation is under-standable. And interestingly, we find a parallel in the case in which we withhold blame from people on a variety of grounds. Kathleen Connelly (in preparation) takes up the case of withholding blame from some who are diagnosed with autism, and writes that:

> In this case and in others, many everyday people and philosophers alike have the
> intuition that someone has been wronged in having blame withheld from him. It
> is disrespectful—it is patronizing—we think, to withhold blame from someone in
> certain ways or for certain reasons.   (Connelly forthcoming)

---

[33] In this way, my view overlaps in some key ways with Eyal's. But where Eyal rejects autonomy as a rationale because it cannot accommodate all that we expect from informed consent, I continue to believe it plays the dominant role.

The similarity of reactions here lends further support for the idea that a similar kind of agency—tied to respect—is at work in both domains.[34]

Second, it is worth noting that unity of a general framework does not entail that the details must be filled in in the same ways in both areas. Though one could treat one domain in a value-laden way and not the other, I have offered reasons to think that the framework should be value laden in both domains.[35] At the same time, there are other ways in which one might treat the two asymmetrically. For example, given the scalarity of quality of opportunity, it is possible to set thresholds in different places. We might have good reason to set the threshold lower for the kind of autonomous agency that generates respect for persons than that of responsible agency that makes one a candidate for justified blame when one acts badly, and in fact, we might set the thresholds at different places within each domain, depending on the stakes.

In fact, these resources are already available in both domains, and point to yet one more intriguing parallel between the debates that so far has only been implicit. On the quality of opportunity account, a more congenial situation can reduce the degree of normative competence that is required for responsibility. And when it comes to the medical context, Buchanan and Brock (1989) have argued that for agents in a simpler decisional situation context less is required in the way of internal capacity. The principled flexibility to adjust requirements for reasons-responsive capacities differently in different situations is an attractive one in both domains and provides additional support for considering the two domains together.

In addition, we have good reason to set the level of justification required for acting on the belief in capacity differently depending on the stakes. This kind of welcome flexibility can also be applied within each of the two domains. Putting the two replies together, I hope to have offered enough reason for taking a shared framework seriously, while also taking there to be flexibility in how it is filled out in each domain.

6.6 Conclusion

I have here tried to show how the quality of opportunity framework for moral responsibility can be fruitfully brought to bear when it comes to understanding

---

[34] One might question the tight link here between respect and responsible agency by noting how differently children's wishes are treated in the clinical and research contexts. (Thanks to Garrett Cullity for pressing this point.) Children's refusal is easily overridden in the clinical setting, but not when they are asked to participate in research. In the latter setting, it seems that although beneficence may be one reason for not overriding their wishes, something akin to respect is also operative. Though this is an important issue in its own right, here I would simply say that one feature of the account of responsible agency put forward here is its scalar nature, allowing that it comes in degrees. This suggests that respect has application in the case of children without undermining the link between respect and responsible agency.
[35] See Knutzen (2020) for consideration of the suggestion that we might go substantive (or value laden) for some purposes and procedural for others.

informed consent, and, in the process, how consideration of the two areas of inquiry together can illuminate both. I have focused primarily on locating a role for a value-laden condition in both areas, which avoids both the Scylla of making outcomes decisive and the Charybdis of arbitrarily excluding value facts from the facts we take it important for people to be able to track. I have also drawn connections to other thriving debates about what else might be required for informed consent, such as the conditions under which ensuring actual uptake of relevant empirical facts might be required. While I have not offered specific suggestions on these matters, my aim is to have contributed by shedding some light on the ways that larger rationales, such as respect for persons and autonomous agency, are in place and where their limitations require supplementation by others.

References

Agule, C. (2016), "Resisting Tracing's Siren Song," in *Journal of Ethics and Social Philosophy* 10: 1–24.

Albuquerque, W. (manuscript), *The Epistemic Condition: A Reckless Mismanagement View*.

American Psychiatric Association (2013), *Diagnostic and Statistical Manual of Mental Disorders, Fifth Edition*.

Arpaly, N. (2004), "Which Autonomy?" in J. Campbell, M. O'Rourke, and D. Shier (ed.) *Freedom and Determinism* (MIT), 173–88.

Blair, J. (2008), "Empathetic Dysfunction in Psychopathic Individuals," in T. Farrow and P. Woodruff (ed.) *Empathy in Mental Illness: Mental Models, Regulation and Measurements of Empathy* (Cambridge University Press), 3–16.

Brink, D. (2021), *Fair Opportunity, Responsibility, and Excuse* (Oxford University Press).

Brink, D. and Nelkin, D. (2013), "Fairness and the Architecture of Responsibility," in *Oxford Studies in Agency and Responsibility* I (Oxford University Press), 284–313.

Buchanan, A.E. and Brock, D.W. (1989), *Deciding for Others: The Ethics of Surrogate Decision-Making* (Cambridge University Press).

Buss, S. and Westlund, A. (2018), "Personal Autonomy," in E. Zalta (ed.) *The Stanford Encyclopedia of Philosophy* Spring 2018 edition. https://plato.stanford.edu/archives/spr2018/entries/personal-autonomy/

Clarke, R. (2014), *Omissions* (Oxford University Press).

Connelly, K. (manuscript), "Blame and Patronizing."

Culver, C. and Gert, B. (2004), "Competence," in J. Radden (ed.), *The Philosophy of Psychiatry: A Companion* (Oxford University Press).

Dworkin, G. (1988), *The Theory and Practice of Autonomy* (Cambridge University Press).

Eyal, N. (2018), "Informed Consent," in A. Müller and P. Schaber (ed.), *The Routledge Handbook of the Ethics of Consent* (Routledge), 272–84.

Eyal, N. (2019), "Informed Consent," in E. Zalta (ed.) *The Stanford Encyclopedia of Philosophy* Spring 2019 edition. https://plato.stanford.edu/archives/spr2019/entries/informed-consent/

Fischer, J., and Ravizza, M. (1998), *Responsibility and Control: A Theory of Moral Responsibility* (Cambridge University Press).

Fischer, J., and Tognazzini, N. (2009), "The Truth About Tracing," in *Noûs* 43: 531–6.

Fistein, E. (2012), "The Mental Capacity Act and Conceptions of the Good," in L. Radoilska (ed.) *Autonomy and Mental Disorder* (Oxford University Press), chapter 8.

Frankfurt, H. (1971), "Freedom of the Will and the Concept of a Person," in *Journal of Philosophy* 68: 5–20.

Freyenhagen, F. (2009), "Personal Autonomy and Mental Capacity," in *Psychiatry* 8: 465–7.

Ganzini, L., Voliser L., Nelson W., Fox, E., and Derse, A. (2005), "Ten Myths About Decision-Making Capcity," in *Journal of the American Medical Directors Association* 6: S100–4.

Grisso, T., Appelbaum, P. S., and Hill-Fotouhi, C. (1997), "The MacCAT-T: A Clinical Tool to Assess Patients' Capacities to Make Treatment Decisions," in *Psychiatric Services* 48, no. 11: 1415–19.

Guerrero, A. (2017), "Intellectual Difficulty and Moral Responsibility," in J. W. Wieland and P. Robichaud (ed.) *Responsibility: The Epistemic Condition* (Oxford University Press), 199–218.

Hawkins, J. and Charland, L. (2020), "Decision-Making Capacity," in E. Zalta (ed.) *The Stanford Encyclopedia of Philosophy* (Fall 2020 edition). URL = <https://plato.stanford.edu/archives/fall2020/entries/decision-capacity/>

Holroyd, J. (2012), "Clarifying Capacity: Value and Reasons," in L. Radoilska (ed.) *Autonomy and Mental Disorder* (Oxford University Press), chapter 7.

Jaworska, A. (1999), "Respecting the Margins of Agency: Alzheimer's Patients and the Capacity to Value," in *Philosophy & Public Affairs* 28: 105–38.

King, M. (2009), "The Problem with Negligence," in *Social Theory and Practice* 35: 577–95.

Knutzen, J. (2020), "The Trouble with Formal Views of Autonomy," in *Journal of Ethics and Social Philosophy* 18: 173–210.

Manson, N., and O'Neill, O. (2007), *Rethinking Informed Consent in Bioethics* (Cambridge University Press).

Meyers, D. (1989), *Self, Society, and Personal Choice* (Columbia University Press).

National Commission for the Protection of Human Subjects of Biomedical and Behavioral Research (1979), *The Belmont Report: Ethical Principles and Guidelines for the Protection of Human Subjects of Research.*

Nelkin, D. (2011), *Making Sense of Freedom and Responsibility* (Oxford University Press).

Nelkin, D. (2015), "Psychopaths, Incorrigible Racists, and the Faces of Responsibility," *Ethics* 125(2): 357–90.

Nelkin, D. (2016), "Accountability and Desert," *Journal of Ethics* 20: 173–89.

Nelkin, D. and Rickless, S. (2017), "Moral Responsibility for Unwitting Omissions: A New Tracing Account," in D. Nelkin and S. Rickless (ed.) *The Ethics and Law of Omissions* (Oxford University Press), 106–29.

O'Neill, O. (2002), "Some Limits of Informed Consent," in *Journal of Medical Ethics* 29: 4–7.

O'Neill, O. (2003), "Autonomy: The Emperor's New Clothes," in *Proceedings of the Aristotelian Society* 77: 1–21.

Owen, G., Freyenhagen, F., Richardson, G., and Hotopf, M. (2009), "Mental Capacity and Decisional Autonomy: An Interdisciplinary Challenge," in *Inquiry* 52: 79–107.

Rosen, G. (2004), "Skepticism about Moral Responsibility," in *Philosophical Perspectives* 18: 295–313.

Rudy-Hiller, F. (manuscript), "Accountability, Reasons-responsiveness, and Narcos' Moral Responsibility."

Savulescu, J. (1994), "Rational Desires and the Limitation of Life-Sustaining Treatment," in *Bioethics* 8: 191–222.

Scanlon, T. (1998), *What We Owe to Each Other* (Harvard University Press).

Shoemaker, D. (2015), *Responsibility from the Margins* (Oxford University Press).

Smith, A. (2015), "Responsibility as Answerability," in *Inquiry* 58: 99–126.

Sripada, C. (2017), "Frankfurt's Unwilling and Willing Addicts," in *Mind* 126: 781–815.

Talbert, M. (2008), "Blame and Responsiveness to Moral Reasons: Are Psychopaths Blameworthy?," in *Pacific Philosophical Quarterly* 89: 516–35.

Talbert, M. (2022), "Attributionism," in D. Nelkin and D. Pereboom (ed.) *The Oxford Handbook of Moral Responsibility* (Oxford University Press).

Tan, J., Hope, T., and Stewart, A. (2003), "Competence to Refuse Treatment in Anorexia Nervosa," in *International Journal of Law and Psychiatry* 26: 697–707.

Tan, J., Stewart, A., Fitzpatrick, R., and Hope, T. (2006), "Competence to Make Treatment Decisions in Anorexia Nervosa: Thinking Processes and Values," in *Philosophy, Psychiatry, and Psychology* 13: 267–82.

Vargas, M. (2005), "The Trouble with Tracing," in *Midwest Studies in Philosophy* 29: 269–91.

Vargas, M. (2013), *Building Better Beings* (Oxford University Press).

Watson, G. (1996/2004), "Two Faces of Responsibility," in *Philosophical Topics* 24: 227–48. (Reprinted in *Agency and Answerability: Selected Essays* (Oxford University Press)).

Watson, G. (2011), "The Trouble with Psychopaths," in S. Freeman, R. Kumar, and R. Wallace (ed.) *Reasons and Recognition: Essays on the Philosophy of T. M. Scanlon* (Oxford University Press), 307–31.

Wolf, S. (1990), *Freedom Within Reason* (Oxford University Press).

Zimmerman, M. (1997), "Moral Responsibility and Ignorance," *Ethics* 107: 410–26.

7

Responsibility for Ill-Health and Lifestyle

Drilling Down into the Details

Neil Levy

7.1 Introduction

Whether agents are morally responsible for their need for scarce resources is a difficult and fraught issue. It's not simply because the question whether agents are morally responsible at all is itself contentious (though it is; moral responsibility scepticism is an increasingly influential view: Caruso 2021; Pereboom 2014; Shaw et al. 2019). It is also because showing that agents are sometimes morally responsible for some of their actions—even for some of the actions that help to cause their ill-health—isn't anywhere near enough to show that they are morally responsible for their need for scarce resources. In this paper, I aim to explore some unappreciated difficulties for the attribution of moral responsibility for needs that arise from the fact that in typical cases, ill-health arises from lifestyle: not, that is, from one bad decision, but from a long-term pattern of actions.

My aim in exploring this issue is twofold. First, I hope to build on Brown and Savulescu's (2019) programmatic exploration of what they call the diachronic condition on moral responsibility for ill-health. I will show that the diachronic condition fractionates in multiple ways, depending on how ill-health is caused as well as on the theory of moral responsibility at issue. Second, I aim to show that it is much harder to satisfy the diachronic condition on moral responsibility for ill-health than is widely assumed. I will argue that we usually cannot be confident that a particular agent (or a class of agents: say alcoholics) is responsible for their ill-health, and that—I will suggest—should make us hesitant to ascribe responsibility to them.

However, there is a feature of responsibility-ascriptions in the health care context that distinguishes it from some other contexts in which we ascribe responsibility, such as the criminal justice system. This feature might render the foregoing, if not quite moot, at least very much muted. In the criminal justice

Neil Levy, *Responsibility for Ill-Health and Lifestyle: Drilling Down into the Details* In: *Responsibility and Healthcare*. Edited by: Benjamin Davies, Gabriel De Marco, Neil Levy, Julian Savulescu, Oxford University Press. © Edited by Benjamin Davies, Gabriel De Marco, Neil Levy, Julian Savulescu 2024. DOI: 10.1093/oso/9780192872234.003.0008

context, we are often deciding whether *anyone* is to suffer a burden.[1] Will someone go to jail or be publicly shamed, for instance? What is at stake is whether someone will be punished. This fact entails that we face a heavy justificatory burden: only if we have shouldered that burden may we proceed to holding the person responsible. But matters are quite different in the context at issue here. When we are dealing with the allocation of scarce resources, the question is not *whether* a burden will be borne, but *who* will bear a burden. Will it be patient A or patient B? Since the question is not whether but who, we may think we need much weaker grounds for settling the question. We needn't meet the threshold the criminal law sets itself in order to be justified in holding agents responsible in contexts like these.

We face, as it were, a different burden of proof in virtue of this difference in the imposition of burdens. In the criminal law, the prosecution has to prove guilt beyond reasonable doubt. The set of conditions just mentioned may be taken to show that we will typically struggle to reach this threshold: there will usually be reasonable doubt whether a particular individual or class of individuals satisfies these conditions. But surely we will often be able to show that *on the balance of probabilities*, the individual or group is responsible? Isn't that enough to show that we can use responsibility as a tie-breaker? Suppose we must allocate a scarce resource (say a kidney) to either patient A or patient B, each of whom has roughly the same medical need for it. Evidence that A is more probably than not responsible for needing the kidney, whereas B is not, might be enough to tip our decision B's way.

It is to this issue I will turn in the final section of this paper. Having shown that we rarely can shoulder a reasonable doubt burden of proof with regard to needs for scarce resources, I will assess the prospects for a balance of probabilities standard. Can we establish that even if we cannot be confident that an agent deserves to bear a burden, we may nevertheless be confident that she deserves to bear it *more* than does some other agent? I will argue that even if we have the right to such confidence, this may not be enough for responsibility to serve as a tie-breaker between agents in the context of the allocation of scarce resources.

7.2 Responsibility and the Allocation of Resources

It is a familiar fact that agents may be causally responsible for their own need for scarce health-related resources. A large and growing proportion of the global

---

[1] Ben Davies has pointed out that this is an oversimplification. There are at least two contexts in which the criminal justice system might be faced with deciding who a burden falls on, rather than whether a burden falls on anyone. One concerns harms to the victims of past wrongdoing: such victims or their families may experience psychological harms stemming from the acquittal of someone they believe (or even know) to be guilty. Another context concerns offenders with a high chance of recidivism: if they are not burdened now, a future victim may be instead. Our unwillingness to relax our standards for a justified finding of responsibility in the criminal law in these contexts may stem in part from a perceived difference in the magnitude and the certainties of the harms that acquittal risks.

burden of disease is lifestyle related (Yoon et al. 2014). For example, it has been estimated that nearly 40 per cent of all cancers diagnosed in the UK could be prevented by changes in lifestyle (Davis 2018). Smoking, which significantly raises the risk for lung cancer, as well as multiple other cancers, is the single biggest lifestyle contributor to cancer risk, but obesity, excessive alcohol consumption, overconsumption of certain foods (like processed meats) and physical inactivity are all risk factors for other cancers. Most of these same factors are also risk factors for other life-threatening conditions like stroke and cardiopulmonary disease. In other words, the big killers—those that account for most deaths in developed countries—are all very sensitive to lifestyle factors, and many deaths are prema-ture as a consequence.

In addition, lifestyle contributes significantly to a range of other problems. Unprotected sex is obviously a risk factor for sexually transmitted disease. Alcohol consumption may lead to kidney disease, and other drugs can lead to a range of health problems ranging from minor to life-threatening. Lifestyle also contrib-utes to mental health problems, in addition to addiction. There is evidence, for instance, that psychosis may be linked to use of cannabis (Di Forti et al. 2014; Volkow et al. 2016).

Health systems therefore very often see patients who need scarce resources as a consequence of their lifestyle. These patients may be chronically ill (for example, with emphysema) and require ongoing care, in their own homes or in hospitals. They may suffer from liver disease and be in need of a transplant. People struggling with addictions and other mental health problems may need services outside the clinical setting, narrowly construed; services ranging from housing through to counselling and job training. Decent societies provide these kinds of care and services to those who need them when they cannot pay for them.

But we may wish to distinguish some potential recipients of these services and care from others. All these resources are, even in the most generous and caring of societies, limited. They cost money, and money is a limited resource. Whatever resources we expend on health care for these patients are not available to others (sick children, say) and the money spent is not available for other ends (funding research; the arts; pensions, and so on). All expenditures have opportunity costs, where the opportunity cost of an expenditure is the best alternatives use of the expenditure. Other resources that are not, or not directly, financial are even more obviously limited: physician's time and attention, for example. Of course many non-financial resources could be expanded given extra financial resources: we could, for example, pay for more hospital beds or for the training of more physicians. But some resources are in inelastic supply: while we could increase supply of kidneys or livers (the latter through living donation) through financial incentives, supply of other organs is harder to increase by increasing expenditure.[2]

---

[2] Of course, many donatable organs are buried or cremated, either because the deceased person or their family is opposed to donation or because they simply failed to express a preference. Opposition of

Because health resources are limited, health care providers are inevitably faced with difficult resource allocation decisions. Every dollar spent on one patient's care might have been spent on another. The subsidy of a class of drug entails that others are not subsidized (or that bed numbers are not increased, and so on). Of course, the need to wrestle with opportunity costs is absolutely pervasive. We all face it, in all domains: going to the cinema entails not going to a restaurant; reading a book entails not reading another book; talking to friends entails not meditating, and so on. The domain of health is not special in entailing opportunity costs nor in entailing difficult decisions (many people buy houses; settling on a particular house is for most extremely stressful, precisely because of the enormous opportunity costs of the biggest expenditure most people will ever make). But in health care, the stakes are literally life and death; which is to say that the opportunity costs of every expenditure are the deaths of other individuals. This is true not only where it is obvious (emergency rooms might have to engage in triage, and as a consequence patients may die who would otherwise have lived) but *everywhere*: while investment in physician numbers or drugs or for that matter public information campaigns have a diminishing marginal utility, it will always be true that had we spent more money *in this way* rather than *that*, people who died might have lived.

It is widely held (even by people who reject consequentialism as the right approach to morality) that the best way to make such life-and-death decisions is via some kind of utilitarian calculus (it is such an approach, for instance, that lies behind the widespread use of QALYs in health allocation decisions and policy; McKie et al. 1998; Neumann et al. 2014). On this view, we should allocate scarce resources to get the best bang for our buck, with "bang" measured in QALYs or via some other mechanism. While all decisions entail opportunity costs, some decisions are better than others. We should choose to expend resources such that opportunity costs are minimized.

While this kind of approach to resource allocation decisions obviously has much to recommend it, even many of its advocates believe that it should be modified. It is at this point that responsibility enters the picture. While we ought to allocate resources to get the best bang for our buck, other things equal, things are not equal when some, but not all, the potential beneficiaries of these resources are responsible for their need for them. On the influential luck egalitarian view, for instance, individuals deserve compensation for the consequences of their bad (brute) luck, when it makes them worse off than others, but not for the consequences of their own decisions (e.g., Cohen 2011). Translated into the context of health care, luck egalitarianism entails that agents who are responsible for their own needs should be given a lower priority than those who are unlucky in needing scarce resources (Segall 2010; in this volume, however, Eyal argues that

the first kind and inertia of the second would both fall given significant financial incentives. It's unlikely that such a measure would eliminate the need to make hard choices between potential recipients of organs, due to the inherent difficulties in transporting organs very far.

luck egalitarianism need not support this conclusion). One need not be a luck egalitarian to believe that health care should be sensitive to responsibility: the view is widely held and may be intuitive for many, and it is enshrined in health policy (see Brown et al. 2019 for discussion).

Importantly, one need not believe that those who are responsible for their ill-health deserve to suffer to find it intuitive that resource allocation should be sensitive to considerations of responsibility. Retributivists believe that the guilty deserve to suffer; the responsibilist might take the analogous line in the domain of health care (retributivism about lifestyle is not a mere thought experiment: think of how medieval Christianity classified gluttony as a deadly sin and depicted gluttons as suffering the torments of hell). But even those who think it is best that *no one* suffers may think it is better that those who are responsible suffer than those who are not. Given that our resources are limited, tough decisions have to be made and someone will always suffer. We may temper mercy with justice, since mercy must be tempered will we or no.

7.3 Obstacles to Responsibility

We temper mercy with justice by ensuring that the burdens of suffering fall on those who are guilty, given they must fall at all, only if it is indeed just that the guilty deserve to suffer to some extent; at least, that their right against such burdens is weaker than the same right of those who are not guilty. I defined basic desert in this kind of minimalist way in *Hard Luck*: to say that someone deserved negative consequences in virtue of their responsibility was not to say that it was good that they suffered, but only to say that such agents "no longer deserve the (full) protection of a right to which they would otherwise be entitled: a right against having their interests discounted in consequentialist calculations" (Levy 2011: 3). I went on to argue that in fact no one is responsible even in this minimalist sense: no one loses the right against having their interests discounted in consequentialist calculations (framed in the language of luck egalitarianism, I argued that no one was responsible for outcomes that are significantly due to luck because all our choices are shot through with luck). Here, I will set my own (idiosyncratic) views aside. Instead, I aim to show that even on more mainstream accounts of responsibility, there are large obstacles to confident attribution of responsibility in the case of lifestyle diseases.

On the consensus view of moral responsibility, for an agent to be morally responsible for an action or a state of affairs they must satisfy two independent conditions: a *control* condition and an *epistemic* condition.[3] To say that the agent exercises (responsibility-level) control over an action or a state of affairs is to say

---

[3] In *Hard Luck*, I argued that these two conditions were not independent (Mele (2010) has independently argued for a similar view). I set aside this idiosyncratic view here; in any case, it complicates the picture without changing it in its main features.

that it is appropriately sensitive to their decisions. While it is notoriously difficult to spell out what exactly is required for the possession of such control, the intuitive idea is clear enough. I control those things I can intentionally alter through my bodily movements (including those movements that consist in verbal behaviour). I am not responsible (in the relevant sense) for those things I cannot affect by acting. To say that the agent satisfies the epistemic condition is to say that she understands (at least implicitly) what the effects of her actions or omissions will be (or, in many versions, that she *ought* to understand). My putting salt in your coffee might be a (mildly) blameworthy act, if I know that the sugar bowl contains salt. But if I don't know that someone has switched the sugar for salt, I fail to satisfy the epistemic condition and don't deserve any blame.

Whether agents who come to need scarce health care resources as the result of their own actions satisfy these conditions has already received extensive discussion. There are serious worries that at least some classes of agent do not satisfy the control condition. For example, the fact that many of the most unhealthy behaviours involve addictive substances—with smoking and excessive consumption of alcohol accounting for the bulk of serious lifestyle-related morbidity and mortality—and that addiction impairs control is well-recognized. Even ordinary self-control problems may undermine responsibility attribution. Elsewhere, I have argued that the best explanation for the correlation between socio-economic status (SES) and unhealthy behaviours (and, correlatively, between SES and ill-health) is that lower SES tends to cause reduced self-control capacities and greater difficulty in resisting temptation (Levy 2019). Lower SES individuals are likely to face more stressors, which are known to reduce self-control, and live in neighbourhoods where temptations are harder to avoid, and so on. For those people most likely to suffer poor health induced by lifestyle, better choices with regard to non-addictive consumption and with regard to inactivity may be just as difficult as (or even more difficult than) better choices for addicts with regard to drugs. Indeed, it has been suggested that overeating may be harder to overcome than substance additions (Persson 2014). In addition, I have suggested that we all face problems with satisfying the epistemic conditions on responsibility for disease, because we live in epistemically polluted environments, in which discovering reliable information is difficult for those who lack the right kinds of networks or training (Levy 2018).

Even in the face of these complications, however, there's little doubt that agents who are *causally* responsible for their own ill-health regularly satisfy standard tests for moral responsibility too. The most influential test is that developed by John Martin Fischer, alone and together with Mark Ravizza (Fischer 2011; Fischer and Ravizza 2000). This tests probes agents' capacities: an agent is morally responsible for an outcome if she's capable of controlling whether it occurs (and she satisfies the epistemic condition with regard to it). She possesses the relevant capacities if she satisfies a counterfactual test: she *would* exercise her control over it given reasons to do so. She must be *receptive* to these reasons (capable of recognizing some of them *as* reasons) and *reactive* to them (adjusting her behaviour in response to some of those she recognizes). Here's not the place to delve into the details of this test. Suffice it to note that even addicts satisfy them. Given a sufficient reason to refrain from smoking, even the heaviest smoker will do so, for a shorter or longer period of time. Despite the myths, the same is true of the heroin addict or the cocaine user. The proverbial police officer at the elbow will motivate every addict to refrain for the time being. But incentives need not be extraordinary to motivate abstinence. Cocaine addicts will refrain for an extended period of time in exchange for low-value vouchers for things like cinema tickets (Higgins et al. 1994; Lussier et al. 2006). In fact, most addicts seem to "mature out" of addiction on their own, and they seem to do in response to ordinary incentives (Heyman 2009). For example, landing a new job, entering a new relationship or having a baby all provide incentives that prove powerful for many addicts.

The counterfactual test for moral responsibility is designed with one-shot cases in mind, and for those purposes may be appropriate. In effect, the test asks how the agent would respond *were something significant at stake*, and when we're concerned with responsibility for crimes, something significant *is* at stake. We think it's perfectly reasonable to enquire after how agents would respond were they attentive and careful, because it's no excuse in contexts like this that we weren't paying attention. But in the kind of cases we're concerned with—smoking, say—much less is at stake on each token occasion, and a test that asks how the agent might respond if they were paying careful attention may not be the right one to apply. There's no doubt that agents who cause their own ill-health are sometimes careful and attentive, but it's an open question how often we satisfy conditions like this with regard to low stakes and often habitual behaviours. Special problems arise for the attribution of moral responsibility in contexts like this.

7.4 Responsibility for Patterns of Behaviour

Brown and Savulescu (2019) argue that holding agents responsible for their ill-health when it arises from lifestyle requires showing that the agents are *diachronically* responsible. Diachronic responsibility consists in being responsible for at least a minimum proportion of the repeated behaviours. As they point out, diachronic responsibility comes in degrees: an agent might satisfy the conditions on being responsible for *all* the instances of behaviour that contribute to ill-health, for the *majority*, or for some minimum threshold. The distinction between diachronic responsibility and what we might call one-shot responsibility is important. I will argue that we need some further distinctions beyond those that Brown and Savulescu introduce. Ill-health may arise from lifestyle in a variety of ways, and different causal relations between lifestyle and ill-health have different implications for diachronic responsibility.

We can distinguish at least three kinds of cases where harm arises from repeated behaviour. The first set consists of cases that fit the *cumulative* model; smoking might be a good example. On the cumulative model, each instance of behaviour independently contributes to the risk of serious ill-health, though the contribution of each might be infinitesimal. When the behaviour is repeated many times (say a pack a day for twenty years), the cumulative risk is raised considerably simply through addition. The second model is the *agglomeration* model. On this model, no individual instance of the behaviour makes an independent contribution to the risk by itself, but excessive consumption significantly raises the risk. Perhaps some illegal drugs are like this: perhaps no single (carefully calibrated) dose of heroin raises the risk of ill-health for an otherwise healthy person, but many doses significantly raises the risk (sugar consumption might be a less contentious example). Substances that agglomerate to be harmful are those of which the old adage "the dose makes the poison" is true. On the agglomerative model, the risk is emergent from the behaviour. On the *stochastic* model, *most—* perhaps almost all—instances of the behaviour carry no negative consequences at all for the person, but each involves a risk of negative consequences. Unprotected sex with multiple partners might be like this: most partners will probably not carry a virus. But repeated behaviour raises the risk considerably.

These distinctions greatly complicate the attribution of responsibility: different patterns of causation may imply different degrees of moral responsibility for the same effects. In what follows, I will work through the different models. As we will see, there are important differences between them in what they imply for moral responsibility (and in particular how confident we have a right to be in attributing moral responsibility to agents for their own ill-health). I will *not* show that agents cannot be responsible for their own ill-health under any of these models. My ambitions are less grand: to sketch the complications that these models entail for the justified attribution of responsibility and to suggest that we often have much less right to confidence about such attributions than we might think.

7.4.1 The Cumulative Model

Let's begin with the cumulative model. On this model, each instance of behaviour carries some independent risk, though it is typically low (if it were not low, most agents would not be willing to run the risk: the fact that smoking *this* cigarette is almost risk-free for me is a necessary condition of my smoking it). Are agents responsible for resulting ill-health?

Once a behaviour is established, it may become habitual. Smoking may easily become habitual because it is widely accepted (or at least was until recently) and doesn't take a great deal of effort in preparation (in these ways, it is quite unlike, say, heroin use, which is unlikely to become habitual). Habitual behaviours are subject to a relatively low degree of control. The habit system may render behaviour "unintentional, robotic, perhaps even unconscious" (Redish et al. 2008: 424). Of course, behaviours that are habitual are not subject to a low degree of control all the time. Smokers give up smoking, obviously. Just as obviously, their behaviour is reasons-responsive (in the sense made famous by Fischer and Ravizza 2000): given sufficient reasons to inhibit the behaviour, the agent will succeed for a shorter or longer period of time. Nevertheless, very many of the behaviours that cumulatively give rise to the risk exhibit a low degree of control.

We don't blame agents for running small risks (the agent who takes a short-cut home knowing there's a one in a million chance of being hit by lightning surely takes a reasonable chance and isn't held responsible if she's subsequently struck). A simplistic framework for attributing responsibility to agents in cases like this would simply add up the token instances on which the agent exercised control for their behaviour and ask whether the total risk arising from those instances combined is sufficient to hold the agent responsible. Given the behaviour is habitual, it's likely that such instances are not sufficiently numerous to underwrite attribution of responsibility.

However, the simplistic framework is too simple. In attributing responsible, we shouldn't treat these decisions to smoke (or to refrain from smoking for some period of time) independently of the agent's other behaviours. The agent may not have exercised responsibility-level control over many token behaviours that together raise the risk to some considerable level, but she's aware that she engages in them, and she ought to take them into account when she makes her attentive decisions. Analogously, I ought to take the behaviour of other agents (over whom I exercise no control) into account in making decisions about how to behave; if I vandalize a painting together with other people, for example, I'm not excused for causing damage on the grounds that my action wasn't sufficient by itself to cause it.

Nevertheless, the fact that control does fluctuate over time and most of the time I may not exercise responsibility-level control over my behaviour should surely reduce my responsibility for the outcome, relative to having caused it in a fully controlled fashion (analogously, the fact that my contribution to the damage wasn't sufficient to cause it might reduce my responsibility for the outcome). Moreover, there are other reasons to think that my responsibility might be reduced on the cumulative model (and on the agglomeration model too) in some instances.

Consider a case like this: an agent causes ill-health by excessive eating. Instances of eating come in two varieties for her. Some feature fine food that she savours; others involve habitual and inattentive gulping of junk for stress relief. The first kind is important to her and are among her most significant pleasures. The second she barely notices. She exercises responsibility-level control only over the first kind, since she is sufficiently attentive only when she savours her food. While there are grounds for thinking she ought to refrain from her controlled eating, in asking her to refrain we ask her to forego a pleasure that is significant to her. Even if she values good health more than these pleasures, their importance to her and the place they occupy in her life seem to constitute grounds for mitigation.

Again, I don't take these considerations to show that she's not responsible. Rather, I take them to complicate attributions of responsibility and to give us reason to be cautious in making such attributions. Of course, we might hope to set all these complications aside by abstracting from the question whether or when the agent exercises control over her behaviour and asking instead about her acquisition of a habit. I set this question aside for the moment.

7.4.2 The Agglomeration Model

I turn now to the agglomeration model. On this model, a risk arises only when the number of behaviours passes a certain threshold. Prior to that threshold, there is no (or no measurable) risk at all. It seems to follow that on such models, an agent might be fully responsible for a great many of her behaviours, but not for a significantly raised risk of ill-health because she is responsible for some number of token behaviours insufficient to cross the threshold required for risk. If that's right, then the kinds of considerations introduced above—about how control fluctuates over time and how the epistemic condition for habitual behaviours may be harder to satisfy than at first seems—may entail that we should not be confident that people are responsible for agglomerative risks.

We've already noted both that this simple way of thinking about risks is too simple, because we shouldn't treat token controlled behaviours as independent of uncontrolled behaviours (at least if the agent is aware of them), and that there may be grounds for mitigating responsibility nevertheless that carry over from the cumulative model. Here I want to note a reason for caution that seems to rise especially pointedly on the agglomeration model. On these models, and depending where the threshold for risk is, attributions of moral responsibility confront a significant worry from moral luck.

Moral luck cases arise when an agent seems responsible for an outcome, but differs from other possible or actual agents who are not responsible for the outcome only in respects that are shot through with luck. The standard illustration is reckless driving: an agent who drives recklessly and injures a pedestrian may differ from another reckless driver who did not cause harm only in experiencing bad luck. Since luck is not a basis for desert, there are strong grounds for treating the agents alike: either blaming both or excusing both (Levy 2016a). Such cases can arise on the agglomeration model.

Consider two agents who engages in behaviour liable to give rise to such harms. One ceases the behaviour just prior to passing the threshold at which there is a significant risk of harm; one continues just beyond that threshold. Is there a difference in their responsibility if they both subsequently develop ill-health? In the case of the first agent, the ill-health is (by stipulation) not a result of their behaviour. Because they ceased the behaviour prior to passing the risk threshold, the ill-health arose by bad luck instead. The other agent may indeed have caused their ill-health (though of course we cannot be sure; we are rarely in a position to be confident that had the person not engaged in certain behaviours, they would not have suffered the problem). But they engaged in the behaviour only slightly more than their counterpart. Moreover, the first is unlikely to have known that they stopped just prior to the threshold (in fact, no one has any clear idea where that threshold is). They did not satisfy the control or epistemic conditions with regard to the fact *that they stopped in time*. So if they escape responsibility, it is in virtue of a difference which is slight and which is little to their credit (both, recall, did stop before developing ill-health). The difference between the two agents seems to be a matter of luck. By hypothesis, the agent who didn't cross the threshold for substantial risk didn't cause their ill-health and therefore isn't liable to bear any additional burdens; since she differs from the other agent only in her luck, there are grounds for excusing *both*.

7.4.3 The Stochastic Model

Before turning to the question whether responsibility can nevertheless serve as tie-breaker between patients, let's briefly consider the stochastic model. On that model, recall, most instances of behaviour do not cause any harm at all but some are likely to cause serious harm. *Epistemically*, token acts may be indistinguishable, but metaphysically they're unalike. We might call this the Russian roulette model. Pointing a gun at one's head and pulling the trigger when the chamber is empty carries no risk, whereas when there's a bullet in the chamber it is almost always fatal. Russian roulette is a gamble because the two cases look identical to players. Russian roulette is one example of a behaviour that fits the stochastic model. Another might be unprotected sex with strangers: on most occasions there may be no risk of a sexually transmitted disease but on some (epistemically indistinguishable) cases, the risk might be high. Sharing needles might be another example.

One way to attribute responsibility on such a model might be as follows: what matters is whether the agent is responsible for the *specific* behaviour that actually caused the harm. Suppose an agent contracts a serious illness from sharing a needle, and the illness was caused by one particular occasion out of eighty times she shared needles. Whether she's responsible for the illness might depend on whether she was responsible for *that* behaviour. If that's right, the stochastic model might be particularly demanding: an agent might be responsible for *almost all* her actions and fail to be responsible for her ill-health. This intuition would be particularly strong with regard to behaviours that are morally innocuous, taken one by one.

However, just as we have reasons not to treat token behaviours independently of one another on the agglomeration and the cumulative models, so we have reason to take actions that didn't result in any harm (directly, at any rate) into account in assessing responsibility on this model. Even though these actions didn't cause ill-health, they may have made the action that *did* cause ill-health more probable. If some of these directly harmless actions are free, then they might underwrite attribution of responsibility to the agent.

Actions may make future (free or unfree) behaviours more likely in the future in one or more of several ways. They may establish habits. They may result in altered assessments of risk ("if sharing needles turned out okay a dozen times, it's probably not very dangerous"). They may alter the environment in which the agent finds herself, and therefore change the temptations or the opportunities (freely installing Tinder on your phone today might make unsafe sex more likely tomorrow, when your self-control is low). Of course, causal responsibility for later unfree behaviours isn't sufficient for moral responsibility for those later behaviours: the agent must satisfy the epistemic condition with regard to them as well. That is, she must know (or be culpable for failing to know) that her behaviour now makes future behaviour more likely. I set this question aside until the next section, which deals with such "tracing" conditions as they pertain to all three models. For the moment, I note only that such assessments may be difficult and at least sometimes agents won't satisfy the epistemic tracing condition sufficiently for their earlier actions to underwrite responsibility for later behaviours.

Just like the agglomeration model, problems of moral luck can arise on the stochastic model. Two agents may have (for example) each shared needles with others on dozens of occasions. Agent A may contract hepatitis on one such occasion, whereas agent B was luckier and did not contract any serious disease. We may hold the tracing condition fixed for both (recall that the problem of moral luck does not require that there is an *actual* luckier counterpart; merely that a counterpart that differed only in their luck could exist[4]). Since the agents differ only in their degree of luck, it is difficult to see how one might deserve a greater burden than the other.

---

[4] It is not sufficient that a counterpart might have existed who did not suffer the harm. There are constraints on the attribution of luck, and probability features, in complex ways, among them. If an event was likely, then the agent is not lucky to have been subject to it. The agent who is hit by a car while walking across the freeway is not unlucky, and the fact that a possible counterpart was not subject to this harm does not provide her with an excuse. See Levy (2011) for discussion.

7.5 Tracing Responsibility

We sometimes want to hold agents responsible for later behaviours in virtue of their previous free actions (Vargas 2005). For example, we may want to blame an addict for her habit in virtue of their actions *prior to* the behaviour becoming habitual—and hence when the low degree of control characteristic of habits was not at issue. Such attributions are subject to a demanding epistemic condition. The epistemic condition entails, among other things, that agents must have some kind of grasp of what is at stake in their behaviour. They must understand that they are at significant risk of developing that pattern of behaviour *and* that if they develop that pattern of behaviour they will run a significant risk of subsequent ill-effects. These conditions are harder to satisfy then is often realized. People may overestimate their resistance to developing a habit or becoming addicted. Overheated drug rhetoric may contribute to agents possessing an inaccurate sense of the degree to which addictions are hard to shake: when they learn that their peers have tried cocaine or heroin without becoming addicted, they may think that popular rhetoric is false and may conclude that the risk is low.

Of course, habits often develop when the person is young, and may have difficulty in assessing risks. A smoking habit developed as a teenager will make later life abstention much more difficult. Young people *also* have weaker control systems: they may fail to satisfy (or fully satisfy) both epistemic and control conditions. While a great many of the instances of behaviours that together give rise to a significant risk of ill-health are sufficiently controlled and sufficiently knowing to count as morally responsible, we often lack grounds for confidence that the agent is responsible for a sufficient number to count as a responsible for the ill-health should it eventuate.

There are also questions of moral luck that arise with regard to the epistemic tracing condition. Agents A and B may have fulfilled the tracing condition to the same extent, with only one developing a harmful habit (perhaps as a result of a genetic vulnerability they could not have known about). Most people who try hard drugs do not in fact become addicts, after all. Since the agents do not differ in the degree to which they satisfy the tracing condition, it seems unfair to hold them responsible to different degrees.

It is also worth emphasizing that some harmful habits develop out of behaviours that are difficult to avoid, or that are innocuous. Early instances of the behaviour might be prudentially neutral at worst; indeed, some instances of some behaviour may be prudentially *required*. The clearest example here is eating. Because eating is prudentially required, it has features that make it peculiarly difficult to control. Eating is unlike smoking or alcohol consumption in that it is not possible to treat the problem by abstinence. The rule "never smoke again" is maximally clear: the contexts to which it is relevant and the behaviours that it prescribes (or proscribes) in these contexts are quite precisely specified. But rules like "do not overeat" are far from clear in their application and require a great deal of interpretation. How much is too much? How should episodes of consumption be balanced against one another and against other activities (e.g., going for a walk)? Is it permissible to eat the birthday cake *now* (perhaps required: my friend's feelings will be hurt if I don't accept) and skip breakfast tomorrow to make up for it? The intrinsic vagueness in dietary rules may entail difficulties in holding agents responsible for the development of harmful habits.

There are also difficulties arising with regard to satisfaction of the epistemic condition in this domain. Some dieticians argue that we should never see some foods as forbidden entirely, because that leaves the dieter feeling deprived and makes them more vulnerable to lapses (which may be seen as devastating failures). But the knowledge that no food is forbidden is itself open to abuse, since once again it is intrinsically difficult to calculate how much is okay. It is cognitively demanding to calculate how much we ought to eat, especially in contemporary environments where food is plentiful and omnipresent. Moreover, the fact that these rules require interpretation opens the way for self-deception in their application (Ainslie 2001; Levy 2016b). All these facts may entail that agents' responsibility for overeating is lower than we might have thought. The agents in question failed at a task that is more difficult than the one facing those who do not experience constant temptation and a subsequent need to engage in cognitively demanding tasks (see Brown, this volume, for further discussion of the complexities in assessing the degree to which agents satisfy the tracing condition).

7.6 Responsibility as a Tie-Breaker

My aim, in the foregoing, has been both to shed light on the structure of behaviour that causes ill-health (in typical cases) and also thereby to suggest we have weaker grounds for concluding that agents are responsible for such ill-health than we might have thought. As we saw at the beginning of this paper, however, there are important differences between responsibility in the health context and responsibility in (say) the context of criminal law.

In the criminal law context, the standard of proof is reasonable doubt. If we have grounds for doubting either that the agent actually committed to crime or that she was responsible for committing it, we should acquit. But we may think we need to satisfy a lower standard of proof in the context of the allocation of scarce resources. Setting consequential considerations aside (as we may often appropriately do in the context of criminal law; in any case, such considerations are irrelevant to questions of basic desert by stipulation), no one is wronged by a guilty agent going free.[5] It is not necessary that a burden fall on anyone. In the resource allocation context, however, matters are crucially different. Resources expended one way are not available for expenditure in others. An agent who is more clearly not responsible for her ill-health may therefore have a claim on these resources that is stronger than another who is less clearly absolved of responsibility. It seems that we may therefore appropriately invoke responsibility, or a reasonable suspicion of responsibility, as a tie-breaker between agents. Other things being equal (each needs the resource to the same degree; the expected benefits to each are more or less identical; QALYs are equalized, and so on), we might prefer to allocate resources to agents who are more clearly absolved of responsibility.

The tie-breaker argument is surely the strongest available to the defender of moral responsibility in the allocation of scarce resources (though it is likely limited in its scope: there may be few cases in which we face a choice between agents who are roughly equal in the expected utility they might derive from a treatment). Those who invoke it may accept that they do not have the right to confidence that a particular agent is morally responsible for their ill-health but note—plausibly–that they have a better reason to think that the agent is morally responsible for their ill-health than another who needs the resource equally badly. If we accept such an argument, however, it will be at the cost of witting unfairness. That is, if we use responsibility as a tie-breaker, we do so in the knowledge that it will rule some people responsible who are not (and perhaps vice versa). That's a moral cost to us and to the individuals wrongly stigmatized as responsible, and it's an open question whether we should pay this cost.

Of course, we must accept that whatever we do, some people will be denied those resources to which they have an equal right. That's a sad entailment of the fact that need exceeds supply. We must allocate resources so some people are not treated as they deserve. But that we should run the additional risk of stigmatizing some in the process surely should not be accepted without further argument. We may prefer to allocate resources on other grounds: a lottery, or a queuing system, for example. While I won't argue for that claim here, the fact that the capacities and opportunities to escape blame are *themselves* unfairly distributed (Levy 2019) entails that when we use these kinds of devices we treat people *more* fairly than when we are sensitive to considerations of desert.

7.7 Conclusion

My aim in this paper has been to draw attention to unappreciated difficulties in assessing the extent to which people may be responsible for their own ill-health.

---

[5] Or at least no one *need* be wronged. Some deontologists may think that if a guilty person goes free, their victims are wronged. But some crimes are victimless, and some victims may be dead. These kinds of effects are potentially important but not essential to the question of basic desert.

These difficulties are unappreciated because ethicists have paid little attention to diachronic responsibility and how it differs from one-shot cases. Further, they have not noticed how different kinds of causal relations between token behaviours and subsequent ill-health entail different sets of complications in the assessment of responsibility. I have not suggested that agents are not responsible for their ill-health under any of the models I have outlined. Rather, I have attempted to delineate the kinds of issues that confront us in attributing responsibility.

Firmer conclusions on when agents are responsible, under any of these models, must await further work which moves beyond the sketch of the landscape I have provided here and focuses on particular kinds of agents in particular contexts. In the meantime, I hope I have not only provided conceptual tools for such further work but also led to us seeing that attributions of responsibility of responsibility are often premature. We are not entitled to the confidence with which many of us blame agents for their health-related needs.

References

Ainslie, G. (2001), *Breakdown of Will* (Cambridge University Press).

Brown, RCH., Maslen, H., and Savulescu, J. (2019), "Against Moral Responsibilisation of Health: Prudential Responsibility and Health Promotion", in *Public Health Ethics* 12: 114–29.

Brown, RCH. and Savulescu, J. (2019), "Responsibility in Healthcare Across Time and Agents", in *Journal of Medical Ethics* 45/10: 636–44.

Caruso, GD. (2021), *Rejecting Retributivism: Free Will, Punishment, and Criminal Justice* (Cambridge University Press).

Cohen, GA. (2011), *On the Currency of Egalitarian Justice, and Other Essays in Political Philosophy* (Princeton University Press).

Davis, N. (2018), "Four in 10 cancer cases could be prevented by lifestyle changes", in *Guardian.* 23 March 2018. https://www.theguardian.com/society/2018/mar/23/four-in-10-cancer-cases-could-be-prevented-by-lifestyle-changes

Di Forti, M., Sallis, H., Allegri, F., et al. (2014), "Daily Use, Especially of High-Potency Cannabis, Drives the Earlier Onset of Psychosis in Cannabis Users", in *Schizophrenia Bulletin* 40: 1509–17. https://doi.org/10.1093/schbul/sbt181

Fischer, JM. (2011), *Deep Control: Essays on Free Will and Value* (Oxford University Press).

Fischer, JM., and Ravizza, M. (2000), *Responsibility and Control: A Theory of Moral Responsibility* (Cambridge University Press).

Heyman, GM. (2009), *Addiction: A Disorder of Choice* (Harvard University Press).

Higgins, ST., Budney, AJ., Bickel, WK., et al. (1994), "Incentives Improve Outcome in Outpatient Behavioral Treatment of Cocaine Dependence", in *Archives of General Psychiatry* 51: 568–76.

Levy, N. (2011), *Hard Luck: How Luck Undermines Free Will and Moral Responsibility* (Oxford University Press).

Levy, N. (2016a), "Dissolving the Puzzle of Resultant Moral Luck", in *Review of Philosophy and Psychology* 7: 127–39.

Levy, N. (2016b), "'My Name is Joe and I'm an Alcoholic': Addiction, Self-Knowledge and the Dangers of Rationalism", in *Mind & Language* 31: 265–76.

Levy, N. (2018), "Taking Responsibility for Health in an Epistemically Polluted Environment", in *Theoretical Medicine and Bioethics* 39: 123–41.

Levy, N. (2019), "Taking Responsibility for Responsibility," in *Public Health Ethics* 12/2: 103–13.

Lussier, JP., Heil, SH., Mongeon, JA., Badger, GJ., and Higgins, ST. (2006), "A Meta-Analysis of Voucher-Based Reinforcement Therapy for Substance Use Disorders," in *Addiction* 101: 192–203.

McKie, J., Singer, P., Kuhse, H., and Richardson, J. (1998), *The Allocation of Health Care Resources: An Ethical Evaluation of the "QALY" Approach* (Routledge).

Mele, A. (2010), "Moral Responsibility for Actions: Epistemic and Freedom Conditions", in *Philosophical Explorations* 13: 101–11.

Neumann, PJ., Cohen, JT., and Weinstein, MC. (2014), "Updating Cost-Effectiveness: The Curious Resilience of the $50,000-per-QALY Threshold", in *New England Journal of Medicine* 371: 796–7.

Pereboom, D. (2014), *Free Will, Agency, and Meaning in Life* (Oxford University Press).

Persson, K. (2014), "Why Bariatric Surgery Should be Given High Priority: An Argument from Law and Morality", in *Health Care Analysis* 22: 305–24.

Redish, AD., Jensen, S., and Johnson, A. (2008), "A Unified Framework for Addiction: Vulnerabilities in the Decision Process", in *Behavioral and Brain Sciences* 31: 415–37.

Segall, S. (2010), *Health, Luck, and Justice* (Princeton University Press).

Shaw, E., Pereboom, D., and Caruso, GD. (2019), *Free Will Skepticism in Law and Society: Challenging Retributive Justice* (Cambridge University Press).

Vargas, M. (2005), "The Trouble with Tracing", in *Midwest Studies in Philosophy* 29: 269–90.

Volkow, ND., Swanson, JM., Evins, AE., et al. (2016), "Effects of Cannabis Use on Human Behavior, Including Cognition, Motivation, and Psychosis: A Review", *JAMA Psychiatry* 73: 292–7.

Yoon, PW., Bastian, B., Anderson, RN., et al. (2014), "Potentially Preventable Deaths from the Five Leading Causes of Death: United States, 2008–2010", *Morbidity and Mortality Weekly Report* 63: 369–74.

8

Obesity and Responsibility for Health

Rekha Nath

8.1 Introduction

Should society adopt health care policies aimed at holding obese individuals responsible for obesity-related health problems?[1] I will argue that it should not. More specifically, holding obese individuals responsible for obesity-related health problems would neither treat people as they deserve to be treated, nor make the distribution of health care resources fairer, nor produce better overall outcomes in the promotion of public health. The arguments to the contrary are based on dubious empirical assumptions, faulty philosophical reasoning, or both.

8.2 The Case for Holding Obese Individuals Responsible in Health Care Domains

Obesity is widely regarded a major public health crisis. Being obese increases a person's risk of heart disease, diabetes, and over a dozen types of cancer, and it is associated with a lower life expectancy. In the United States, over 40 percent of the adult population is obese, and nearly one in five children—some 14 million—are obese (Hales et al. 2020; Centers for Disease Control and Prevention 2021). Since the 1960s, the obesity rate among American adults has nearly tripled. Similar trends of sharply rising obesity rates have been seen in many other high-income nations in recent decades too. More than one of every four adults in Canada and the United Kingdom is obese, and about one of every three adults in Australia and New Zealand is obese (Statistics Canada 2019; Baker 2021; Australian Institute of Health and Welfare 2020; Ministry of Health NZ 2021). What is more, obesity rates now seem to be on the rise in low-income and middle-income nations. According to the World Health Organization (WHO 2021), as of 2016, around 13

---

[1] "Obese" is often regarded as a pejorative label, and many scholars and activists who discuss weight from a social justice perspective prefer the term "fat" to "obese." Centrally, the label "obese" is rejected on the grounds that it medicalizes having a larger body, presuming that obesity is a problem in need of fixing. See, e.g., Wann (2009). I do not endorse the medicalized approach to obesity. Nevertheless, in this chapter, I use the terms "obese" and "obesity" because the health care policy matters I set out to address concern those to whom that clinical designation applies (that is, those persons who have a body mass index (BMI) of at least 30).

Rekha Nath, *Obesity and Responsibility for Health* In: *Responsibility and Healthcare*. Edited by: Benjamin Davies, Gabriel De Marco, Neil Levy, Julian Savulescu, Oxford University Press. © Edited by Benjamin Davies, Gabriel De Marco, Neil Levy, Julian Savulescu 2024. DOI: 10.1093/oso/9780192872234.003.0009

percent of all adults worldwide are obese. In a 2018 report, the WHO (2018) deems childhood obesity "one of the most serious global public health challenges of the 21st century."

Besides the immense toll that obesity can take on obese individuals themselves—threatening to prematurely shorten their lives and cause chronic conditions that diminish their quality of life—high obesity rates can also produce significant costs for society at large. For one thing, it is expensive to treat obesity-related medical conditions. In the United States, the annual health care expenditures of treating those conditions have been estimated in recent years to exceed $172 billion (Ward et al. 2021). For another, health care resources—not just money, but hospital beds, time spent with doctors, the numbers of surgeries performed, and much else—are in limited supply. Addressing the health needs of some patients means there are fewer health care resources available to treat others. As such, attending to the considerable, and by all accounts ever-growing, obesity-related disease burden has a steep opportunity cost.

Lifestyle factors are an important contributor to the high incidence of obesity. That is, obesity isn't something that merely happens to people, but rather people's choices and behaviors are implicated in making them obese. The two familiar lifestyle factors at issue are poor dietary habits—taking in too many calories, especially an excessive consumption of junk food—and insufficient physical activity. It is against this backdrop that the question arises of whether people should be held responsible for being obese. Holding someone responsible for being obese could take different forms. It might involve getting obese persons to accept that their high weight is the consequence of their own poor dietary choices. Or, it might involve morally reproaching obese individuals for failing to embrace a healthier lifestyle. In what follows, the specific sense of holding people responsible for their weight that will be my focus concerns whether obese individuals should be made to bear some (or even all) of the higher health care costs associated with being that way.

To investigate this matter, it will be helpful to first consider what holding obese people responsible for the higher health care expenditures associated with their weight might entail in practice. Different sorts of policies could be adopted toward this end. One approach would be to hold all obese individuals responsible for their weight *ex ante*, making them personally bear the higher health care costs associated with being obese. Here the reasoning is that whether or not a particular obese individual experiences poor health, her weight puts her at significantly heightened risk of poor health, and she should be held to account for the costs associated with that risk, a risk which she could avoid through lifestyle changes. Based on this rationale, obese individuals might be charged more than nonobese individuals for health care—for instance, through higher premiums or co-pays, or they might be required to purchase supplementary insurance. In the United States, this sort of policy approach has been adopted in the case of tobacco use. Employers may charge smokers more for health insurance than nonsmokers.

A different policy approach would involve holding obese individuals responsible for poor health outcomes *ex post*, by making it costlier or otherwise more difficult for those who experience obesity-related health problems to access health care. On this basis, obese patients might be denied access to certain medical treatments altogether, they might have to pay more for certain treatments, or they might be given lower priority than nonobese patients for receiving some forms of medical treatment. Alternatively, their access to health care might be made conditional on their adoption of certain lifestyle changes (e.g., dietary modifications) or their achievement of certain outcomes (e.g., losing a specified amount of weight to be eligible for knee surgery). This approach is embodied in the long-standing clinical practice in the United States and elsewhere of requiring alcoholics to abstain from alcohol for at least six months before they qualify for a life-saving liver transplant.

Those are some of the sorts of policies that might be adopted to hold obese persons responsible for their weight or for obesity-related health problems. In what follows, I set out to answer the question of whether, in general, there is a sound basis for adopting any such policies, which I will refer to as "obesity-penalizing policies."[2] Obesity-penalizing policies might be construed as serving to remedy the problem of *moral hazard* that arises without some measures in place aimed at making obese individuals bear health-related costs associated with their higher weight. A moral hazard arises when people are shielded from negative conse-quences of their risky choices, which, in the case at hand, would seem to occur if individuals aren't made to personally shoulder a non-negligible share of the added health care costs that their obesity risks producing. It might seem quite plausible that we ought to try to reduce, if not eliminate, this moral hazard by making obese individuals bear (some of) the health-related costs associated with being that way.

Let us consider the basis for that claim. For one thing, it might seem plainly unfair that those individuals who make prudent lifestyle choices concerning their health and weight (e.g., the slim, salad-eating joggers) should have to subsidize the significant health care expenditures that wouldn't arise if not for the poorer choices made by some of their fellow citizens (e.g., the fat, Cheetos-eating couch potatoes). In addition, it might seem unfair that individuals who make prudent lifestyle choices would receive inferior health care, or face barriers to accessing affordable health care, because others avoidably utilize scarce health care resources on account of their imprudent choices. Health policies targeting the obese would mitigate both forms of unfairness. By reducing obese individuals' access to subsidized health care, nonobese individuals wouldn't have to bear as much of the costs generated by the less responsible lifestyle choices made by the obese, and the freed-up health care resources could be redirected toward better

---

[2] In pursuing this query, my focus is on the American context. However, there are good reasons to expect my main conclusions on whether such policies should be adopted to apply more broadly. In particular, I take it that similar conclusions on that matter likely hold for other high-income nations.

addressing the medical needs of more deserving recipients, namely, those with conditions that aren't the product of poor lifestyle choices. We can call this basis for holding obese persons responsible in health care domains *the fairness argument*.

Besides promoting fairness, a different reason for adopting obesity-penalizing policies has to do with the better public health outcomes they might bring about. Here the rationale is that making obese individuals personally bear more of the costs of obesity-related health problems would generate a stronger incentive for all persons subject to such policies to make healthier lifestyle choices regarding their weight—choices that would, in turn, reduce obesity rates. Such policies, the reasoning continues, would both nudge those who are currently obese to work harder to lose weight and encourage nonobese individuals to make a greater effort to stay that way. This would be a win-win for all members of society: for many persons who are obese or are at risk of becoming obese, obesity-penalizing policies could help them make better choices to keep their weight down and thereby avoid serious health problems. Moreover, lower obesity rates would translate into improved health care for individuals across the board, as the reduction to the overall disease burden implied by lower obesity rates would result in freeing up health care resources that could be redirected toward other, nonobesity-related medical issues. Call this *the public health promotion argument*.

Both the fairness argument and the public health promotion argument might seem initially compelling.[3] But I will argue that on reflection they are not. Contrary to the fairness argument, policies aimed at holding obese individuals personally responsible for their weight or for weight-related health problems would neither treat people as they deserve to be treated nor make the distribution of health care resources fairer. And contrary to the public health promotion argument, those policies would not produce better health outcomes for society at large. Indeed, they would tend to do just the opposite. Moreover, I will argue, reflection on the shortcomings of those arguments reveals that their conclusion is false: society should *not* adopt policies aimed at holding obese individuals personally responsible for their weight or for weight-related health problems.

8.3 Are Obese Individuals a Drain on the Health Care System?

The claim that obese individuals should be made to bear higher costs in health care settings is predicated on the supposition that these individuals, in virtue of their weight, avoidably impose a burden on the health care system. The claim here is not merely that obese persons, on average, use more health care resources than

---

[3] For further assessment of one or both of these arguments for holding individuals responsible for unhealthy lifestyle choices (not just pertaining to weight), see also Wikler (1987); Feiring (2008); Friesen (2018); Brown et al. (2019).

others (although proponents of the reasoning that obese people should pay more than others often take that to be the case too). After all, we do not generally take issue with the fact that some individuals—for example, a child with leukemia or someone with a congenital heart defect—use considerably more health care resources than most others do on account of their expensive-to-treat conditions. The intuition that we shouldn't as a matter of fairness make those persons bear more of the costs of their higher health care seems to be explained by it not being their fault that they have those medical conditions. The charge against obese persons, then, is that by being obese they produce, or risk producing, higher health care costs that would not obtain if they made better lifestyle choices—choices they can be reasonably expected to make.

What is the evidence that obesity leads to higher health care expenses? Over the past few decades, numerous studies have been conducted on this matter. The broad consensus is that obese people use more health care resources than others. Obese persons are more likely than their nonobese counterparts to have chronic diseases, to be hospitalized, and to have ongoing pharmaceutical drug needs (Musich et al. 2016). The per capita annual health care costs associated with being obese are estimated to be higher than those of normal-weight individuals, although estimates of just how much higher are wide-ranging—with some as low as $732 and others as high as $3508 (Finkelstein et al. 2003; Cawley et al. 2015).[4]

However, even if obese individuals use more health care resources per annum than nonobese individuals, it does not follow that they incur greater health care expenses over the course of their lifetimes. Indeed, if we instead compare the average *lifetime* health care costs of obese and nonobese persons, it may well be that those of the former tend to be lower than those of the latter. A parallel observation has been made about the comparative health care expenses of smokers and nonsmokers. That is, over their lifetimes, on average, smokers appear to cost the health care system quite a bit less than nonsmokers do (Barendregt et al. 1997).[5] This has to do with just how smoking typically harms a person's health. Smoking greatly increases the risk of contracting dangerous medical conditions (e.g., lung cancer) that once contracted tend to dramatically worsen a person's health (and raise one's health care expenses) before bringing one's life to an end relatively swiftly. This is where, to put it rather crudely, the cost-savings come about: by cutting people's lives prematurely short in this way, smoking-related diseases significantly reduce the number of years that smokers tend to use health care resources.

---

[4] These estimates are based on the US health care context. Not surprisingly, cross-country estimates of obesity-attributable health care costs are more wide-ranging yet.

[5] Based in part on this observation, Resnick (2013) advances a challenge to the case for charging smokers more for health care parallel to the one I raise here for obesity.

The medical conditions linked to obesity differ from those linked to smoking in a couple of respects that are salient to the current analysis: first, obesity-related diseases tend to be chronic conditions that affect individuals for many more years than the conditions that afflict smokers, and second, obesity-related diseases don't generally reduce life expectancy as dramatically as smoking does. Both these reasons would suggest that being obese may not imply reduced lifetime health care costs in the way that being a smoker does. However, in spite of those key differences, some researchers conclude that obese individuals have lower lifetime health care costs than their nonobese counterparts on the basis of a counterfactual assessment of the so-called "substitute" diseases obese individuals would be likely to get if they weren't obese and thus were to live somewhat longer lives (van Baal et al. 2008).[6]

The widespread tendency to focus on the narrower construal of obesity-related costs, in per annum terms, is understandable. That obese patients use more health care resources than others within, say, a given financial year will be immediately apparent to those in health care settings. For example, health care providers and insurers will tend to notice that obese patients make more annual medical visits, require more surgeries, have longer hospital stays, and so on. Accordingly, it will be easy enough for such parties to quantify the greater expenses incurred by this group within a fixed timeframe. By comparison, the health care savings that are linked to lower life expectancy of a group will be less visible. Those same parties, it would seem, would be less likely to take notice of the savings implied by the considerable financial costs and other health care resources that would have been used by people had they lived longer lives. Based on methodological complexities that arise in trying to calculate lifetime health care costs, it's also far easier for researchers to measure annual rather than lifetime comparative health care costs of different groups (Schell et al. 2021). Even so, it does not follow that we would be justified in relying on the narrower, per annum assessment of the comparative health care expenditures of these two groups to arrive at a reasoned verdict as to whether by being obese a person will tend to, all things considered, risk burdening the health care system.

I do not claim to have settled the empirical issue of whether obese people use more health care resources than they would if they weren't obese. I have merely sought to show that the widely touted claim that they do isn't as well-substantiated as we might suppose. That claim, as it's typically advanced, relies on a narrow construal of the relevant costs. An argument is needed to show why that construal is preferable to a broader construal, which might not deliver that same verdict. Unless such an argument is provided, the widely touted claim may rest on a

---

[6] This assessment of the lifetime medical costs associated with obesity is not uncontested. Indeed, others conclude just the opposite, namely, that obese individuals have higher average lifetime health care expenses than nonobese individuals. See, e.g., Allison et al. (1999).

dubious assessment. That's because if we instead focus on the lifetime health care costs of obesity, which can be harder to accurately estimate, it is far from obvious that obese individuals have greater health care expenditures than nonobese individuals.

If it is true that obese people, on average, do not use more health care resources over their lifetimes than they would were they not obese, how would that claim bear on the two arguments under consideration? It would have different implications for each of those arguments. The motivating concern of the fairness argument, to recall, is the perceived unfairness of the status quo in which it is presumed that obese persons tend to use more health care resources than they would if they weren't obese—an outcome it's supposed that they could, and, moreover, should, avoid. And, on that basis, it is thought that obese individuals use more than their fair share. Yet, if we consider their lifetime health care costs, it is not clear that obese persons do, in fact, typically have higher health care expenses as a result of being obese. And, from a budgeting standpoint in which we are seeking to account for who uses how much of a given resource, it is hard to see why our ultimate concern shouldn't be people's lifetime health care costs rather than their per annum health care costs.[7]

Let's turn to the public health promotion argument. On that argument, part of the rationale for policies holding obese persons responsible is that those policies would help reduce obesity rates and thereby free up considerable health care resources that could be used to attend to other health needs. But that rationale is flawed if it turns out that many people spared of obesity by such policies would live longer and require costlier end-of-life care that they wouldn't have otherwise required. Still, even if obesity-penalizing policies wouldn't for this reason reduce health care spending, they could improve overall public health. That's because by enabling people to live longer and healthier lives than they otherwise would, these policies might well produce net gains in public health, notwithstanding the increases in health care expenditures that they might also produce. Consequently, even it turns out that obese people don't use more health care resources than they would if they weren't obese, that wouldn't necessarily undermine the public health promotion argument.

The claim that obese individuals are a burden on the health care system faces other challenges as well. Let me briefly mention a few.[8] For one thing, the relationship between a person's weight (or, more precisely, their BMI) and their health status is far from straightforward. A number of variables—such as whether an obese person's high weight is owing to greater muscle mass or fat, where on

---

[7] Granted, in some cases, it might make sense to focus on annual rather than lifetime health care costs. For instance, in the United States in which health care is often employer-funded, employers might have to pay more each year for health care for their obese employees. Even so, it is unclear that in general per annum measures are more relevant.

[8] For a more detailed treatment of these issues, see Nath (2019); Nath (forthcoming).

one's body one carries excess fat, and, perhaps surprisingly, demographic factors such as race—can significantly affect whether, and if so to what extent, being obese may adversely affect a person's health. In addition, lifestyle factors and obesity don't always neatly correspond. Some obese individuals regularly exercise and eat healthfully (e.g., eating plenty of fruits and vegetables and mostly avoiding junk food), and some thin people don't do those things. Diet and fitness may be better predictors of a person's health status than weight, especially as pertains to a person's risk for major diseases linked to obesity. So, the relationship between weight, lifestyle choices, and health is quite a bit more complicated than we are usually led to believe. In some cases, it would be a mistake to regard obesity as a proxy for either poor choices or poor health. And if that is so, then we have further reason to take issue with policies penalizing obese individuals for their weight.

8.4 Do Obese Individuals Deserve to Pay More?

For the sake of argument, let us set aside the empirical disputes from the previous section and assume that a person would tend to require more health care resources if she were obese than if she were not. On the fairness argument, it is claimed that obese individuals themselves should be made to shoulder some or all the costs associated with their higher weights because it is their fault for being that way. Indeed, only if it is obese individuals' fault for being obese would it seem plausible to conclude that those individuals deserve to bear the costs at issue (and, conversely, that it would be unfair to impose those costs on undeserving others who, after all, play no role in avoidably and knowingly producing them).

The notion of fault that this reasoning relies on ascribes to obese individuals *moral responsibility* for their being obese. To be morally responsible for something means two things. First, to be morally responsible for how one is means that being the given way is a matter of moral concern. The moral complaint about obesity is that in virtue of being obese a person risks unfairly burdening others by avoidably imposing added costs on the health care system.[9] Second, to say that an obese person is morally responsible for her weight implies that her obesity is *attributable* to her in some meaningful sense. This isn't just a matter of one's choices or actions playing a role in causing one's obesity (here, we can take "actions" to encompass that which a person does and doesn't do). It is also a matter of a person's choices or actions that make her obese reflecting her *agency*. Just what it means for a person to exercise agency over some choices or actions is an extensively debated issue. As I employ the term, a person exercises sufficient agency over her conduct to qualify as morally responsible for the same if, and only if, she: (a) has suitable

---

[9] For a qualified defense of the view that individuals might owe it to their fellow citizens to look after their health to avoid burdening them in this way, see Buyx (2008).

control over behaving in the given manner, and (b) is appropriately well-informed about the nature of the relevant course of action.[10]

So, what basis might there be for supposing that obese individuals tend to have requisite agency over their weight via their choices and actions to bear moral responsibility for the same? The usual explanation goes like this: people are obese because of their poor lifestyle choices, specifically, choices concerning diet and exercise that they both make freely and know better than to have made. But is that convincing? No doubt, lifestyle factors such as diet and physical activity often play an important role in explaining why some people are obese. However, the causes of obesity are multifaceted and varied. And based on a more sophisticated analysis of these causes—an analysis to which we will turn presently—it is questionable that obese persons in general are morally responsible for their weight as would be required on the fairness argument to justify policies that would make them bear (at least some of) the actual or probable costs of obesity-related health problems.[11]

People's lifestyle choices that affect their weight are significantly influenced by factors that lie outside of their control. Genetics are one such key factor. These days, there is scientific consensus that obesity is a highly heritable trait. Numerous studies from the latter half of the twentieth century reveal that genes strongly predict obesity, and in the decades since geneticists have been able to identify multiple biological pathways that can help explain why that is so (Bouchard 2021). Another crucial such factor is social environment. Public health experts describe the environment of contemporary Western societies as "obesogenic," by which they mean an environment that drives people toward obesity (Swinburn and Egger 2002). The two hallmarks of this obesogenic environment are, first, an increasingly sedentary lifestyle, enabled in large part by technological advance-ments and, second, the ubiquity of inexpensive, high-calorie, highly processed foods (or "junk food" for short). Genes and environment interact in important ways, which can go some way toward explaining why some persons, but not others, may be highly susceptible to becoming obese against the backdrop of an obesogenic environment. It's not that the interplay of both such factors that lie outside of people's control cause some to become obese regardless of the choices that they make. Rather, a growing body of evidence suggests that a critical means by which genes drive obesity in some individuals when they live in an enabling

---

[10] In the philosophical literature on moral responsibility, it is widely supposed that something along the lines of these two conditions must be satisfied for a person to be morally responsible for her behavior. See, for instance, the theory of moral responsibility proposed by Fischer and Ravizza (1998). Clearly, my formulation above leaves unanswered a number of key questions (e.g., "What constitutes *suitable* control over one's behavior?" and "*Which* aspects of how a person behaves must one be appropriately well-informed about?").

[11] Dixon (2018) advances a similar argument to the one I put forth in this section. She explores in much greater depth than I do here the matter of how we might fruitfully theorize about obese individuals' moral responsibility for unhealthy dietary choices connected to their weight. Although Dixon doesn't engage with the policy question that is this chapter's focus, her conclusions dovetail with those I defend on that matter in this section.

environment is precisely by conditioning their choices (O'Rahilly and Farooqi 2006; Mark 2008).

Let's consider in some detail how individual choices are influenced by the latter of the two aforementioned environmental factors. Junk foods have been deliberately engineered by the food industry with a just-so ratio of salt, sugar, and fat to render them "hyperpalatable."[12] Their textures too are carefully crafted to be optimally pleasing to consumers. We are increasingly gaining a more sophisticated neuroscientific understanding of how intense cravings for these foods come to be formed. Upon taking the first bite of a junk food, a person gets nearly instant gratification, with the brain's so-called pleasure receptors lighting up within seconds—much quicker than it takes for the flow of positive feelings to wash over a person upon taking a sip of an alcoholic beverage or taking the first drag of a cigarette. How quickly these pleasurable feelings arise upon ingesting or otherwise using a substance seems to be a key part of how human beings form cravings and, relatedly, how we can become addicted to substances. Food corporations have been wildly successful in securing an ample consumer base by packaging their products enticingly and bombarding consumers with relentless marketing. And of course, junk food is cheap, readily available, and convenient to eat on the go, making it all the more likely that people will overindulge.

Against this backdrop, it is no surprise that many people acquire a taste for junk foods. And once in the habit of regular junk-food consumption, people can find it extremely difficult to resist their cravings for these foods. Many such individuals desperately wish to change their eating habits, and they try to do just that. Yet they usually don't succeed in the long term. Of course, it is not impossible for such individuals to adopt healthier dietary habits. But doing so can prove to be very hard. Fighting intense cravings in the face of nearly ever-present opportunities to indulge them tends to be a losing battle. To the extent that a person's environment is characterized by such factors that render one less capable of making healthier dietary choices—choices that one might ultimately prefer to make—we can say that one's agency over those choices is thereby diminished. And if a person's ability to make better dietary choices that would enable one to avoid being obese is *significantly* compromised, then one should not be held responsible for being obese by being made to bear penalties of the sort associated with the policies under consideration.[13]

---

[12] The empirical characterization in this paragraph draws on two recent books by Michael Moss (2013, 2021), as well as Gearhardt et al. (2011).

[13] The reasoning here is that in the face of significant agential constraints, the responsibility an individual bears for being obese may be diminished to such an extent that one shouldn't be held responsible for being that way via obesity-penalizing policies. In such cases, some find it helpful to distinguish between how such agential constraints bear on whether one *is* responsible for a given poor health outcome and on whether one *should be held* responsible for the same. See, e.g., Schwan (2021).

In light of the foregoing, one might grant that we shouldn't hold obese individuals responsible for their failure to change poor dietary habits once those habits have been firmly established. Still, one might find it plausible to hold obese individuals responsible for the "original sin" of their initial foray down the path of junk-food eating. Indeed, given the difficulty of breaking poor eating habits and the well-known health risks of eating too much junk food, one might reason that people ought to be vigilant about not forming dangerous dietary habits in the first place and should be made to personally bear whatever negative consequences result from a failure to do so.

But just how much agency do people tend to have over forming the poor dietary habits at issue? For some, arguably many, this too will be a matter that they turn out to have little, if any, meaningful control over. For one thing, it is not at all uncommon for individuals to form the habit of regularly indulging in junk food in childhood. In forming these tastes and habits, some children will be so young that they do not yet possess the basic agential capacities that are required for ascriptions of moral responsibility. Those basic capacities include the capacity to understand the long-term consequences of their choices and to rationally deliberate about how they should act. Plainly, this will be so for toddlers who eagerly accept soda, candy, or chips offered to them. What is more, the food industry spends hundreds of billions of dollars annually advertising its products to children, and some 98 percent of the food advertising on television targeting American children features unhealthy foods (Harris et al. 2009). For many older children and adolescents who possess the requisite deliberative capacities to reflect on the consequences of their dietary choices, they may not have much meaningful control over the foods that are generally available to them. They might well be at their parents' (or other caretakers') mercy concerning what is served at mealtimes as well as what foods are kept around the house.[14] It seems doubtful that many individuals who form poor dietary habits under such circumstances have thereby engaged in a suitably robust exercise of agency—of the sort that would be needed to justify policies that would hold them responsible for their weight.

Even individuals who have the capacity to rationally deliberate about their dietary choices and who possess effective means to avoid junk food may be significantly constrained in their abilities to make better initial dietary choices. Not infrequently, people face epistemic obstacles to making better initial dietary choices; choices that may align with their own values, desires, and goals. For many an ordinary person putting forth a good-faith effort to make healthy dietary choices, it won't be clear just what one should and shouldn't eat. To be sure, it

---

[14] These observations naturally invite the following queries: Should parents be held responsible for the unhealthy diets and obesity of their children? And, if so, is there a case for adopting policies that would penalize the parents of obese children? I am not able to address these questions here, but they have been explored by others. See, e.g., Holm (2008).

is common knowledge in our society that eating too much junk food can cause weight gain and poor health. But it is not clear what follows, practically speaking, as to what sort of junk-food consumption carries unacceptable health risks and thus should be avoided.[15] For instance, is a person to refrain from ever tasting any junk food? Or from eating junk food more than a dozen times ever? Or from allowing junk food to become a part of one's daily diet? At what point do dietary choices that prove to be innocuous if made only occasionally (e.g., having a donut with one's colleagues on an afternoon coffee break or grabbing a drive-through fast food meal) run the risk of producing unhealthy and hard-to-break habits if made more frequently? The answers to these questions turn partly on empirical issues that have yet to be resolved. And even where there is a scholarly consensus on such issues, those findings would hardly qualify as common knowledge among most ordinary individuals at present.

Further, we are regularly subject to conflicting messages about the risks of junk-food indulgence. On the one hand, public health authorities tell us that junk food is bad for us. On the other hand, the food industry deploys strategic marketing and advertising tactics portraying junk-food indulgence as a normal, acceptable every-day behavior. Sometimes consumers can be fooled into judging products with low nutritional merits to be reasonably healthy (for instance, when unregulated terms like "natural" are rather indiscriminately slapped onto food labels) (Dixon 2018: 624). One might protest that we shouldn't be so gullible as to fall for misleading advertising and marketing strategies aimed at downplaying the risks of junk food. But that overestimates the degree of rational control we have in resisting such ploys. We know that continual exposure to such messages subtly influences people's beliefs and judgments, and it does so by bypassing our capacities for rational engagement with their content.

In any case, it's not just unsavory corporate tactics that contribute to this tendency. The message that there's nothing wrong with junk-food consumption seems to find plenty of support in our everyday experiences. We need only look around us, at what so many others regularly eat, to confirm that there's nothing wrong with junk-food indulgence. Junk food is everywhere: at schools, in the workplace, at social gatherings, and so on. What is more, most of us regularly observe some people around us eating chips, cookies, and the like, without becoming obese or suffering serious health problems. So, in the status quo, powerful forces *obscure* the gravity of the risks of junk-food consumption—some by design and others not. It would thus seem quite reasonable for a person to suppose that regularly eating junk food doesn't pose a great danger to her health. And to the extent that a person's ability to determine which dietary choices

---

[15] Of course, *some* dietary choices, on the face of it, seem patently unhealthy (for example, most of us can judge that it's plainly inadvisable to eat nothing but Twinkies and Doritos for a full year).

would and would not be health-promoting is compromised in such ways, her responsibility for making poor dietary choices would thereby be mitigated.

In high-income societies, it can be especially difficult for poor persons to make lifestyle choices that are conducive to avoiding obesity.[16] For one thing, low-income individuals tend to be less knowledgeable about the nutritional merits of different common foods (Cluss et al. 2013). Poor individuals are also more likely to live in geographical areas characterized by the twin curses of being "food deserts" (offering limited access to affordable and nutritious food, especially to fresh produce) and "food swamps" (having a high concentration of junk-food availability). Indeed, such ready access to low-cost junk food is thought to be an important part of what explains the higher obesity rates sometimes observed among the poor (Cooksey-Stowers et al. 2017; Drewnowski and Specter 2004). Furthermore, in low-income neighborhoods, residents are less likely to have access to outdoor spaces in which to safely engage in recreational activities (e.g., walking or cycling paths, parks, and sports fields). Other dimensions of living in poverty can make it harder still for a person to regularly make healthy lifestyle choices. Take a cash-strapped and time-poor individual who works several jobs, doesn't have a car or access to decent public transportation, and lacks access to cooking facilities at home. Compared to someone with considerable disposable income and ample free time, it will tend to be much more difficult for such a person to routinely opt for healthier meals over fast food, or to regularly exercise.

Further, experiencing poverty can reduce the cognitive resources needed for regularly resisting the temptation to engage in various unhealthy behaviors (Mullainathan and Shafir 2013). Living in poverty can take a steep mental toll on a person by, for instance, reducing her capacity to engage in long-term planning or to exert impulse control, or causing chronic stress, which in turn can hinder her ability to pay attention. These and other cognitive effects of poverty, some of which emerge in one's early childhood years, can severely constrain a person's capacities to form and to act upon goals.[17] Taken together, these factors would seem to go some way toward explaining why socioeconomically disadvantaged individuals are more prone than others to making certain types of unhealthy choices and to being obese as a result.

[16] Although poor individuals are more likely to be obese than those who aren't poor, the relationship between poverty and obesity is far from straightforward. For reasons that aren't entirely clear, that relationship intersects with gender. So, for instance, in the United States, obesity rates are markedly lower among rich women than among both poor and middle-class women. For men, however, a weak, inverse relationship between class and weight is observed such that an American man's chances of being obese *increase* slightly as his socioeconomic standing rises. Race and ethnicity further complicate the relationship between poverty and obesity. On these associations, see Ogden et al. (2010).

[17] For further discussion of such phenomena and, in particular, of how they bear on the moral plausibility of holding individuals responsible for poor lifestyle choices and associated adverse health outcomes, see Brown (2013); Levy (2019).

To be clear, that factors lying outside of a person's control compromise one's ability to avoid being obese to any degree *whatsoever* would not imply that one shouldn't be held responsible for being obese. Rather, it is when a person's ability to avoid obesity is significantly compromised that it would be implausible to conclude that she deserves to be penalized for being that way, as the sorts of policies under consideration would seek to do. It is doubtful that all obese persons experience sufficiently serious impairments to their agential capacities to avoid obesity such that they shouldn't be held responsible for being that way. Given that, one might wonder about the prospects of crafting obesity-penalizing policies that would seek to impose penalties on those, and only those, obese individuals who exercise suitable agency over their relevant lifestyle choices to warrant holding them responsible for the same.

Consider one way this might be done. We could adopt an obesity-penalizing policy from which individuals who, say, are in a low-income bracket or lack a college education, would be exempt. On the face of it, such a policy might seem morally attractive. On the one hand, it wouldn't unfairly penalize socioeconomically disadvantaged persons, who are more likely to face distinctive hardships in avoiding obesity.[18] On the other hand, it would hold accountable relatively privileged individuals, who would seem to be comparatively well equipped to make the necessary lifestyle choices to avoid being obese. Upon reflection, however, such a policy would still objectionably impose penalties on many individuals who don't deserve to be penalized. That is because even if agential constraints to obesity-avoidance are especially pronounced for those living in poverty, the constraints endured by numerous others in society who are not poor are arguably serious enough to tell against subjecting those persons to obesity-penalizing policies as well.[19]

An alternative policy that we might consider would be one on which obese individuals were assessed on a case-by-case basis to determine their respective deservingness of penalties. Doing so, however, would be neither practical nor morally desirable. It's just not feasible for policies that would apply to large numbers of people to be so fine-grained to adequately distinguish between those

---

[18] Indeed, an obesity-penalizing policy that disproportionately imposed burdens on members of already socially disadvantaged groups would invite two distinct complaints of unfairness. Besides unfairly penalizing many poor persons who don't deserve to be penalized, such a policy also seems unfair for compounding (often unjust) systematic, group-based social inequalities. On the pressing need for public health initiatives to be sensitive to the latter moral concern, see Braveman and Gruskin (2003); Powers and Faden (2006).

[19] Although my focus has been on environmental factors, this case is bolstered considerably by what is known about the influence of genetic factors. See, e.g., Mark (2008). Crucially, people's ability to avoid being obese can be significantly constrained by the interplay between their genes and their environment, regardless of their socioeconomic status. Although it's hard to say, even roughly, what proportion of obese individuals who make poor lifestyle choices experience sufficiently serious agential constraints to tell against holding them responsible for that failure, available evidence suggests that many do.

individuals who should and shouldn't be held responsible for their weight. Indeed, the costs involved in carrying out those assessments could be so high as to outweigh any savings they might generate (Levy 2019: 109). Practical difficulties aside, attempts to procure the sort of information needed to make such assessments would threaten to be unduly intrusive and disrespectful toward the individuals subject to them.[20] Even if we could find some way of identifying individuals who bear responsibility for being obese that managed to avoid the concerns just raised, it is far from obvious that such individuals should be penalized in the sorts of ways at issue. Depriving a person of affordable access to a medical intervention that stands to significantly improve one's health, for instance, hardly seems a fitting response to the transgression of embracing poor lifestyle habits.[21]

So far in criticizing the fairness argument, I have focused on the problem that many obese persons may lack requisite agency over their weight-affecting choices for them to be plausibly held responsible for the same, and attempts to identify and hold to account just those individuals who satisfy the relevant agential requirements would give rise to other concerns. I will briefly mention two other challenges the fairness argument faces. First, up until now, we have presumed the truth of a key moral premise of that argument, namely, the claim that individuals shouldn't impose on their fellow citizens (at least some) health care costs they could avoid producing by making different lifestyle choices. But why accept this? No doubt, there are all sorts of actions that we shouldn't be free to undertake based on the detrimental consequences those actions risk producing for others (e.g., dropping large rocks from an overpass onto a busy freeway below). However, it seems just as plain that people should be free to act in a variety of other ways notwithstanding the risk of harming others associated with those actions. For instance, that painting my house bright pink and filling my front lawn with kitschy garden gnomes causes a substantial decline in the property value of my next-door neighbor's home doesn't clearly imply that I shouldn't be free to do those things. What is more, we do not generally suppose that people are to avoid acting in any ways at all that stand to worsen their health (e.g., failing to regularly get at least seven hours of sleep a night) based on the health care costs that might produce for others. What is needed, then, is a principled explanation of why the specific burden at issue—that is, raising health care costs for one's fellow participants of a shared health care system by being obese—is a burden that individuals don't have the right to impose on others.[22]

---

[20] For discussion of similar concerns about attempts to distinguish between responsible and irresponsible parties that have been raised against luck egalitarian theories, see Wolff (1998); Anderson (1999).

[21] This concern could be escaped by an obesity-penalizing policy that imposed only relatively minor burdens on those it sought to hold responsible (e.g., a very slight increase in health care costs).

[22] For further discussion of this sort of challenge, see Wikler (1987: 18–21).

Second, even if we find it plausible that individuals should be held responsible for their unhealthy lifestyle choices, it might be thought premature to institute policies that would do so until society has met *its* obligation to its members to meaningfully enable them to make healthier choices.[23] Overall, then, we have ample reason to reject the fairness argument.

8.5 Would Holding Obese Individuals Responsible Improve Public Health?

Let's turn to the public health promotion argument. Unlike the fairness argument, this argument isn't about adopting obesity-penalizing policies so as to treat people as they deserve. Rather, it is about the expected consequences of such policies: that they will tend to improve public health. On this argument, to recall, there are two main ways in which obesity-penalizing policies are expected to produce better health outcomes. First, by helping reduce obesity rates, many people would benefit from them by being spared from obesity-related medical conditions that they otherwise would be more likely to endure. Second, by decreasing the incidence of obesity-related diseases, such policies would reduce society's overall disease burden, which, in turn, would translate into more health care resources being available to treat other health needs of the population.[24]

This argument in favor of adopting such policies, like the previous argument, does not succeed. I will focus on its main empirical premise: penalizing obesity in health care contexts would prompt many people to adopt lifestyle changes that they would need to make to avoid becoming or remaining obese. That premise might seem plausible initially. But is it really so plausible that obesity-penalizing policies would have the desired effect? It is already the case that most people who are obese don't want to be that way. And given how our society treats people who are obese, that is hardly a surprise. Obese individuals endure pervasive and wide-ranging social and economic harms based on how most people regard them in virtue of their weight. To name just a few: obese children are bullied and teased; weight-based workforce discrimination is rampant; and being obese negatively affects people's interpersonal relationships (Puhl and Heuer 2009).

In any event, the empirical evidence strongly indicates that the premise is false: penalizing obesity in health care contexts would *not* provide many people with just the nudge they need to succeed in their obesity-avoidance efforts. Most obese

---

[23] For different arguments along these lines, see Daniels (2011); Levy (2019).

[24] One might point to another potential health benefit of such policies: reducing obese individuals' (affordable) access to health care would translate into nonobese individuals enjoying lower health care costs and/or expanded access to health care resources. However, that outcome wouldn't tend to amount to a *net* gain in public health since producing those benefits for nonobese persons would come at the cost of worse health outcomes for obese persons.

persons report having tried (usually multiple times) to lose and keep off large amounts of weight. In the vast majority of cases, what is stopping people from obesity-avoidance isn't a lack of motivation on their part but rather a lack of a feasible means by which to fulfill that goal. At present, with the possible exception of bariatric surgery, we don't know what works as a safe and practical method to enable most obese people to sustain anything more than a modest weight loss in the long term. In the past few decades, this has emerged as the consensus view among scientists and public health scholars who study this matter. For instance, in a recent review of the current state of knowledge on long-term maintenance of weight loss for obese individuals, obesity experts Kevin Hall and Scott Kahan (2018: 191) write: "numerous studies show that diet, exercise, and behavioral counseling, in the best of cases, only leads to 5% to 10% average weight loss, and few patients with significantly elevated initial weights achieve and maintain an 'ideal' body weight." The claim that we don't know what would work to get the majority of obese individuals to stop being obese is deeply at odds with what most people believe. As most see it, it is obvious what is to be done: to stop being obese, people need to eat less and move more.

To be sure, strictly speaking, it is possible for most obese people to lose enough weight to no longer be obese and to sustain a much lower weight through changes in their dietary and exercise habits. Indeed, if we were to confine most any individual to a laboratory setting in which one were forced to abide by a severely restrictive low-calorie diet for months without any opportunities to stray from it, one would almost certainly lose a lot of weight and keep that weight off while being made to stick to that regimen. But the significance of that thought experiment is limited. One might suppose that achieving that same result outside of a lab setting simply requires obese persons to make a good-faith effort to embrace better habits. The trouble is that in the real world, obese individuals who do this almost never succeed.

Study after study reveals a strikingly similar pattern in people's weight-loss attempts through behavioral changes (Mann et al. 2007). Individuals attempting to slim down typically take off the greatest amount of weight about three to six months into their weight-loss endeavors. Then, despite continued efforts, for almost all of these individuals, some of the initially lost weight starts creeping back on. And as more time passes, more of the lost weight returns. After a few years, many dieters end up back at their starting weights, or sometimes heavier. The best-case scenarios that emerge from the pooled results of rigorous studies on this matter reveal that about three years after embarking on a diet, the most successful dieters manage to sustain weight losses of around 5 percent of their starting weights (Franz et al. 2007). For most obese individuals—indeed, for all but those at the very lowest end of the mild-obesity range—to stop being obese, they would need to sustain weight losses that are substantially greater than that which is achieved by the most successful subjects in those studies. So, we know that most obese people who try to avoid being obese fail in the long term.[25] And if it is the case that most who try fail, there is little reason to conclude that all that those individuals need to succeed is a nudge of the sort that the obesity-penalizing policies at issue would provide.[26]

Besides being unlikely to help many people avoid obesity, obesity-penalizing policies may actually harm obese individuals' health. There are several ways this might happen. To begin with, such policies might drive people to embrace patently unhealthy tactics that can facilitate or sustain more dramatic weight losses than can otherwise be achieved. People may resort to crash-dieting, dangerous fad diets, purging, and other forms of disordered eating. Or, to keep their weight down, they might turn to smoking, diet pills, laxatives, or illegal drugs such as cocaine. Obesity-penalizing policies might also make it less likely that people would do things that improve their health. For some time now, it has been recognized that even very modest weight reductions for obese individuals can produce clinically significant health improvements. For instance, one study found that for obese individuals at high risk of contracting diabetes, losing just 5 kilos (about 11 pounds), cut their risk of getting the disease by over one-half (Hamman et al. 2006).

Furthermore, some of the standard behavioral modifications that obese individuals are counseled to make to help them lose weight—specifically, eating more nutritious foods and becoming more physically active—have been shown to yield important health benefits independently of any weight losses they might help produce (Gaesser et al. 2011). *Those* health improvements are achievable for many obese people. By contrast, obesity-penalizing policies emphasize the goal of substantial weight reduction, well known to be unattainable in real-world settings for the vast majority who are obese. Such policies could inadvertently discourage people from adopting achievable healthy lifestyle changes. Suppose you have been instructed to eat more vegetables and to take regular walks *as a means* to drastically reduce your weight, and you fail to see the sought-after results despite diligently adhering to these healthier habits for months. It would hardly be surprising if you felt frustrated and abandoned those habits—habits that are, in fact, likely to improve your health in the long run.

Obesity-penalizing policies run the risk of worsening obese persons' health in other ways as well. For one thing, such policies might help reinforce, if not

---

[25] Parallel reflections seem to hold true for obesity prevention efforts. The consensus view among experts is that we do not know at present what works in real-life settings to help prevent many individuals from becoming obese. On this matter, see Nath (forthcoming).

[26] It is commonly supposed that the majority of people's weight-loss attempts fail because they don't exert sufficient willpower. As I discuss elsewhere (Nath, forthcoming), available evidence does not support that contention. However, even if it were true that weakness of will explained the high failure rates of people's weight-loss attempts, that would do little to rescue the public health promotion argument from the challenge under discussion. For the purposes of that argument, what matters is whether obesity-penalizing policies would help meaningfully lower obesity rates. And if most obese people who desperately want to lose weight fail in their efforts to do so owing to weakness of will, then why should we expect these policies to circumvent that obstacle?

exacerbate, weight stigma. In health care settings, weight stigma is already pervasive, and the great harm it does to obese persons has been well documented (Phelan et al. 2015). Doctors and nurses report high levels of anti-fat bias, and that can translate into inferior health care for obese patients. Doctors, for instance, spend less time with obese patients and may be less likely to explore plausible diagnoses for their medical complaints. Feeling negatively judged in clinical settings, in turn (understandably) makes obese people more likely to delay and cancel medical appointments, behaviors that are linked to worse health outcomes. By emphasizing the connections between obesity, unhealthy lifestyle choices, and poor health, obesity-penalizing policies threaten to reinforce negative beliefs and attitudes about obese individuals that are already widely held by medical practitioners as well as by the public at large.[27] These policies might also promote the internalization of such beliefs and attitudes by obese persons themselves. Those who experience internalized anti-fat bias, which is already widespread among obese individuals, are at heightened risk of suffering poor mental health and of engaging in a variety of disordered eating behaviors (Pearl and Puhl 2018).

In addition, making health care more costly or otherwise harder to access for obese individuals, as obesity-penalizing policies do, would almost certainly translate into worse health outcomes for this group. Charging people more to use health care will make them more likely to avoid and delay medical appointments. The health of obese individuals would also be worsened by such policies delaying, if not outright preventing, their (affordable) access to medically necessary treatments (e.g., if their access to treatments is conditional on their losing weight, or if a treatment is made prohibitively expensive for them). Obese individuals already endure significant disadvantages in health care settings that negatively impact their health. From a public health standpoint, we should not be adopting policies that could be reasonably expected to further compromise obese individuals' access to decent health care.

Now, it is not the case that obesity-penalizing policies wouldn't have any beneficiaries. Most likely, those policies would prompt some people to make better lifestyle choices and to be healthier as a result. But it's doubtful that the benefits produced by these policies would outweigh their likely harms. As we have seen, there's little empirical basis for expecting obesity-penalizing policies to bring about substantial reductions in obesity rates. Indeed, we have good reason to believe that no more than a relatively small proportion of the population would stand to benefit from them in that way. And given the reasonable expectation that high obesity rates would persist under such policies, the seriously detrimental effects those policies would threaten to have on obese individuals' health and well-being would likely be endured by a great many people.

---

[27] On this concern, see Wikler (2004: 128).

A further concern about the expected consequences of obesity-penalizing policies has to do with how the benefits and burdens they produce would be distributed across different social groups. If such policies were to help decrease obesity rates over time, that benefit would tend to be disproportionately enjoyed by members of relatively privileged social groups. And individuals who are poor or otherwise socially marginalized would tend to be disproportionately harmed by these policies. This projection is based, in part, on the fact that in high-income nations, socially disadvantaged individuals are more likely to be obese in the first place (Devaux and Sassi 2013). In the United States, obesity rates are much higher among the poor compared to the wealthy, as well as among Blacks and Hispanics as compared to individuals belonging to other racial and ethnic groups (independently of socioeconomic class). What is more, socially disadvantaged individuals aren't as likely as others to adopt the sorts of lifestyle changes that obesity-penalizing policies seek to elicit in those they target. Indeed, a body of evidence suggests that public health campaigns aimed at modifying individual behavior tend to disproportionately benefit society's already well-off members while disproportionately burdening the already disadvantaged (Goldberg 2012). For instance, we have seen this happen in the decades-long fight to reduce tobacco use in the United States (Bell et al. 2010). In practice, then, obesity-penalizing policies would likely worsen the health and well-being of already badly-off segments of the population without providing the majority of those individuals with any compensating benefits, and they would do this for the sake of producing benefits primarily for already better-off individuals.

Taken together, the reflections canvassed in this section strongly suggest that obesity-penalizing policies wouldn't improve public health. For one thing, improving public health is, arguably, an endeavor that ought to be concerned with *equitably* promoting better health outcomes for a population. Although there is much debate over just what it means to promote health in an equitable manner, on each of the main competing views defended in the public health ethics literature on this matter, policies holding obese persons responsible would fail to do so.[28] But putting aside considerations of equity, if we were only concerned with producing net gains in health for a population at large, still we should oppose obesity-penalizing policies. That's because it is doubtful that such policies would be conducive to advancing even that narrower aim.[29]

---

[28] Here, the views that I have in mind are the following: *egalitarian* views, which are concerned with reducing health inequalities; *prioritarian* views, which assign added weight to promoting better health for society's worst-off; and, *sufficientarian* views, on which all individuals should have, or be positioned to attain, some threshold level of health. See Faden et al. (2020).

[29] The public health promotion argument for obesity-penalizing policies may face other challenges too. For instance, even if such policies were reasonably expected to improve people's health by reducing obesity rates, they might be opposed on the grounds of being objectionably paternalistic. For discussion of this concern, see Wikler (1987: 12–13).

8.6 Why Single Out Obesity?

Considerable cherry-picking occurs as concerns which segments of the population and which behaviors are targeted by calls for individuals to take greater responsibility for their health. Society demands that those who smoke, drink excessively, and eat too much should make themselves do better. But where are the exhortations, say, of those working high-powered jobs and subjecting themselves to avoidable health risks associated with regular sleep deprivation and high levels of stress? Or, of extreme athletes engaging in risky behaviors? Drawing attention to this disconnect, Daniel Wikler (2004: 129) observes the rather neat overlap between, on the one hand, those "lifestyles deemed burdensomely expensive" and, on the other hand, those "lifestyles deemed sinful, or... people deemed unworthy." That is, we seem prone to singling out as irresponsible those individuals that are a certain way, or do a certain thing, that many in society disapprove of or otherwise look down on. This tendency for our biases to color our judgments about which lifestyles we criticize as irresponsible and unhealthy is especially pertinent in the case at hand. Living in societies in which anti-fat bias is prevalent, we are all the more susceptible to embracing punitive policies against obese people. Based on the trends surveyed above, we also have good reason to suspect that anti-fat bias intersects with classism and racism, thereby compounding this problematic tendency.[30] Given these concerns, we should be cognizant of the possibility that arguments advanced in favor of policies holding obese people responsible might seem more convincing than they in fact are.

8.7 Conclusion

In sum, we have ample reason to be wary of policies aimed at holding obese people responsible in health care contexts. The case for such policies relies on a questionable empirical assessment of the health care costs associated with obesity. It also relies on overly simplistic assumptions about the relationship between weight, lifestyle choices, and health. Based on those reasons alone, it seems ill-advised to adopt policies grounded in the supposition that an obese individual's weight tells us something meaningful about one's choices or health status.

Even putting aside those concerns, there are other serious problems with obesity-penalizing policies—problems that emerged in examining the two arguments that may be offered in their defense. The fairness argument presumes that

---

[30] Although I haven't discussed the phenomenon, anti-fat bias also importantly interacts with sexism. In wide-ranging domains including the workplace, women endure much harsher penalties than men do for being heavier. See, e.g., Fikkan and Rothblum (2012). For further discussion of the intersections between weight and these other social inequalities, see Farrell (2011); Saguy, (2013); Strings (2019).

obese individuals bear moral responsibility for making poor choices that account for their weighing too much. But for many individuals (especially, but by no means exclusively, for those who live in poverty), their abilities to consistently make better choices might be significantly constrained by factors that lie outside of their control. The influence of such factors may well render them susceptible to making poor choices, to being obese, and to experiencing ill health. As such, those hardly seem to be outcomes that those individuals should be held responsible for, certainly not by penalizing them in the health care domain.

The public health promotion argument for those policies fares no better. Available evidence doesn't support its key contention that to avoid being obese, many people just need the sort of nudge such policies would provide. Obesity isn't a simple problem that admits of such a simple fix. Obesity experts have not, to date, identified a single prescription for how individuals are to go about modifying their behavior that, when put into practice in real-world contexts, produces an impressive long-term reduction in obesity rates. So, we shouldn't expect ramping up the already immense pressure on obese individuals to stop being that way to work. What is more, obesity-penalizing policies can be reasonably expected to do harm: by worsening obese individuals' health and by disproportionately imposing social penalties on those who are already badly off.[31]

References

Allison, D. B., Zannolli, R., and Venkat Narayan, K. M. (1999), "The Direct Health Care Costs of Obesity in the United States," in *American Journal of Public Health* 89/8: 1194–9.

Anderson, E. (1999), "What is the Point of Equality?" in *Ethics* 109/2: 287–337.

Australian Institute of Health and Welfare (2020), "Overweight and Obesity," July 23, 2020. https://www.aihw.gov.au/reports/australias-health/overweight-and-obesity

Baker, C. (2021), *Obesity Statistics: Briefing Paper Number 3336* (House of Commons Library). https://dera.ioe.ac.uk//37120

Barendregt, J. J., Bonneux, L., and van der Maas, P. J. (1997), "The Health Care Costs of Smoking," in *New England Journal of Medicine* 337/15: 1052–7.

Bell, K., Salmon, A., Bowers, M., Bell, J., and McCullough, L. (2010), "Smoking, Stigma and Tobacco 'Denormalization': Further Reflections on the Use of Stigma as a Public Health Tool," in *Social Science & Medicine* 70/6: 795–9.

Bouchard, C. (2021), "Genetics of Obesity: What We Have Learned Over Decades of Research," in *Obesity* 29/5: 802–20.

[31] I am very grateful to Torin Alter, Ben Davies, and Neil Levy for their helpful comments on earlier drafts.

Braveman, P., and Gruskin, S. (2003), "Defining Equity in Health," in *Journal of Epidemiology & Community Health* 57/4: 254–8

Brown, R. C. H. (2013), "Moral Responsibility for (Un)Healthy Behaviour," in *Journal of Medical Ethics* 39/11: 695–8.

Brown, R.C.H., Maslen, H., and Savulescu, J. (2019), "Against Moral Responsibilisation of Health: Prudential Responsibility and Health Promotion," in *Public Health Ethics* 12/2: 114–29.

Buyx, A. (2008), "Personal Responsibility for Health as a Rationing Criterion: Why We Don't Like it and Why Maybe We Should," in *Journal of Medical Ethics* 34/12: 871–4.

Cawley, J., Meyerhoefer, C., Biener, A., Hammer, M., and Wintfield, N. (2015), "Savings in Medical Expenditures Associated with Reductions in Body Mass Index Among US Adults with Obesity, by Diabetes Status," in *PharmacoEconomics* 33/7: 707–22.

Centers for Disease Control and Prevention (2021), *Childhood Obesity Facts*. https://www.cdc.gov/obesity/data/childhood.html

Cluss, P. A., Ewing, L., King, W. C., et al. (2013), "Nutrition Knowledge of Low-Income Parents of Obese Children," in *Translational Behavioral Medicine* 3/2: 218–25.

Cooksey-Stowers, K., Schwartz, M. B., and Brownell, K. D. (2017), "Food Swamps Predict Obesity Rates Better Than Food Deserts in the United States," in *International Journal of Environmental Research and Public Health* 14/11: 1366

Daniels, N. (2011), "Individual and Social Responsibility for Health," in C. Knight and Z Stemplowska (ed.) *Responsibility and Distributive Justice* (Oxford University Press), 266–86.

Devaux, M., and Sassi, F. (2013), "Social Inequalities in Obesity and Overweight in 11 OECD Countries," in *European Journal of Public Health* 23/3: 464–9.

Dixon, B. (2018), "Obesity and Responsibility," in A. Barnhill, M. Budolfson, and T. Doggett (ed.) *The Oxford Handbook of Food Ethics* (Oxford University Press), 614–33.

Drewnowski, A., and Specter, S. E. (2004), "Poverty and Obesity: The Role of Energy Density and Energy Costs," in *The American Journal of Clinical Nutrition* 79/1: 6–16.

Faden, R., Bernstein, J., and Shebaya, S. (2020), "Public Health Ethics," in E. Zalta (ed.) *The Stanford Encyclopedia of Philosophy*. https://plato.stanford.edu/archives/spr2022/entries/publichealth-ethics/

Farrell, A. E. (2011), *Fat Shame: Stigma and the Fat Body in American Culture* (New York University Press).

Feiring, E. (2008), "Lifestyle, Responsibility and Justice," in *Journal of Medical Ethics* 34/1: 33–6.

Fikkan, J. L., and Rothblum, E. D. (2012), "Is Fat a Feminist Issue? Exploring the Gendered Nature of Weight Bias," in *Sex Roles* 66/9–10: 575–92.

Finkelstein, E. A., Fiebelkorn, I. C., and Wang, G. (2003), "National Medical Spending Attributable to Overweight and Obesity: How Much, And Who's Paying?," in *Health Affairs* 22/11: 219–26.

Fischer, J. M., and Ravizza, M. (1998), *Responsibility and Control: A Theory of Moral Responsibility* (Cambridge University Press).

Franz, M. J., VanWormer, J. J., Crain, A. L., et al. (2007), "Weight-Loss Outcomes: A Systematic Review and Meta-Analysis of Weight-Loss Clinical Trials with a Minimum 1-Year Follow-Up," in *Journal of the American Dietetic Association* 107/10: 1755–67.

Friesen, P. (2018), "Personal Responsibility within Health Policy: Unethical and Ineffective," in *Journal of Medical Ethics* 44/1: 53–8.

Gaesser, G. A., Angadi, S. S., and Sawyer, B. J. (2011), "Exercise and Diet, Independent of Weight Loss, Improve Cardiometabolic Risk Profile in Overweight and Obese Individuals," in *The Physician and Sportsmedicine* 39/2: 87–97.

Gearhardt, A. N., Davis, C., Kuschner, R., and Brownell, K. D. (2011), "The Addiction Potential of Hyperpalatable Foods," in *Current Drug Abuse Reviews* 4/3: 140–5.

Goldberg, D. S. (2012), "Social Justice, Health Inequalities and Methodological Individualism in US Health Promotion," in *Public Health Ethics* 5/2: 104–15.

Hales C. M., Carroll, M. D., Fryar, C. D., and Ogden, C. L. (2020), "Prevalence of Obesity and Severe Obesity Among Adults: United States, 2017–2018," *NCHS Data Brief* (National Center for Health Statistics).

Hall, K. D., and Kahan, S. (2018), "Maintenance of Lost Weight and Long-Term Management of Obesity," in *Medical Clinics of North America* 102/1: 183–97.

Hamman, R. F., Wing, R. R., Edelstein, S. L., et al. (2006), "Effect of Weight Loss With Lifestyle Intervention on Risk of Diabetes," in *Diabetes Care* 29/9: 2102–7.

Harris, J. L., Pomeranz, J. L., Lobstein, T., and Brownell, K. D. (2009), "A Crisis in the Marketplace: How Food Marketing Contributes to Childhood Obesity and What Can Be Done," in *Annual Review of Public Health* 30/1: 211–25.

Holm, S. (2008), "Parental Responsibility and Obesity in Children," in *Public Health Ethics* 1/1: 21–9.

Levy, N. (2019), "Taking Responsibility for Responsibility," in *Public Health Ethics* 12/2: 103–13.

Mann, T., Tomiyama, A. J., Westling, E., et al. (2007), "Medicare's Search for Effective Obesity Treatments: Diets Are Not the Answer," in *American Psychologist* 62/3: 220–33.

Mark, A. L. (2008), "Dietary Therapy for Obesity: An Emperor With No Clothes," in *Hypertension* 51/6: 1426–34.

Ministry of Health NZ (2021), *Obesity Statistics*. https://www.health.govt.nz/nz-health-statistics/health-statistics-and-data-sets/obesity-statistics

Moss, M. (2013), *Salt Sugar Fat: How the Food Giants Hooked Us* (Random House).

Moss, M. (2021), *Hooked: How Processed Food Became Addictive* (Random House).

Mullainathan, S., and Shafir, E. (2013), *Scarcity: The New Science of Having Less and How it Defines Our Lives* (Picador).

Musich, S., MacLeod, S., Bhattarai, G. R., et al. (2016), "The Impact of Obesity on Health Care Utilization and Expenditures in a Medicare Supplement Population," in *Gerontology and Geriatric Medicine* 2: 1–9.

Nath, R. (2019), "The Injustice of Fat Stigma," in *Bioethics* 33/5: 577–90.

Nath. R. (forthcoming), *Why It's OK To Be Fat* (Routledge).

Ogden, C. L., Lamb, M. M., Carroll, M. D., and Flegal, K. M. (2010), "Obesity and Socioeconomic Status in Adults: United States, 2005–2008," in *NCHS Data Brief* 50: 1–8.

O'Rahilly, S. and Farooqi, I. S. (2006), "Genetics of Obesity," in *Philosophical Transactions of the Royal Society of London. Series B, Biological Sciences* 361/1471: 1095–105.

Pearl, R. L., and Puhl, R. M. (2018), "Weight Bias Internalization and Health: A Systematic Review," in *Obesity Reviews* 19/8: 1141–63.

Phelan, S. M., Burgess, D. J., Yeazel, M. W., et al. (2015), "Impact of Weight Bias and Stigma on Quality of Care and Outcomes for Patients with Obesity," in *Obesity Reviews* 16/4: 319–26.

Powers, M., and Faden, R. (2006), *Social Justice: The Moral Foundations of Public Health and Health Policy* (Oxford University Press).

Puhl, R. M., and Heuer, C. A. (2009), "The Stigma of Obesity: A Review and Update," in *Obesity* 17/5: 941–64.

Resnick, D. B. (2013), "Charging Smokers Higher Health Insurance Rates: Is It Ethical?" in *The Hastings Center* (blog), September 19, 2013. https://www.thehastingscenter.org/charging-smokers-higher-health-insurance-rates-is-it-ethical

Saguy, A. C. (2013), *What's Wrong with Fat?* (Oxford University Press).

Schell, R. C., Just, D. R., and Levitsky, D. A. (2021), "Methodological Challenges in Estimating the Lifetime Medical Care Cost Externality of Obesity," in *Journal of Benefit-Cost Analysis* 12/3: 441–65.

Schwan, B. (2021), "Responsibility amid the Social Determinants of Health," in *Bioethics* 35/1: 6–14.

Statistics Canada (2019), *Overweight and Obese Adults, 2018* (Government of Canada) June 25 2019, https://www150.statcan.gc.ca/n1/pub/82-625-x/2019001/article/00005-eng.htm

Strings, S. (2019), *Fearing the Black Body: The Racial Origins of Fat Phobia* (New York University Press).

Swinburn, B., and Egger, G. (2002), "Preventive Strategies against Weight Gain and Obesity," in *Obesity Reviews* 3/4: 289–301.

van Baal, P. H. M., Polder, J. J., de Wit, G. A., et al. (2008), "Lifetime Medical Costs of Obesity: Prevention No Cure for Increasing Health Expenditure," in *PLoS Medicine* 5/2: 242–9.

Wann, M. (2009), "Fat Studies: An Invitation to Revolution," in E. Rothblum and S. Solovay (ed.) *The Fat Studies Reader* (New York University Press), ix–xxv.

Ward, Z. J., Bleich, S. N., Long, M. W., and Gortmaker, S. L. (2021), "Association of Body Mass Index with Health Care Expenditures in the United States by Age and Sex," in *PloS One* 16/3: e0247307.

Wikler, D. (1987), "Who Should Be Blamed for Being Sick?," in *Health Education Quarterly* 14/1: 11–25.

Wikler, D. (2004), "Personal and Social Responsibility for Health," in S. Anand, F. Peter, and A. Sen (ed.) *Public Health, Ethics, and Equity* (Oxford University Press), 109–34.

Wolff, J. (1998), "Fairness, Respect, and the Egalitarian Ethos," in *Philosophy & Public Affairs* 27/2: 97–122.

World Health Organization (2018), *Taking Action on Childhood Obesity*. https://apps. who.int/nutrition/publications/obesity/taking-action-childhood-obesity-report/en/ index.html

World Health Organization (2021), *Obesity and Overweight*. July 9, 2021. https://www. who.int/news-room/fact-sheets/detail/obesity-and-overweight

9

Habitual Health-Related Behaviour and Responsibility

Rebecca Brown

9.1 Introduction

Many health-related behaviours are, to a significant extent, habitual. This means, roughly, that they are repeated behaviours, which occur in similar contexts, and whose initiation and maintenance requires little (if any) conscious control. They include behaviours like diet, physical activity, smoking and alcohol consumption which influence the likelihood of developing non-communicable diseases like cardiac disease, stroke, lung disease and type II diabetes. The aim of this chapter will be to lay some of the groundwork for considering whether or not (or to what extent) people are responsible for these kinds of behaviours. My aim will be first to clarify some of the key concepts here: habitual, health-related and behaviour. I will indicate what features of habitual health-related behaviour seem important for considering whether or not agents are typically responsible for these behaviours. I point to two approaches to establishing responsibility for habitual health-related behaviour: the first assumes that agents have control-in-the moment over their behaviour, and the second assumes agents lack such control but possessed it at some earlier point (the tracing approach). I point out some challenges to establishing responsibility through these routes before discussing the relevance of this analysis for health care provision and policy.

9.2 Why Does Responsibility for Habitual Health-Related Behaviours Matter?

First, responsibility for habitual health-related behaviour might be of ethical importance. Responsibility plays an important role in most moral theories. Praising, blaming, rewarding and punishing on the basis of judgements of responsibility are commonplace across cultures, although differ in their form and implications (Weiner 1995; Maciel 2015; Miller, Goyal et al. 2017; Hannikainen, Machery et al. 2019). Responsibility can modify and regulate personal relationships and is central to institutions such as the criminal justice system (although

Rebecca Brown, *Habitual Health-Related Behaviour and Responsibility* In: *Responsibility and Healthcare*. Edited by: Benjamin Davies, Gabriel De Marco, Neil Levy, Julian Savulescu, Oxford University Press. © Edited by Benjamin Davies, Gabriel De Marco, Neil Levy, Julian Savulescu 2024. DOI: 10.1093/oso/9780192872234.003.0010

the role of responsibility in such contexts isn't free of criticism (Caruso 2021)). Unless there is some reason for thinking that habitual health-related behaviours are an exception, then knowing whether or not people are responsible for these behaviours is likely to be of ethical relevance.

Some have argued that responsibility (either individual responsibility or social responsibility or some other kind) should play an explicit role in determining the provision and receipt of health care resources (Glannon 1998; Buyx 2008; Callahan 2012; Cavallero 2019; Davies and Savulescu 2019; Feng-Gu, Everett et al. 2021). Elsewhere, I have written about the ways in which implicit responsibility judgements appear to infiltrate policy (Brown 2018, 2019). The legitimacy of allowing judgements of people's responsibility to influence activity in these ways can be illuminated by an account of responsibility for habitual health-related behaviours.

A second use for an account of responsibility for habitual health-related behaviour is to identify (or rule out) practical solutions for addressing health harms. For instance, if an assessment of responsibility suggests that people lack responsibility due to a lack of control over these behaviours, then health promotion approaches which require individuals to exercise such control are unlikely to be successful. Clearly, such a discussion must be informed by empirical research which describes the ways in which people can exert control over their behaviour.

9.3 What are Habitual Health-Related Behaviours?

In this section I will seek to shed light on what I mean by the terms *habitual, behaviour* and *health-related*. Whilst these terms raise few problems in everyday contexts, a little more precision will be helpful if we are to put them to work in better understanding responsibility. None of these terms have precise, uncontested definitions. I will not attempt to provide a set of necessary and sufficient conditions which mark out all and only cases of "habitual health-related behaviour", but instead will seek to provide a working definition that captures the majority of cases. In order to describe what counts as habitual health-related behaviour I will refer to work in psychology which is where these terms have, for the most part, been used.

9.3.1 Habitual

Paradigmatically habitual behaviours are repeated overt behaviours that are cued by features of the environment (external stimuli) (Gardner 2015). They include daily routines like brushing one's teeth, making a cup of tea or cycling a particular route to work. In psychology, habitual behaviour is distinguished from goal-directed behaviour (Lingawi, Dezfouli et al. 2016). The former involves learning to perform a certain behaviour in response to a particular stimulus (e.g., a hungry rat presses a lever in an operant chamber to access food). This learning occurs through repeated pairing of the response behaviour to a valued outcome (e.g., food). For habitual behaviours, once the stimulus and response have been paired, the animal will perform the response behaviour in the presence of the stimulus even if the valued outcome is degraded (e.g., if acting habitually, the rat will press the food lever when in the chamber even if it has been fed and is no longer hungry). Goal-directed behaviours are performed to access a valued outcome. Thus, if the valued outcome is degraded, the animal will no longer perform the behaviour that will achieve that outcome (e.g., if acting in a goal-directed manner, the rat will not press the lever if it has already been fed).

Thus habits,[1] once established, are more or less insensitive to the devaluation of their outcomes. That is, habitual behaviours may still be performed in response to environmental stimuli even when those behaviours no longer bring about desirable outcomes. The extent to which this is true can be influenced by how well established the habit is, as described by correlation theory (Baum 1973; Dickinson 1985). So a rat that has only just learned to associate the operant chamber + food lever + hunger with the behaviour of pressing the lever and the valued outcome of food will stop pressing the lever in the absence of hunger quite quickly. In contrast, a rat that has been "overtrained" on these stimuli will continue pressing the lever regardless of hunger for much longer (Adams 1982).

Performing behaviours habitually is energy conserving. The mechanisms governing habitual behavioural responses to environmental stimuli can be regulated through automatic processes. There remains a question as to the extent to which automatic behaviours are conscious (I will say more on this), but there is agreement that they at least require much less conscious involvement than goal-directed behaviours. Controlling behaviours through conscious mechanisms is, in contrast, energetically costly.

At least as I will use the concepts here, habitual behaviour and automatic behaviour are not one and the same. Habitual behaviours may include quite complex behaviours (such as cycling a specific route to work), large portions of which are initiated and maintained by automatic processes. "Pure" habitual behaviour (as in the case of the overtrained rat who presses a lever for food regardless of hunger) may be entirely regulated through automatic processes. Unfortunately, life rarely lines up neatly with conceptual boundaries and most of the health-related behaviours we are interested in will not look like this (Gardner 2015). Dietary behaviour, physical activity, smoking and drinking alcohol are commonly described as "habitual" behaviours or "habits". Although they will often be largely composed of learned associations between environmental

---

[1] I will use "habit" and "habitual behaviour" interchangeably in this chapter.

stimuli and behavioural responses, and feature high degrees of automaticity, it would by incorrect to act as if they are *entirely* controlled through such mechanisms. The "habitual" behaviours we are interested in from a health perspective are rarely completely insensitive to changes in outcome value as in the "pure" habitual behaviour case. It is helpful, however, to be able to refer to such behaviours that are *largely but not entirely* "habitual" in the stricter sense. Without wishing to be revisionary I will, therefore, use "habitual" more loosely, to include complex behaviours which depend upon learned stimulus response mechanisms and are largely automatic, but which may feature some (minimal) degree of goal-directedness and reflectivity.

As mentioned, habitual behaviours (here understood) will feature high degrees of automaticity but need not be entirely composed of automatic processes. It is this automaticity that prompts special consideration of how people come to be responsible for habitual health-related behaviours. Despite the prevalence of habitual behaviours, most discussion of moral responsibility has focused on discrete behaviours with clear moral valence. Questions of how agency affects responsibility often focus on exceptional events or individuals: controlling forces such as manipulation and coercion, or agents with mental disorders. Automaticity, in contrast, is ever-present and mundane. Behaviours initiated and maintained automatically do not involve deliberative agency at the time they are performed (i.e., they are not directly under conscious control). Reflexes are paradigmatic automatic processes. They form stimulus-response mechanisms that can initiate, for instance, muscle contraction in the presence of some stimulus (light, pressure, etc.) extremely quickly and without the involvement of the brain. Some reflexes can be modified and controlled consciously. For example, one can (sometimes) consciously keep hold of a hot object, inhibiting the reflex to drop it.

Automaticity can be learned (or conditioned) such that agents perform some behavioural response automatically in the presence of some stimuli without conscious control (Bargh and Chartrand 1999). This learned stimulus–response association can be acquired intentionally, for example in the case of skill acquisition (such as practicing a piece of piano music). Automaticity can also arise unintentionally, simply through the repetition of a particular behaviour in certain circumstances, as may happen with someone who always has a cigarette immediately upon waking. The extent to which habitual health-related behaviours are automatic is likely to depend in part on how frequently they have been performed (under relevantly similar conditions) in the past.

9.3.2 Behaviour

It is more common for philosophers to talk of actions rather than behaviours. Actions are generally taken to require agency, and so to relate to reasons and intentions (Davidson 1980; Frankfurt 1988); actions are things done *by* an agent, rather than things that happen *to* an agent. The latter include bodily processes like digestion, or movements the agent makes whilst unconscious. Reflexes, to the extent that they are not under the control of the agent, would not normally be considered actions. I use the term habitual health-related *behaviour* rather than *action* for two reasons. First, it is consistent with the language used elsewhere in the literature discussing the kinds of health-related activities of interest. Second, it is neutral with regards to the role of agency (and reasons and intentions) in bringing about the activity.

On this understanding of behaviour, we are *behaving* some way or another all the time, even if we are not *acting*. People can (most likely will) engage in multiple behaviours at any given time (sitting whilst talking on the phone and eating a snack). Questions of how to individuate behaviours, as with actions, may arise here. Often discussed habitual health-related behaviours such as eating, exercising, smoking and drinking could be understood to combine multiple behaviours (going to the pub, meeting with friends, spending money). The fact they are habitual also means that they tend to be repeated (*regularly* going to the pub; smoking a cigarette *every morning upon waking*) (Brown and Savulescu 2019). Questions of responsibility could be applied to different aspects of dietary behaviour over different time frames (for instance, buying food, taking a single bite, having chips every Friday). Talk of "responsibility for diet", then, is ambiguous and could refer to responsibility for a number of things.

9.3.3 Health-Related

As discussed, there is interest in whether or not people are responsible for health-related behaviours (Brown, Maslen et al. 2019; Cavallero 2019; Davies and Savulescu 2019). Health is an important political issue, commanding large budgets. Health, or at least access to adequate health care, is often considered a basic human right (Daniels 1985; Sen 2008). It is now frequently grouped together with well-being, acting as a "super value"—a dominant value to be pursued to the exclusion of (almost) all other values (Crawford 1980; WHO 2023). Further, current concern with chronic (non-communicable) diseases such as cancer, heart disease and lung disease has led to a focus on risk factors, which can be seen as diseases themselves (e.g., obesity (Wilding, Mooney et al. 2019)) or used to identify people as being in a state of "pre-morbidity" (Crawford 1980) and thus not truly "healthy".

The tendency to construe health very broadly means almost any behaviour can be considered health-related. This includes not only behaviours directly concerned with health (such as taking medication, undergoing surgery or attending medical appointments) but almost anything a person does. From the perspective of considering whether or not an agent is responsible for some behaviour, its health impact seems irrelevant. We can simply[2] specify the behaviour (buying some cigarettes, attending the pub every Friday, routinely adding salt to one's meals) and set about considering the extent to which it meets the conditions of responsibility.

This might not be the full picture. The aim of considering people's responsibility for health-related behaviour is, ultimately, to inform practical decisions about how we should act. This includes how we attempt to change behaviour, and whether or not it is appropriate to blame (punish, disadvantage) or praise (reward, advantage) people on the basis of their health-related behaviour. I have argued elsewhere that people lack a state-enforceable moral obligation to adopt healthy behaviours (Brown, Maslen et al. 2019). Yet one might reject this and insist that, for instance, solidarity-based health care systems can legitimately demand of people that they not jeopardise their health. Or one might accept the absence of a *moral* responsibility to remain healthy but nonetheless think the state would be justified in disincentivising unhealthy behaviours, so long as people possess responsibility for those behaviours, since it has a special and legitimate interest in the health of its citizens.

I happen to think it would be wise to constrain what gets to be considered a "health-related behaviour" rather more than is typical. Often the evidence base that certain behaviours have some effect (often small, delayed and uncertain) on health is weaker than is apparent, and the benefits of changing one's behaviour may be small. But I do not wish to argue for a specific conception of "health-related" here since I do not think it matters for present purposes. I shall assume that the example behaviours I frequently refer to—diet, physical activity, smoking and drinking alcohol—can be considered health-related on most definitions of that term.

9.4 Are Agents Responsible for Habitual Health-Related Behaviours?

It is common to assume that, in order for an agent to be either directly or indirectly responsible for her behaviour, she must, at some point, be able to foresee the moral significance of her behaviour and be able to control that behaviour. These can be referred to the *epistemic* and *control* conditions of responsibility. Such an approach is broadly popular, with the conditions being specified in different ways (for instance, different degrees and kinds of control needed; different depth of knowledge required).[3] The requirements of these conditions will determine how frequently and to what extent agents are responsible.

---

[2] As mentioned, this might, in fact, not be so simple.
[3] For discussion of varieties of the epistemic and control conditions, see Robichaud and Wieland (2017); Fischer and Ravizza (1998); Mele (2010).

Some might take any specification of the epistemic and control conditions that finds that habitual behaviour does not meet these conditions (i.e., finds that agents are not responsible for habitual behaviour) to be a reductio of those conditions. Accounts of responsibility might thus be developed in order to ensure that habitual behaviour is behaviour for which agents are responsible. I will assume that it is plausible that agents are not responsible for behaviour performed habitually and/or that features of behaving habitually mitigate an agent's responsibility.

I will focus on the significance of automaticity and repeatedness in judging responsibility for habitual health-related behaviours. Most authors who have discussed responsibility for habitual behaviours assume that agents are typically responsible for those behaviours. Few authors have discussed responsibility for habits directly and at length. An exception is Pollard (2003, 2006, 2010) who argues that, unlike "reflexes, bodily processes, phobias or compulsions", habits exhibit *intervention control*, which means that:

> [I]f somebody does something out of habit, it is perfectly within their power to have done otherwise were they to choose to do so ... This is to say that the agent could intervene at any point before or during the behaviour in question, and do something else, or nothing at all. The suggestion is that since it is the agent herself who has this sort of control over the exercise of her habits, it is she who is responsible for them. It would therefore be rational to hold various attitudes towards her, say of blame or praise, since she could have done otherwise.
>
> (Pollard 2006)

Vargas (2013) discusses responsibility for habits briefly in his book *Building Better Beings*. Rather than assume that agents have "intervention control" (or some other form of control sufficient to fulfil the control condition) when they behave habitually, Vargas uses a tracing argument to establish indirect responsibility for habitual behaviour:

> On a tracing theory, one way we can be responsible for what we do is by being responsible for who we are. This capacity is important, as much of what we do is a product of habits, policies, and character traits. It is by being responsible for the formation of these habits, policies, and character traits that we come to be responsible for much of what we do. That is, we can trace our responsibility for actions that derive from habits, policies, and character traits back to our antecedent choices that led to those aspects of ourselves.   (Vargas 2013)

The first (Pollardian) approach assumes there's nothing special about habitual behaviour; agents are responsible for habitual behaviour in the way they are responsible for goal-directed, non-habitual behaviour, since they have the right sort of control over their habitual behaviour whilst performing it. The second

(Vargasian) approach suggests there may be something special about habitual behaviour which means agents are only responsible for it insofar as it is possible to trace back to some previous time where they were able to exercise control over the development of the habit that led to the behaviour.[4] I'll say a little more about both of these approaches to establishing responsibility for habitual behaviour since they both face some challenges.

9.4.1 Control-in-the-Moment

On Pollard's account, agents have control-in-the-moment over their habitual behaviour. He calls this "intervention control": at any point whilst behaving habitually the agent could pay attention to what she is doing and do something else instead.

This is true for at least some habitual behaviours. Agents who are behaving habitually often do intervene in their behaviour to modify it in some way. The agent who habitually cycles a particular route to work but, on this occasion, wishes to take an alternative route is often successful in following this non-typical route. But what are we to make of those times when the agent fails to take the non-typical route? Or those instances of agents who express a desire to reduce snacking and yet constantly eat peanuts from the bowl placed in front of them? Presumably, Pollard would still want to say that such agents possess intervention control: they *could* change their behaviour if they were attentive to it. The mere fact that they fail to attend to their behaviour (and hence fail to change it) does not show them to lack control.

The problem for such an approach to establishing responsibility, if there is one, seems to lie in part with the agent's ability to be attentive. As discussed, habitual behaviours feature high degrees of automaticity (paradigmatic habitual behaviours are entirely automatic). Automatic behaviours can be performed without attention and even without the agent being consciously aware of what she is doing. If behaviour lacks conscious awareness, it may be that the agent is not responsible for it. Levy, for instance, argues that consciousness is needed for agents to be sensitive to reasons in a sufficiently broad and consistent manner so as to enable their behaviour to reflect them in a meaningful way (Levy 2014). This is because consciousness provides the capacity to integrate information across different domains.

A review of non-conscious processes in health-related behaviour describes how difficult it is to determine whether or not agents are conscious of factors influencing their behaviour. The authors describe how:

---

[4] Alternatively, Vargas might want to adopt a hybrid approach, where responsibility stems from some combination of (direct) control agents have over their behaviour at the time it is performed, and (indirect) control they have at some time prior to performing the behaviour.

Determining whether any given behaviour can be described as conscious is a complex task. Each behaviour and its activation is a composite of many conscious and non-conscious processes and may arise from an array of internal and external cues and their interaction. This is further complicated by the fact that behaviours can be analysed and described at a number of levels and so the admixture of conscious and non-conscious processing may be different depending on the level at which the analysis is applied.

(Hollands, Marteau et al. 2016)

Hollands et al. argue for the view that behaviours sit on a spectrum of consciousness, with agents more or less aware of external influences and their own responses to them. The authors also provide examples from intervention testing where it seems people lack awareness (as measured by ability to report) of influences on their behaviour, such as portion size and posters promoting particular diet recipes (Papies and Hamstra 2010; Rolls, Roe et al. 2010; Hollands, Marteau et al. 2016).

Moors and De Houwer (2006) offer an overview of automaticity, describing how automatic processes only deplete very minimal (cognitive) resources by minimising attentional input. Automatic behaviour "once started . . . runs to completion with no need for conscious guidance or monitoring" (Moors and De Houwer 2006). This is consistent with Levy's description of ballistic "action scripts" which, once initiated, can run to completion without conscious input (Levy 2014).

Habitual health-related behaviours will involve varying degrees of conscious awareness and the opportunity for conscious control. Those who are sympathetic to the claim that consciousness is necessary for agents to be responsible must, therefore, consider the extent to which habitually behaving agents are consciously aware of what they do. Where behaviours are largely or entirely automatic (initiated and maintained through automatic processes) there will be less scope for responsibility, at least on the control-in-the-moment approach. This could include behaviours like eating chocolate from a bowl placed in front of you (Allan, Johnston et al. 2010) or brushing your teeth.

This doesn't allow any firm conclusions regarding whether or not agents are responsible for habitual health-related behaviours. But it suggests where further attention could be directed: elucidating the degree to which habitual health-related behaviours are controlled through automatic processes, and the degree to which automatically controlled processes are subject to conscious intervention.[5] Alternatively, one might adopt an account of responsibility which rejects the idea that consciousness is needed (for instance, by modelling responsible agency in a far less reflective manner, as Doris (2015) proposes).

---

[5] This could be achieved through regulatory mechanisms which might themselves operate automatically, for instance, a "vigilance" mechanism. Thanks to Gabriel De Marco for this point.

9.4.2 Tracing

One might accept that, to the extent that habitual behaviours are automatic, we should consider agents to lack control-in-the-moment. This need not, however, mean agents are not responsible for habitual behaviours. It is common to use a tracing approach where agents lack control-in-the-moment but possessed control at some prior point, sufficient to ground indirect responsibility. A common example is drink driving: drunk drivers lack the control to drive safely, and on a control-in-the-moment account would not be responsible for any accidents they cause whilst driving. Yet we typically *do* think they are responsible for such accidents, and this is based on the idea that they should have taken steps to avoid a situation where they would be driving drunk in the first place: not driving to the pub, giving their car keys to the barman to look after, and so on. We can trace back to some point in time where the agent could have taken steps to avoid driving whilst drunk and thus hold them responsible for their failure to do so.

The same might be said of habitual health-related behaviours: agents should have avoided developing the habit of smoking or eating unhealthily in the first place (or taken steps to modify this habit once developed) and then they would not be in a position where they automatically perform these behaviours with little or no control. As mentioned, Vargas (2013) thinks this is how we come to be responsible for habits.

This looks promising and avoids the difficulty of needing to establish the degree of automaticity and control-in-the-moment for habitual behaviours. There are, however, some remaining challenges with showing agents to be responsible for habitual health-related behaviours on a tracing approach.

First, assuming agents can exercise control over their future behaviour (e.g., by avoiding developing a habit in the first place) it must also be established that they are aware of this capacity and the need to use it. As discussed, habitual health-related behaviours consist in learned stimulus–response pairings where behaviours are triggered automatically by contextual cues. These develop through learning where particular behaviours are performed in response to particular cues repeatedly. Habits thus develop by repeatedly doing the same thing in the same context. To avoid developing habits the agent must avoid doing the same thing in the same context, and avoid allowing her behaviour to become automatised.

Agents may not realise they are establishing a habit, and so may acquire habits unintentionally. Someone who cycles the same route to work everyday may not realise that their behaviour is becoming automatised. Alternatively, they may establish a habit under conditions where doing so is not harmful, and where they did not (perhaps could not) foresee that the habit would be harmful in the future. Consider here those who became habitual smokers before it was widely recognised that cigarettes were extremely damaging for one's health. Other stimulus–response pairings are acquired extremely easily, such as that between having tasty snacks within reach and eating those snacks. One need only eat salted peanuts a few times to acquire an automatic mechanism to eat peanuts when they are within reach.

The development of many habitual health-related behaviours is unintentional (Bargh and Chartrand 1999). This is quite different from, for instance, skills like piano playing, where agents intentionally and effortfully practice certain behaviours in order to establish automaticity. Although unintentional, the acquisition of particular habitual health-related behaviours isn't necessarily unavoidable: agents may, prior to automaticity developing, be able to take steps to inhibit this process. But, in considering whether agents can be responsible for habits by tracing back to a time before the habit became established it is necessary to consider whether the epistemic condition for responsibility was fulfilled.

One might think that, even once a habit is established, agents still have opportunities to exercise control over that habit, since it is (presumably) not always operative. In such a case, in order to establish responsibility via tracing, it must be shown that the agent really was able to exercise control over her future behaviour in the way required. Further, it must be shown that the agent knew how to do this. For example, in order to refrain from smoking, the habitual smoker may need to avoid buying cigarettes in the first place, or facilitate a new habit such as vaping.

This will be possible for some habitual health-related behaviours, but for others it will be trickier. A notoriously hard behaviour to change is dietary behaviour. Many people who are overweight or obese proclaim a desire to lose weight, yet despite multiple attempts and numerous different techniques, they often fail to do so. One reason why changing dietary behaviour is so difficult is that people are frequently exposed to environmental cues for eating that it is difficult to avoid. If food is available, people may eat it (more or less automatically). They lack control (to some extent) over their eating behaviour once food is present, but they also lack the ability to control (again, to some extent) whether food is present.

The fact that people find it hard to control their exposure to environmental cues for eating and other behaviours does not mean they lack control to a sufficient degree to undermine responsibility. This will depend on the particular formulation of the control condition that one accepts. A relatively undemanding formulation such as Fischer and Ravizza's (1998) guidance control, requiring only that the mechanism which produced the behaviour was moderately reasons responsive, is likely to be fulfilled much of the time.[6]

There is also a necessary epistemic component to altering established habits. In order to change habits, it is sometimes recommended that people substitute them with a different habit, or try to change habits in the context of bigger environmental changes (Wood, Tam et al. 2005; Bouton 2011). For instance,

---

[6] On Fischer and Ravizza's account, guidance control is typically fulfilled for habitual behaviours anyway, without the need to refer to tracing.

swapping cigarettes for e-cigarettes, snacking on rice cakes rather than cookies, or establishing a jogging routine after you've just moved house. Yet people may well not know such techniques for exercising willpower and discarding undesired habits, nor the kind of mental resolve needed in order to conform their behaviour over time to a particular plan (Holton 2003). Such epistemic requirements might need to be fulfilled if agents are to exercise control over their habitual behaviour.

9.5 Concluding Remarks: Implications of Responsibility for Habitual Health-Related Behaviour

In the above sections I have discussed two approaches to establishing responsibility for habitual health-related behaviour. The first is to argue that agents have control-in-the-moment over their habitual behaviour, and hence will ordinarily fulfil the control and epistemic conditions for responsibility as a result. I have suggested that, to the extent that habitual behaviours feature high degrees of automaticity, agents may lack conscious awareness of their behaviour, and may not exert conscious control over that behaviour. For those who think, in order for an agent's behaviour to reflect her agency it must be subject to her conscious control, automaticity could diminish the range of habitual behaviours for which an agent is responsible. Where behaviour is initiated and maintained through automatic processes it may not reflect an agent's values and preferences. Without the ability to integrate information across different domains, automatically produced behaviours will not be sensitive to reasons in the way that consciously produced behaviours will be.

Habitual behaviours, in their "pure" form, depend only on the presence of the appropriate stimulus to be cued. Even if the desired outcome such behaviours initially achieved disappears, people acting habitually may still perform the response behaviour (and this could be true even when that response behaviour becomes inappropriate and undesirable). What might appear to be an "irrational" behaviour, then, may be the continued performance of a learned response.

The degree of automaticity, and its impact on responsibility, will vary across habitual health-related behaviours. Those which are highly automatised (like eating high calorie foods that are readily available, or smoking a cigarette once it is lit and in one's hand) are likely to leave less room for conscious control, especially if undertaken whilst performing other, cognitively demanding activities. Behaviours which have been performed more frequently (under the relevant conditions) are likely to become more automatised. This means that people who have long-standing dietary or exercise habits are likely to find them harder to alter than those who have recently established habits (Verplanken and Orbell 2003; Orbell 2013).

The second approach discussed traces an agent's responsibility to some previous point in time where she was able to control her behaviour, either before she has established a habit (and automatised stimulus–response pairings), or if she has already established a habit, then at some point prior to that habit being performed (before automatic behaviour is initiated). I have suggested that there remains a possibility that, first, agents unintentionally acquire habitual behaviours and are unaware of their opportunities to control automatic behaviours in the way needed for responsibility, and second, that the opportunity to control environmental cues (and thus avoid stimulating automatic processes) may be limited.

It is important to consider the fact that some stimulus–response pairings will develop during childhood, and/or develop very swiftly, such as habitually overeating calorie dense foods. In such cases, it is hard to see how this is avoidable for the agent, and means the agent must have some later opportunity to exercise control over the habit if she is to be considered responsible for it.

When it comes to agents who have established habits, but could theoretically control their behaviour by avoiding the relevant environmental cues, it will be necessary to consider how much people are aware of such tools and techniques for controlling their behaviour. Whilst behavioural scientists are aware of the importance of environmental stimuli for behaviour, non-experts may not recognise how automatic processes and environments can influence their actions.[7] Levy (2017) discusses evidence that suggests some people *do* seem able to pursue long-term goals by changing their environment in order to avoid temptations. This suggests awareness of techniques to weaken the role of automatic mechanisms of behavioural control. But it is unclear how representative individuals who successfully employ such strategies are: if those who fail to use such techniques do so because they are unaware of them, such ignorance could undermine responsibility for failing to take steps to avoid the triggering of habitual behaviours. This could, however, be rectified by drawing agents' attention to such features of behaviour and presenting them with opportunities to change it. "Golden opportunities" have been identified as a tool for ensuring agents have the skills (including knowledge) and reasonable, realistic opportunities to change health-related behaviours (Savulescu 2018). Under such conditions—including appropriate support—responsibility may be restored.

Even when agents are aware of the role environmental stimuli play in their behaviour, it may be much more challenging for some than others to avoid those stimuli. For instance, environmental cues for impulsive behaviours (such as gambling and eating fast food) are more prevalent in areas of high socio-economic deprivation than areas of low deprivation (Macdonald, Cummins et al. 2007; Wardle, Keily et al. 2014). The effort required to avoid such stimuli is thus much greater for some than others. Whilst it might nonetheless be *possible* for

---

[7] See, for instance, the fundamental attribution error, where people tend to overestimate the importance of an individual's personal characteristics in explaining some action, whilst underestimating the contribution of situational factors.

agents to avoid cues for automatic behaviour (and thus they can be considered responsible), the fact that they will frequently have to exert and sustain significant effort in order to do so means their responsibility and blameworthiness for any failure should be mitigated.

None of this discussion suggests that agents are not, in general, responsible for habitual health-related behaviour. Instead, it indicates that in order to establish responsibility, attention must be paid to the degree to which behaviours are automatically cued and maintained; the extent to which agents are aware of the development of automatic (habitual) behaviours; and the opportunities agents have for shaping their environment in the ways necessary to avoid cueing undesired automatic behaviours.

Habitual health-related behaviours are significant contributors to non-communicable disease, which cause a huge amount of suffering worldwide and is economically costly (WHO 2018). If agents are responsible for habitual health-related behaviours then they might be responsible for their consequences (i.e., poor health) and it might be justifiable to use responsibility-sensitive health care policies to tackle disease. This could involve, for example, deprioritising for treatment those who are responsible for their poor health. This idea is contentious. Some argue, on principle, that health care just isn't the kind of thing that should depend on (personal) responsibility (Wikler 2002; Waller 2005). Others assert that systemic inequality means those who are already worse off due to no fault of their own would be unfairly punished by policies designed to "hold people responsible" for their behaviour (Brownell 1991; Cavallero 2019). Still others think that attempts to make health care responsibility-sensitive will have net negative effects on health and well-being, or be practically unachievable, or unacceptably intrusive (Waring 2006; Eyal 2013; Friesen 2017). This discussion shouldn't be taken as a response to any of these concerns. There could be independent reasons for thinking that responsibility ought not to play a role in health care policy. My intention has only been to consider the challenges features of habitual health-related behaviour *specifically* create for grounding responsibility. If responsibility is frequently mitigated or absent due to some of these challenges, then we should ensure that health policy and practice reflects this and does not incorporate mistaken assumptions about responsibility.

References

Adams, CD. (1982), "Variations in the Sensitivity of Instrumental Responding to Reinforcer Devaluation", in *Quarterly Journal of Experimental Psychology Section B* 34/2b: 77–98.

Allan, JL., Johnston, M. and Campbell, N. (2010), "Unintentional Eating. What Determines Goal-Incongruent Chocolate Consumption?" in *Appetite* 54/2: 422–5.

Bargh, JA., and Chartrand, TL. (1999), "The Unbearable Automaticity of Being", in *American Psychologist* 54/7: 462–79.

Baum, WM. (1973), "The Correlation-Based Law of Effect 1", in *Journal of the Experimental Analysis of Behavior* 20/1: 137–53.

Bouton, ME. (2011), "Learning and the Persistence of Appetite: Extinction and the Motivation to Eat and Overeat", in *Physiology & Behavior* 103/1: 51–8.

Brown, RC. (2019), "Irresponsibly Infertile? Obesity, Efficiency, and Exclusion from Treatment", in *Health Care Analysis* 27/2: 61–76.

Brown, RC, and Savulescu, J. (2019), "Responsibility in Healthcare Across Time and Agents", in *Journal of Medical Ethics* 45/10: 636–44.

Brown, RCH. (2018), "Resisting Moralisation in Health Promotion", in *Ethical Theory and Moral Practice* 21/4: 1–15.

Brown, RCH., Maslen, H. and Savulescu, J. (2019), "Responsibility, Prudence and Health Promotion", in *Journal of Public Health* 23/1: 31–5.

Brownell, KD. (1991), "Personal Responsibility and Control Over Our Bodies: When Expectation Exceeds Reality", in *Health Psychology* 10/5: 303–10.

Buyx, AM. (2008), "Personal Responsibility for Health as a Rationing Criterion: Why We Don't Like It and Why Maybe We Should", in *Journal of Medical Ethics* 34/12: 871–4.

Callahan, D. (2012), "Obesity: Chasing an Elusive Epidemic", in *Hastings Center Report* 43/1: 34–40.

Caruso, GD. (2021), *Rejecting Retributivism: Free Will, Punishment, and Criminal Justice* (Cambridge University Press).

Cavallero, E. (2019), "Opportunity and Responsibility for Health", in *Journal of Ethics* 23/4: 369–86.

Crawford, R. (1980), "Healthism and the Medicalization of Everyday Life", in *International Journal of Health Services* 10/3: 365–88.

Daniels, N. (1985), *Just Health Care* (Cambridge University Press).

Davidson, D. (1980), *Essays on Actions and Events* (Oxford University Press).

Davies, B. and Savulescu, J. (2019), "Solidarity and Responsibility in Health Care", in *Public Health Ethics* 12/2: 133–44.

Dickinson, A. (1985), "Actions and Habits: The Development of Behavioural Autonomy", in *Philosophical Transactions of the Royal Society of London. B, Biological Sciences* 308/1135: 67–78.

Doris, J. (2015), "Doing Without (Arguing about) Desert", *Philosophical Studies* 172/10: 1–10.

Eyal, N. (2013), "Denial of Treatment to Obese Patients: The Wrong Policy on Personal Responsibility for Health", in *International Journal of Health Policy and Management* 1/2: 1–4.

Feng-Gu, E., Everett, J., Brown, RCH, et al. (2021), "Prospective Intention-Based Lifestyle Contracts: Health Technology and Responsibility in Healthcare", in *Health Care Analysis* 29/3: 189–212.

Fischer, JM and Ravizza, M. (1998), *Responsibility and Control: A Theory of Moral Responsibility* (Cambridge University Press).

Frankfurt, HG. (1988), *The Importance of What We Care About: Philosophical Essays* (Cambridge University Press).

Friesen, P. (2017), "Personal Responsibility Within Health Policy: Unethical and Ineffective", in *Journal of Medical Ethics* 44/1: 53–8.

Gardner, B. (2015), "A Review and Analysis of the Use of 'Habit' in Understanding, Predicting and Influencing Health-Related Behaviour", in *Health Psychology Review* 9/3: 277–95.

Glannon, W. (1998), "Responsibility, Alcoholism, and Liver Transplantation", in *Journal of Medicine and Philosophy* 23/1: 31–49.

Hannikainen, IR., Machery, E., Rose, D., et al. (2019), "For Whom Does Determinism Undermine Moral Responsibility? Surveying the Conditions for Free Will Across Cultures", in *Frontiers in Psychology* 10/2428: 1–13.

Hollands, GJ., Marteau, TM., and Fletcher, PC. (2016), "Non-Conscious Processes in Changing Health-Related Behaviour: A Conceptual Analysis and Framework", in *Health Psychology Review* 10/4: 381–94.

Holton, R. (2003), "How is Strength of Will Possible", in S. Stroud and C. Tappolet (ed.), *Weakness of Will and Practical Irrationality* (Oxford University Press): 39–67.

Levy, N. (2014), *Consciousness and Moral Responsibility* (Oxford University Press).

Levy, N. (2017), "Of Marshmallows and Moderation", in *Moral Psychology* 5: 197–214.

Lingawi, NW., Dezfouli, A., and Balleine, BW. (2016), "The Psychological and Physiological A. Mechanisms of Habit Formation", in RA. Murphy and RC. Honey (ed.), *The Wiley Handbook on the Cognitive Neuroscience of Learning* (Wiley-Blackwell): 411–40.

Macdonald, L., Cummins, S., and Macintyre, S. (2007), "Neighbourhood Fast Food Environment and Area Deprivation: Substitution or Concentration?" in *Appetite* 49/1: 251–4.

Maciel, R. (2015), *Relative Justice: Cultural Diversity, Free Will, and Moral Responsibility* (Taylor & Francis).

Mele, A. (2010), "Moral Responsibility for Actions: Epistemic and Freedom Conditions", *Philosophical Explorations* 13/2: 101–11.

Miller, JG., Goyal, N., and Wice, M. (2017), "A Cultural Psychology of Agency: Morality, Motivation, and Reciprocity", in *Perspectives on Psychological Science* 12/5: 867–75.

Moors, A. and De Houwer, J. (2006), "Automaticity: A Theoretical and Conceptual Analysis", in *Psychological Bulletin* 132/2: 297–326.

Orbell, S. (2013), "Habit Strength", in MD. Gellman and JR. Turner (ed.), *Encyclopedia of Behavioral Medicine* (Springer New York): 885–6.

Papies, EK. and Hamstra, P. (2010), "Goal Priming and Eating Behavior: Enhancing Self-Regulation by Environmental Cues", *Health Psychology* 29/4: 384–8.

Pollard, B. (2003), "Can Virtuous Actions Be Both Habitual and Rational?", in *Ethical Theory and Moral Practice* 6/4: 411–25.

Pollard, B. (2006), "Explaining Actions with Habits", in *American Philosophical Quarterly* 43/1: 57–69.

Pollard, B. (2010), "Habitual Actions", in T. O'Connor and C. Sandis (ed.), *A Companion to the Philosophy of Action* (Wiley-Blackwell): 74–81.

Robichaud, P. and Wieland, JW. (2017), *Responsibility* (Oxford University Press).

Rolls, BJ., Roe, LS. and Meengs, JS. (2010), "Portion Size Can Be Used Strategically to Increase Vegetable Consumption in Adults", in *American Journal of Clinical Nutrition* 91/4: 913–22.

Savulescu, J. (2018), "Golden Opportunity, Reasonable Risk and Personal Responsibility for Health", in *Journal of Medical Ethics* 44/1: 59–61.

Sen, A. (2008), "Why and How is Health a Human Right?" in *The Lancet* 372/9655: 2010.

Vargas, M. (2013), *Building Better Beings* (Oxford University Press).

Verplanken, B. and Orbell, S. (2003), "Reflections on Past Behavior: A Self-Report Index of Habit Strength", in *Journal of Applied Social Psychology* 33/6: 1313–30.

Waller, BN. (2005), "Responsibility and Health", in *Cambridge Quarterly of Healthcare Ethics* 14: 177–88.

Wardle, H., Keily, R., Astbury, G., and Reith, G. (2014), "'Risky Places?': Mapping Gambling Machine Density and Socio-Economic Deprivation", in *Journal of Gambling Studies* 30/1: 201–12.

Waring, R. (2006), "Imposing Personal Responsibility for Health", in *New England Journal of Medicine* 355/8: 1–4.

Weiner, B. (1995), *Judgments of Responsibility: A Foundation for a Theory of Social Conduct* (Guilford Press).

WHO. (2018), "Noncommunicable Diseases", https://www.who.int/news-room/fact-sheets/detail/noncommunicable-diseases.

WHO. (2023), "WHO Constitution", Retrieved 29/08/23, 2020, from https://www.who.int/about/governance/constitution.

Wikler, D. (2002), "Personal and Social Responsibility for Health", in *Ethics & International Affairs* 16/2: 47–55.

Wilding, JPH., Mooney, V. and Pile, R. (2019), "Should Obesity be Recognised as a Disease?" in *BMJ* 366: l4258.

Wood, W., Tam, L., and Witt, MG. (2005), "Changing Circumstances, Disrupting Habits", in *Journal of Personality and Social Psychology* 88/6: 918–33.

PART IV

BEYOND PATIENT RESPONSIBILITY: HEALTH CARE PROFESSIONALS

10

Taking Responsibility for Uncertainty

Richard Holton and Zoë Fritz

10.1 Introduction

We start with the realities of contemporary medicine. When a patient consults a doctor about a set of symptoms, the doctor will typically form not a definitive diagnosis, but one that is, in the jargon, 'differential': that is, a list of possible diagnoses, with some rating of the likelihood and severity of each, and with some indication of what would need to be done to narrow down these possibilities.[1]

But the patient will not typically be given the full differential diagnosis. Instead they will typically be given the most likely possibility, perhaps two, or else a 'putative' or working diagnosis of the most likely cause. Why is this? Part of the reason stems from the complexity. Understanding a single diagnosis can be hard enough; understanding a whole list of possible diagnoses will be much harder. But in addition, the notion of uncertainty, and of the probabilities in which it is typically framed, brings a whole set of further difficulties. How seriously should one take a 2 per cent probability? Or a 2 per cent probability of something really bad? How should one weigh a test with a 10 per cent chance of false positives? How should one factor in the relevant base rates? We know from many studies that even doctors are not good on many of these questions (especially the last) (Gigerenzer 2008: ch. 9). For patients they risk becoming unmanageable. We know that people typically react in particular ways, over-weighting small probabilities, underweighting large, ignoring base rates and so on. Some of these difficulties can be minimized by a sensitive presentation of probabilistic information: natural frequencies ('of 100 people who test positive, 7 will have the disease, 93 will not') are easier to understand than probabilities and base rates ('the rate of the disease in the population is 1 per cent; the proportion of false negatives in the test results is 0; the proportion of false positives in the test results is 10 per cent') (Gigerenzer 2008: ch. 12). For others though there is no technical fix. There is no neutral answer to when we should worry about a possibility, and when we should ignore it as too small to worry about.

---

[1] This is a common use of the term among practicing doctors. The next stage, the act of distinguishing one cause from the other possibilities, is colloquially referred to as 'working through the differential' or 'narrowing down the differential'.

Richard Holton and Zoë Fritz, *Taking Responsibility for Uncertainty* In: *Responsibility and Healthcare*. Edited by: Benjamin Davies, Gabriel De Marco, Neil Levy, Julian Savulescu, Oxford University Press. © Edited by Benjamin Davies, Gabriel De Marco, Neil Levy, Julian Savulescu 2024. DOI: 10.1093/oso/9780192872234.003.0011

More broadly, we know that people do not simply find uncertainty intellectually challenging—it is emotionally challenging too, unsettling and aversive. People do not, in general, like being in a state of uncertainty, and consequently, of indecision, especially when this involves their own fate. A good illustration of this comes from the experiences of a doctor, Franz Ingelfinger (1980). He was diagnosed with a carcinoma in the gastroesophageal junction, an illness about which he was an expert. Nevertheless, worrying about the best treatment, he became, in his own words, "increasingly confused and emotionally distraught. Finally, when the indecision had become nearly intolerable, one wise physician friend said: 'What you need is a doctor.'"

If even an expert like Ingelfinger can react to the uncertainty of what to do in this way, how much greater the risk for someone who finds it hard to understand what is going on. There has been some discussion of the 'diagnostic moment': the point at which a patient comes to know what they have, and can reconceptualize themselves and their future accordingly (Jutel 2014; Heritage and McArthur 2019). A differential diagnosis presents no such moment. Rather than leaving appropriately reassured, the patient may leave with heightened anxiety. Rather than deciding on a reasonable treatment, they may ask for further unwarranted and potentially dangerous investigations; or, conversely, they may be scared into refusing treatment altogether. We have heard anecdotally from quite a number of clinicians of outcomes in both of these directions. So it is not surprising that clinicians who are concerned for the well-being of their patients might want to restrict the information that they give them. In fact we are sure that almost all already do, if only because of the time constraints on the typical consultation. Possible diagnoses that are very unlikely are typically not reported to patients; nor are possible but very unlikely outcomes of treatment. 'I rarely mention anything with a chance of less than 5 per cent' one consultant told us.

But of course, this raises worries. Withholding information so as not to alarm a patient looks like paternalism at its most overt. Perhaps more worrying, it looks to be incompatible with two of the standard ethical requirements on modern medicine: (i) informed consent, the idea that the patient's consent must be given for any medical intervention, and that to properly give it the patient must fully understand what they are consenting to; and (ii) shared decision-making, the idea that clinicians and patients should share the best available evidence when deciding on a treatment (Elwyn et al. 2010). For if information is withheld, the patient cannot be fully informed; and if they are not fully informed they cannot give informed consent, and decisions cannot be truly shared.

Of course informed consent and shared decision-making do not require that the patient know everything about their treatment, down to the provider of the oxygen. A reasonable criterion of what they need to know might be: everything that, were they to know it, would reasonably affect their choice of whether to have the treatment. Something like that was articulated by the UK Supreme Court in *Montgomery* (2015), specifically in relation to treatment decisions:

> The doctor is therefore under a duty to take reasonable care to ensure that the patient is aware of any material risks involved in any recommended treatment, and of any reasonable alternative or variant treatments. The test of materiality is whether, in the circumstances of the particular case, a reasonable person in the patient's position would be likely to attach significance to the risk, or the doctor is or should reasonably be aware that the particular patient would be likely to attach significance to it.

But even with this reduced standard for what a patient needs to know, the problem remains. If the doctor's decision to withhold information is based on a concern about alarming the patient, then that is exactly to think that the information does meet this test of materiality: the patient would attach significance to the risk it were they informed of it.

We might interpret the judgement as proposing a straightforwardly objective test: it is not whether the patient would in fact be alarmed, but whether they would reasonably be. But as we have said, it is very hard to know what is a reasonable response to, say, a small chance of a bad outcome. And besides, it is unclear that that is how we should interpret the passage. While the talk of the 'reasonable person in the patient's position' might encourage that interpretation, the requirement that the test places on the doctor—that they 'should reasonably be aware that the particular patient would be likely to attach significance to it'—makes no requirement of the reasonableness of the patient's response. It is clear from the passage cited that the judgement requires a doctor to inform a patient of a risk that they reasonably think the patient would be alarmed by, whether or not they in turn think that alarm reasonable (and they must in addition inform the patient if they think that they would have reason to be alarmed, even if they do not in fact think they would be).

Nevertheless two paragraphs earlier (§ 85) in the very same judgement we read:

> A person can of course decide that she does not wish to be informed of risks of injury (just as a person may choose to ignore the information leaflet enclosed with her medicine); and a doctor is not obliged to discuss the risks inherent in treatment with a person who makes it clear that she would prefer not to discuss the matter. Deciding whether a person is so disinclined may involve the doctor making a judgment; but it is not a judgment which is dependent on medical expertise.[2]

---

[2] For a general discussion of how to apply Montgomerty to the issue of the communication of diagnosis, see Liddell, Skopek, Le Gallez, and Fritz (2022).

How do we fit these two ideas together? Sometimes there will indeed be an explicit request not to be told. But this will not remove the discretion from the doctor, for even patients who don't want to know certain things will rarely want absolute ignorance. Which things won't they want to know? At best they might be able to outline this in general terms—'I don't want to know the details'; 'Don't tell me about what might go wrong'—but the doctor will have to make decisions about how to understand such a request. Where the patient has not explicitly requested not to be told, things are even more complicated. Should the doctor ask them what they want to know? Questions themselves convey information, via their framing and presuppositions. To ask whether someone wants to know about other possibilities or risks is to communicate that there are other possibilities or risks. Even if the patient knows that there are going to be risks, to ask a question like this is to draw attention to them. If the patient had been wilfully ignoring them up till now, that stance is no longer available to them.

In the background of these concerns is what we will call *the irreversibility of knowledge*. Many treatments can be given on a trial basis: see if it works, and if it doesn't, discontinue. Knowledge is not like that. You cannot tell someone something, and then withdraw the knowledge if it distresses them, or if they don't want to know if for some other reason. You can perhaps tell someone not to worry about something that they have not been able to understand—tell them, that is, that they do not need to persist in trying to understand it—and that will sometimes be reassuring (though sometimes not). But that is a very different idea. You cannot simply tell someone to forget something that they have understood.

There is a second complication concerning the 'test of materiality' that we outlined above: the test of what information is relevant, and hence what people should normally be told. A natural way of spelling that test out in more details makes use of a hypothetical retrospective test: if, at the end of the treatment, one were to ask the patient what, in retrospect, they would have wanted to have been told, and what not, then that which they say they would have wanted to know is material. The problem with this is that it is likely to be highly contingent on how things turn out. To see this, consider a parallel proposal for deciding on diagnostic procedures. Suppose there is a remote chance—0.01 per cent say—that a patient has something very nasty. There is a test for revealing whether or not they do, but it is itself intrusive and unpleasant. Suppose, therefore, that the standard procedure is not to test. The vast majority of patients, those for whom the test would have been negative, might, in retrospect, be pleased not to have been tested. But the unfortunate one in 10,000 who actually has the condition is likely, in retrospect, to be very regretful that they did not have the test. We can call this *the contingency of regret*.[3] It

---

[3] A related issue arises in discussions of *ex post* and *ex ante* construals of contractualism; see, for instance, Frick (2015) and Suikkanen (2019). Our concern is related to some of the concerns about the *ex post* approach.

is a feature that can frequently be observed in debates about diagnostic testing. Very often patient groups are all in favour of such testing; but they, of course, are the ones who, suffering from the condition, would regret not having been tested. We tend not to hear from those who turn out not to have the condition, and would rather never have been tested. What is relevant to our concerns is that a similar consideration applies to knowledge: Would a patient want to have known about some remote possibility of a bad outcome? That might well depend on the outcome: if the possibility is not realized, then no, if it is, yes. This is important, because the cases that reach the courts are typically those—like Montgomery, cited above—in which the bad but unlikely outcome has come about. There is a risk that we will build our policy primarily to avoid such unexpected but unlikely bad outcomes. We need to ensure that in doing so we do not come up with a policy that is ill-suited to the vast majority of cases.

10.2 The Realities of the Consultation

Imagine a consultation in which a patient has presented with some symptoms, and the doctor has elicited their history and examined them. With this information, the doctor has determined a differential diagnosis, and they have reported (some of) it to the patient. The doctor has outlined the three most likely explanations, with different degrees of probability; and has told the patient that there are further tests that could be done, some of which carry risks, all of which would reduce the level of uncertainty. The patient listens carefully, nodding. Then they say:

But doctor, what do you think I have?

Most clinicians will recognize this as a familiar response. What is the patient after? They have just heard the differential diagnosis. Should the doctor just repeat it, stressing their uncertainty? That might stop the patient persisting the question, but it doesn't seem that it will provide them with what they were after.

To throw some light on what the patient was really after, consider a slightly different case. Instead of further investigations to be undertaken, the patient is faced with three treatments which could be started, though still without a definitive diagnosis in place. Again the patient listens carefully to the three options, and then says:

But doctor, what would you do?

Again most clinicians will be very familiar with such questions. In this second case the request is more straightforward; the patient is after practical, not epistemic, advice. Our suggestion is that in fact the two cases are very closely related, and that even in the first it is something like practical advice that is wanted. The core thought is this: in the face of all the uncertainty, the patient is looking for a more secure path; they want to be on the other side of the diagnostic moment. This is so at the practical level, since the patient wants to decide on a treatment option. But it may also be so at the theoretical level. When the patient asks, in the face of differential diagnosis, what the doctor really thinks they have, they are asking for something like advice on what to believe. They want a diagnosis to use as a working hypothesis. For a patient who wants this epistemic simplification, the theoretical and the practical will typically go together, since the working hypothesis that they adopt will be that they have the condition on which the treatment is to be targeted.

Let us suppose then that the doctor says what they would do, and what, in that sense, they think the patient has; and that the patient is happy to adopt the treatment, and the corresponding working hypothesis, as the path forward. Does this mean that the other possibilities are ignored? In a sense that is so for the patient: the other possibilities are put out of mind. But a thoughtful patient will know that this is not how the doctor is thinking. More test results are pending; and there is of course no information yet on how the patient will respond to the treatment. The doctor needs to keep check on these, and the patient trusts them to do so.

So what we have is a kind of mental division of labour. The patient can focus on the treatment, on what they need to do for it to have a chance of working, and on what they need to keep an eye on and report ('Be sure to tell me if you have any new symptoms, especially if you feel ... '). And they have a diagnosis that they can research, discuss with friends and family, and so on. They know that this diagnosis is not definitive: they know that there is a chance that what they have is something else, but that is left in the background. So they do not have to face the unsettling prospect of keeping more than one possibility live in their mind. The doctor, in contrast has a much more open epistemic field. The other elements of the differential diagnosis remain live, in that they are still being actively investigated. And if evidence comes in that starts to make one of them look somewhat more likely, the doctor will have to decide whether, and if so when, to tell the patient. This may be only if a change of treatment is called for, but it may be well before. That is a decision to be made in the light of the interests, needs, and capacities of the patient. We can say, in short, that the doctor takes responsibility for the uncertainty.

What we have sketched is a fairly explicit case of such taking responsibility. We suggest though that this is happening implicitly all the time. In a consultation a doctor will constantly be assessing how much of their own uncertainty they should communicate to the patient; and the patient will likewise be indicating how much they want to take on. At one extreme the doctor may convey very little of the uncertainty: they may give suggest a line of treatment, outline the diagnosis on which this would be based, mention that there are other possibilities that remain under investigation, and ask the patient if they have any questions. At the other extreme they may share all of the uncertainty, and keep the patient fully apprised of every bit of evidence as it comes in. In between are many possibilities.

Patient and doctor need a shared understanding of who will take responsibility for the uncertainty. This is delicate because of the irreversibility of knowledge that we mentioned at the outset. So it must be done with care. Given that they have more experience in these areas, much of the onus will fall on the doctor. Asking:

Do you have any questions?

provides an opening for questions, but scarcely encourages them; much more encouragement is provided by an approach that presupposes that there will be questions:

So which questions would you like to ask me?[4]

There is even more encouragement if the patient is given time to prepare, and the expectation that they will:

When you get home I'd like you to get some paper and jot down any questions you might have, and we can go through them when I see you tomorrow.

Similarly:

There are some other possibilities that the tests might show.

invites less inquiry than:

There are some other possibilities that the tests might show. Would you like me to go over them now, or shall we wait and see if the tests throw anything up?

Kerr and Read, in the Supreme Court ruling cited above, claimed that judgements about the issues involved here are 'not dependent on medical expertise'. We disagree. There are skills involved in these negotiations, and they are a central part of effective medical treatment. They are hard won. They are rarely taught

---

[4] For a wide-ranging discussion of the way that questions can invite certain sorts of responses see Boyd and Heritage (2006, esp. 177–8). Very often open-ended questions of the kind we are considering will elicit quite new concerns, so-called 'doorknob concerns'. For discussion see Robinson et al. (2015).

explicitly (though perhaps they could be); most doctors learn them as part of their medical experience. Some doctors are undoubtedly better at making them than others.[5]

10.3 Taking Responsibility

We have spoken of *taking* responsibility for uncertainty; here we want to say a little about what we mean by this. The expression is crucially ambiguous. One way to take responsibility is to let oneself be held responsible, and correspondingly to hold oneself responsible. A second way is to be given charge, or to take charge. We are primarily concerned with the second sense here.

In contrast, the familiar notion of moral responsibility, which has come to hold a very central place in moral philosophy,[6] makes use of the first notion—that of being held responsible. This central understanding of moral responsibility has come to be framed in terms of accountability—indeed much of the recent debate has been over whether we can get by with just this, or whether we need in addition what are, by our lights, the fairly closely related notions of attributability or answerability.[7] This understanding is sometimes buttressed with a bit of etymology: to be responsible, on this understanding, is to be able, or liable, to respond; the responsible individual needs to be ready to give an account of themselves. Responsibility is response-ability, as John Gardner (2007: 182) once put it.

In spelling out the alternative conception that we have in mind, we'll start with that etymology, since there is something rather curious about it (to be fair to Gardner, it's not clear how seriously he meant it, but we think it's worth pursuing). The English suffixes 'ible' and 'able' are typically passive: the drinkable thing is something that can be drunk, not something that can drink, and so on.[8] In its earliest usages, 'responsible' followed the same path: a responsible person in Anglo-Norman law was not someone who had to respond; on the contrary, they were the person to whom a response had to be given. The responsible were those who issued writs; others then responded to them.[9] The responsible person was thus passive when it came to responding, but of course they were active in a deeper

---

[5] For discussion of the claim that these should be thought of as medical skills even though they are not explicitly taught, see Liddell, Skopek, Le Gallez, and Fritz (2022).

[6] The most widely used introductory philosophical textbook is entitled *Reason and Responsibility* (Shafer-Landau and Feinberg 2016); desires for alliteration notwithstanding, 'Responsibility' is clearly seen as so central that it can here stand for the whole of ethics. This centrality is really a twentieth-century phenomenon; Hume and Mill scarcely discussed it; Sidgwick talks about it only a little when talking about free will.

[7] See Watson (2004); Shoemaker (2011); Smith (2012).

[8] We see the force of this clearly in new coinages: a writeable disc, for instance, is a disc upon which one can write, not a disc that can write; someone is Googleable if they can be found using Google, not if they are able to use Google; and so on.

[9] See, for instance Pike (1900: 158–9). Pike translates 'responsible' as 'entitled to an answer'.

sense, in that of demanding the answer. It was the responsible person who was calling the shots.

This deeply active notion of responsibility has continued on in the rather different notion of responsibility we are after. If someone is given responsibility for the drinks at the party, it is true that you might go to them to complain if it turns out that the drinks ran out. But to focus there is to focus in the wrong place. The core of the issue is that they were in the position of managing the drinks, and that would be so even if no one were ever in a position to hold them to account—if no one were in a position to *hold* them responsible. One way of putting the distinction we have in mind is thus to contrast (i) being *given responsibility* for something; with (ii) being *held responsible* for something.

Being given responsibility though is only one way to get it: one might assume responsibility oneself. So to give a canonical usage that is broad enough to cover the cases, from now on we will contrast

being responsible for something

with

being held responsible for something.

Drawing the distinction this way, though somewhat stipulative, is close enough to normal usage. The first is the notion we are after; the second is the notion that dominates the philosophical discussions of moral responsibility.

There have been a few philosophical discussions of this notion of being responsible. Hart terms it 'role responsibility', but we find this somewhat misleading since it suggests that it comes automatically with a given role, whereas it is much more plastic than that—plastic in what is required, and plastic in the situations in which it can arise.[10] It has though received far less attention that the idea of being held responsible.

We do not deny that these two states can go together, and typically they do: if one has responsibility for something, one is typically held responsible if it goes wrong, and conversely. But they can come apart.

One can be held responsible, or take responsibility, for an outcome, even if one was not responsible for the events: one could take responsibility for events that happened before one was born, or for the doings of one's subordinates, for example. The classic British example of the latter is of a government minister

---

[10] Hart (1968: 212–13). Hart did want to generalize the idea to a task 'assigned by agreement or otherwise', although most subsequent discussion has kept the idea of an institutional role; see, for instance, Cane (2016). The other recent discussion has been in the works of hard determinists, who want to use it in lieu of the more traditional accounts of responsibility; see e.g., Waller (2011) where it is termed 'take-charge responsibility'. We want to retain both notions.

taking responsibility for things done in his or her department, even without, quite legitimately, having had any knowledge of them.[11]

Conversely, one can have responsibility without being held responsible; for instance if one was too young, too inexperienced, or out of one's depth. To take another British example, the initial responsibility for handling the disastrous Grenfell Tower fire in West Kensington in 2017 fell to a relatively junior watch officer, who had no experience of high-rise fires and had been poorly informed about the particular circumstances of that building. As became clear from reactions to his heartfelt testimony in the Public Enquiry that followed, it is very plausible that though he certainly had responsibility for the fire-brigade's initial response, he shouldn't be held responsible for it, perhaps because he shouldn't have been given that responsibility in the first place.

Even when someone both has responsibility and is properly being held responsible, there is no guarantee that the two will be proportional. Both come in degrees and those degrees are independent. If I have given responsibility for you to you, the two of us together may be held responsible for the outcome. The link between the two notions is not an analytic one, but a substantial moral one: it a substantial moral claim that someone should be held responsible to the degree that they are responsible. While we accept that there is much that is attractive in that claim, the examples that we have given should show that it is far from obvious that it is true.

So when we talk of responsibility for uncertainty in medicine we mean it primarily in terms of having responsibility, rather than of being held responsible for it. To take responsibility for the uncertainty is to manage it. This involves both an epistemic dimension, and a practical one. The person who manages the uncertainty needs to keep abreast of the information they have, and to handle it properly—no ignoring of base rates. And they need to act on this information, and the inferences they draw from it in the right way. And in addition there is also an emotional load. As we saw earlier, uncertainty can be emotionally draining. Doctors are typically better placed than their patients to cope with this: better informed and less involved. But we need to be aware of the costs.

Should the doctor also be held responsible for these decisions? We remain uncommitted, since there are difficult issues here. Some concern the issue of who is to hold them responsible—the patient? the family? the authorities? Others concern what we earlier called the contingency of regret. Suppose that a doctor does take responsibility for uncertainty, decides not to follow up a remote

---

[11] Even in the UK, this seems to be little honoured nowadays: the standard example concerns Austen Chamberlain's resignation after the disastrous siege of Kut in WWI; somewhat more recent (though less clear cut) is the resignation of Lord Carrington, the British Foreign Secretary, after the invasion of the Falkland Islands. More recently UK ministers have tended to follow the practice of assuming that if they can show they did not know what was going on, this absolves them of being held responsible—that is, in our terminology, they think that if they did not have responsibility, they cannot be held responsible. In this they follow most other jurisdictions.

possibility, and things go horribly wrong because of it. It is the stuff of medical nightmare. No wonder there is an emotional cost to taking responsibility. Should the doctor be held responsible for the outcome? They can of course justify their decision in terms of the expected outcomes, and that is certainly what they would be likely to do if there is any malpractice issue. Suppose though there was no malpractice: a reasonable decision was made, and the patient was unlucky.

In retrospect, of course both doctor and patient will wish that the remote possibility had been followed up. The doctor will doubtless feel agent regret—we would be concerned if they did not.[12] Designing policy to avoid such cases leads to defensive medicine with all its costs, which might suggest that the doctor simply not be held responsible. But that in turn brings concerns about accountability, and the incentive structure that such a system would produce. We have, in short, a difficult tangle of both moral and policy issues that we do not seek to untangle here. How to respond in such circumstances—who, for instance, should pay to ameliorate the damage—and whether this should be understood following a broadly 'civil' or 'criminal' model, is a question for a very different paper.[13]

10.4 Justification

We have tried to describe what it is to take responsibility for uncertainty; but we have said little about the circumstances under which it might be defensible. The question here is tied to the broader one of when a doctor can be justified in not telling a patient something, a topic that has generated a fair bit of discussion.[14] We start with some cases, keeping the discussion focused as far as possible on the specific issues around uncertainty.

Case 1: A woman in her late 30s presents to the hospital with a cough and some chest pain. She is worried that she might have a clot on her lung—she is in the early weeks of a pregnancy. She is fit and active, and has walked up the stairs to the clinic without difficulty. On further discussion, it transpires that she has had some general aches and pains and a slight fever; her husband has had similar symptoms. Her blood tests are consistent with a viral infection (and she is Covid negative). The consultant simply tells her that she does not need a scan, since it is very unlikely that she has a clot, and the scan carries a risk for both her and the foetus. The consultant reassures her that if she herself were in this position, she would not be having a scan. She tells her to come back if it gets worse, and gives her the number of the clinic to call.

---

[12] Williams (1981).   [13] For a good discussion of the distinction see Cane (2002).
[14] See especially the discussion around nocebos, for instance, Wells and Kaptchuk (2012); Cohen (2014).

Note what is happening here. The consultant does not just *offer* a course of action to the patient, or make a *suggestion* as to what to do. Rather she makes what Stivers et al. call a *pronouncement*: she tells the patient that she does not need a scan.[15] There is a reason for this: the patient is anxious, the consultant seeks to reassure, and by taking charge in this way she does so. A suggestion or an offer— 'What if we send you home without a scan?'; 'We could skip the scan; what do you think?'—would be less effective as assurance. It remains open to the patient to insist that she wants a more thorough investigation, but that would require an active intervention, taking back the responsibility. The consultant is responding to clinical considerations, but, if she is doing her job well, she is responding to interactional ones too. She has judged that the patient needs the kind of guidance and reassurance that her pronouncement provides, and that it would be welcome, not an unwanted imposition. Such judgements are subtle, and it is all too easy to get them wrong, but they are the stuff of day to day medicine.

Case 2: A 90-year-old patient enters a hospice on their GP's referral. They have a cancer that they have chosen not to have treated. They expect this to be their last few weeks, and make their plans and set their mind accordingly. The consultant, who is less convinced than the GP that their cancer is so advanced, tells the patient that, unless their condition deteriorates rapidly, they will be sent home after a couple of weeks of observation. The patient spends the week in a state of anxiety— fearful that they are a fraud, and fearful that they will be sent home—before it becomes clear that the cancer is well advanced, and there is no question of them being sent home.

This case brings more complications. Clearly the patient had built up an understanding with the GP, on the basis of which they had moved into the hospice. The consultant has sought to construct a rather different relationship, one based on a more explicit presentation of the possibilities. We do not argue that they were obviously wrong to do so; there might have been something important to be gained from the greater transparency. But given the predictable distress that ensued from it, we do suggest that it would have been at least permissible to have maintained something like the GP's attitude, at least until the course of the disease was clearer. That would have been to take on the responsibility for uncertainty from the GP, rather than giving it back to the patient.

Case 3: A 27-year-old man is admitted to hospital after a motorcycle accident. He has suffered a catastrophic brain injury; while he is not 'brain dead' there is very little brain activity that can be picked up on the many scans that have been taken. He is on a ventilator, which is keeping him breathing. The neurosurgeon looking

---

[15] Stivers et al. (2017).

after him cannot reliably put a number on his chances of meaningful recovery. He doubts they would be as good as 1/100; he would hazard a guess at 1/1000. From past experience he knows that if he were to give any such figure to the family they would be more likely to want to keep him on the ventilator than if he were simply to say that he thinks it is almost certain that he won't make a meaningful recovery; and he knows what existence is like for the 999 who do not. So he has a blanket policy of not giving probabilities when the chances are small. He simply tells the family that he is very sorry, but he doesn't think the patient will recover.

Here again the neurosurgeon has taken responsibility for the uncertainty. Here though, we are less convinced of the rightness of the approach, and others have shared our concern. The issue is the blanket policy. It is one thing to negotiate an understanding with a patient or with their family, and on the basis of this to take on the responsibility for the uncertainty. It is quite another to have a unilateral policy of doing so, even if this is motivated by legitimate beliefs about the best outcome.

We tentatively propose then that the key requirement for taking responsibility for uncertainty stems from a relationship with the patient, or, if they are in no position to form it, with those close to them. We think it is best characterized as a relationship of trust. To start to spell it out, let us return to the issue of informed consent.

10.5 Informed Consent, Informed Trust

At this point, there are two obvious approaches that we might take. One is to say that, in some of the cases described above, a doctor can be morally justified in taking responsibility for uncertainty; this is incompatible with informed consent, since by hypothesis the patient isn't properly informed; and hence informed consent must give way as a necessary condition. The other is to say that taking responsibility for uncertainty is really quite compatible with informed consent, once the latter is properly understood: it is just that the patient needs to be informed at some more abstract level, and then to consent to that. In fact our approach runs somewhere between the two. We do think that there is an understanding of what it is to be informed that justifies the claim that there is informed consent in these cases. But rather than being the touchstone of good medical practice, the corresponding notion of informed consent must find its place amidst a welter of different requirements that need to be suitably balanced. Most centrally, we think that we need a notion of *informed trust*.

We start with Manson and O'Neill's influential discussion in *Rethinking Informed Consent*. One of their targets is the 'conduit/container model' of information flow: the idea that, in a consultation, information held by the doctor flows to patient. The patient needs to understand the information for the flow to work;

but once it has worked, the patient is able to provide informed consent, and this is the basis for legitimate treatment.[16]

Manson and O'Neill think that the conduit model is hopeless: it cannot be universally applied, and even where it can it doesn't give the right results. In its place they propose an 'agency model'; their stress is on the need to recognize not just what is said (the speech content) but also what is done (the speech act). We want to explore this idea further. A consultation is indeed not just a transfer of information. Like any conversation it involves a set of illocutionary acts (Manson and O'Neill's speech acts), and a set of perlocutionary consequences. Assertion is one such illocutionary act; a change in the hearer's belief—a coming to be informed—is a typical perlocutionary consequence. Consenting is also an illocutionary act; the creation of an agreement, or a contract is a typical perlocutionary consequence. Focusing just on those gives a single model for a successful consultation: the patient informs the doctor; the doctor informs the patient; this continues until all the necessary information has been transferred. The doctor then proposes a treatment. The patient is then in a position to give, or to withhold, informed consent.

But now consider some of the huge range of other illocutionary actions, and other conversational interactions, that occur in a typical consultation, and the perlocutionary states that result. These have been well documented in sociolinguistics. They include:

- questioning
- advising
- reassuring
- warning
- offering
- bargaining.

A co-operative conversation involves the attempt by all parties to understand each other, and from this to build an agreed understanding: a common ground. If things go well, from all of this comes a growth in trust: each of the parties comes to trust the other. Inviting trust is not typically a direct illocutionary act: it is possible to say 'Trust me', but this will often be either redundant or ineffectual. Instead trust grows through the variety of other interactions. If a patient and doctor have worked together for years, then it can be largely grounded in a history of effective treatment. But if no treatment has happened yet, the trust must be grounded in the conversations around diagnosis: in the kinds of speech interactions that we have been discussing up till now.

---

[16] Manson and O'Neill (2007, esp. ch. 4).

Our basic contention is this: for the trust to be legitimate, both patient and doctor need to be relevantly informed. That is, they need to be informed to the point that can legitimate that trust. Call the resultant trust 'informed trust'. Clearly you cannot build legitimate trust on self-serving lies. Nor can you build it by withholding information on self-serving grounds. But the whole point of trust is that it grants a degree of discretion to the one you trust, and that discretion will mean that very often there will be important things that they know and you don't. This happens all the time, in perfectly routine cases. You trust a professional to fix some electrical thing in your house whose workings you do not understand. You trust your partner to organize a surprise event for your birthday; telling you would spoil the surprise. Or you trust them to take the children rock-climbing without telling you quite what they are doing, because the whole idea of the children exposed on the mountain terrifies you.

Do you have informed consent in these cases? Not really. You consent, but your consent is not fully informed. Because you trust the person, you consent to them doing things about which you are not informed. But the trust itself has to be informed: you need to know that your electrician knows what they are doing; that your partner will organize a party that, even if it doesn't delight you, will not appal you when it happens; and that they will be responsible on the mountains, even if they do things with the children that you would never dream of doing.[17] Note though that there are differences between the cases: when you trust the electrician, you are primarily trusting them to be technically proficient. When you trust your partner you are trusting them to make decisions that are much broader than that.

We think that in good medical practice things will be similar with the relationship between patient and doctor; here again we have been much influenced by Manson and O'Neill's discussion (2007: ch. 7). The doctor trusts the patient: trusts them to be honest in their reporting of symptoms, trusts them to keep to the treatment, to make sensible decision on when to call if things start to go wrong. Equally, the patient trusts the doctor. This has been much more discussed in the medical literature, but too often the trust has been talked about as though it is just trust in technical proficiency: the patient trusts the surgeon to operate on the right organ, complete the procedure, not leave their instruments inside them, and so on. But as we hope our discussion has shown, much of the trust is more like that shown between partners. The patient will often accord much discretion to the doctor: to decide on treatment, to handle the uncertainty, perhaps to act as a custodian of information, selecting and sharing that which is in their interests, and withholding information which they think will not be in their interests to know. Here there should be consent, but again it will not always be fully informed consent. What needs to be informed is the trust.

---

[17] Of course, if things go wrong it might turn out that your views on what was responsible will come apart—but this, we think, is another instance where we shouldn't understand things in terms of what happens in case of failure. It may well be that the failure will change your attitude to the whole thing.

As with relations between partners, there is no uniform level of discretion that will suit all cases. Some patients will want more control than others, as indeed will some doctors; one of the functions of the verbal interactions we have been talking about is to arrive at the right level here.

10.6 Shared Decision-Making, Shared Understanding

We turn finally and briefly to shared decision-making. There has been a great deal of enthusiasm to involve patients in the decision-making process.[18] As that approach has been developed, the idea has been that to truly share a decision, both parties must be equally well informed. Sometimes that is achievable, and desirable, and many of the proponents of shared decision-making have convincingly argued that we often give up on it too early: patients are put off by a lack of self-confidence, or a culture of deference.

But in many of the cases we have been talking about it is neither possible nor desirable for all of the parties to be equally well-informed; that has been our theme. So in so far as full shared decision-making requires full information, it will not be possible. What is achievable, and desirable, is a shared understanding of how the labour will be divided: of who will manage the uncertainty, who will make the decisions, and so on. At that point there should be shared decision-making—a shared decision on the division of labour. But that may equally be a decision about where decisions will not be shared.[19]

10.7 Conclusion

We have stressed here the role of informed trust in legitimating a doctor's taking responsibility for uncertainty. We could equally have stressed the role of informed trust in many other medical contexts; we do not mean to suggest that it only has application here. But this will do for now.

References

Boyd, E., and Heritage, J. (2006), 'Taking the History: Questioning During Comprehensive History-Taking', in J. Heritage and D. Maynard (ed.), *Communication in Medical Care* (Cambridge: Cambridge University Press), 151–84.

Cane, P. (2002), *Responsibility in Law and Morality* (Hart).

---

[18] See for instance Elwyn et al. (2012); Elwyn, Edwards, and Thompson (2016).
[19] We are grateful to Richard Lehman for discussions here.

Cane, P. (2016), 'Role Responsibility', in *Journal of Ethics* 20: 279–98.

Cohen, S. (2014), 'The Nocebo Effect of Informed Consent' in *Bioethics* 28/3:147–54.

Elwyn, G., Edwards, A., and Thompson, R. (2016), *Shared Decision-Making in Health Care: Achieving Evidence-Based Patient Choice* (Oxford University Press).

Elwyn, G., Frosch, D., Thomson, R., et al. (2012), 'Shared Decision-Making: A Model for Clinical Practice', in *Journal of General Internal Medicine* 27: 1361–7.

Elwyn, G., Laitner, S., Coulter, A., et al. (2010), 'Implementing Shared Decision-Making in the NHS', in *BMJ* 341: c5146.

Frick, J. (2015), 'Contractualism and Social Risk', in *Philosophy and Public Affairs* 43: 176–223.

Gardner, J. (2007), *Offences and Defences* (Oxford University Press).

Gigerenzer, G. (2008), *Rationality for Mortals* (Oxford University Press).

Hart, HLA. (1968), *Punishment and Responsibility* (Oxford University Press).

Heritage, J., and McArthur, A. (2019), 'The Diagnostic Moment: A Study in US Primary Care', in *Social Science & Medicine* 228: 262–71.

Ingelfinger, FJ. (1980), 'Arrogance', in *New England Journal of Medicine* 303/26: 1507–11.

Jutel, A., (2014), 'When the Penny Drops: Diagnosis and the Transformative Moment', in A. Jutel and K. Dew (ed.), *Social Issues in Diagnosis: An Introduction for Students and Clinicians* (Johns Hopkins University Press), 78–92.

Liddell, K., Skopek, JM, Le Gallez, I., and Fritz, Z. (2022), 'Differentiating Negligent Standards of Care in Diagnosis', in *Medical Law Review* 30/1: 33–59.

Manson, N. and O'Neill, O. (2007), *Rethinking Informed Consent in Bioethics* (Cambridge University Press).

Montgomery v. Lanarkshire Health Board [2015] UKSC 11 §87, per Kerr and Read.

Pike, LO. (1900), *Year Books of the Reign of Edward III, Trinity Term of the 16th Year* (1342), trans. and ed. Luke Owen Pike (Cambridge University Press).

Robinson, JD., Tate, A., and Heritage, J. (2015), 'Agenda-Setting Revisited: When and How do Primary-Care Physicians Solicit Patients' Additional Concerns?', in *Patient Education and Counseling* 99/5: 718–23.

Shafer-Landau, R, and Feinberg, J. (2016), *Reason and Responsibility: Readings in Some Basic Problems of Philosophy*, 16th edn. Cengage Learning.

Shoemaker, D. (2011), 'Attributability, Answerability, and Accountability: Toward a Wider Theory of Moral Responsibility', *Ethics* 121: 602–32.

Smith, A. (2012), 'Attributability, Answerability, and Accountability: In Defense of a Unified Account', in *Ethics* 122: 575–89.

Stivers, T., Heritage, J, Barnes, RK., et al. (2017), 'Treatment Recommendations as Actions', in *Health Communication* 33/11: 1335–44.

Suikkanen, J. (2019), 'Ex Ante and Ex Post Contractualism: A Synthesis', in *Journal of Ethics* 23/1: 77–98.

Waller, B. (2011), *Against Moral Responsibility* (MIT).

Watson, G. (2004), *Agency and Answerability* (Oxford University Press).

Wells, RE., and Kaptchuk, TJ. (2012), 'To Tell the Truth, the Whole Truth, May Do Patients Harm: The Problem of the Nocebo Effect for Informed Consent', in *American Journal of Bioethics* 12/3:22-9.

Williams, B. (1981), *Moral Luck* (Cambridge University Press).

11

Physician, Heal Thyself

Do Doctors Have a Responsibility To Practise Self-Care?

Joshua Parker and Ben Davies

11.1 Introduction

Burnout among health care professionals is at epidemic proportions.[1] Global rates of burnout are high, although they vary by country and speciality (Lemaire and Wallace 2017). In one study of UK specialists in Obstetrics & Gynaecology, 35 per cent of respondents met the criteria for burnout. Comparable findings have come from Europe and the USA (Bourne et al. 2019). A 2019 survey by the UK's General Medical Council (GMC) suggests that over a third of junior doctors find their work emotionally exhausting, while nearly a quarter feel burnt out. More starkly, the British Medical Association (BMA) has found that 80 per cent of doctors are at high or very high risk of burnout (BMA 2019). Predictably, burnout worsened during the Covid-19 pandemic, with Azoulay et al. (2020) finding that just over half of 1000 respondents worldwide showed signs of "severe" burnout.

Although not classified as a medical issue, burnout was included in the most recent revision of the World Health Organisation's International Classification of Diseases, which defines it as a syndrome resulting from chronic workplace stress, characterised by:

- feelings of energy depletion or exhaustion;
- increased mental distance from one's job, or feelings of negativism or cynicism related to one's job; and
- reduced professional efficacy. (World Health Organisation 2019)

Burnout has implications for patients since it is associated with poorer care (Linzer 2018); greater error (Fahrenkopf et al. 2008); lower patient satisfaction (Vahey et al. 2004); and lower physician empathy (Brazeau et al. 2010). Health

---

[1] Thanks to Gabriel De Marco and Neil Levy for helpful comments on this chapter.

Joshua Parker and Ben Davies, *Physician, Heal Thyself: Do Doctors Have a Responsibility To Practise Self-Care?*
In: *Responsibility and Healthcare*. Edited by: Benjamin Davies, Gabriel De Marco, Neil Levy, Julian Savulescu,
Oxford University Press. © Edited by Benjamin Davies, Gabriel De Marco, Neil Levy, Julian Savulescu 2024.
DOI: 10.1093/oso/9780192872234.003.0012

care systems also feel the effects of burnout through staff absence and problems with retention in addition to consequences for productivity (Iliffe and Manthorpe 2019). Staff sickness rates in the UK's National Health Service (NHS) are double those in the private sector (ONS 2018).

One response to doctors' work-related suffering has been to emphasise strategies focusing on what individuals can do, including self-care training, stress management, mindfulness-based approaches, and small support groups (West et al. 2016). These approaches share a focus on tackling mental health problems, including burnout, by bolstering individual resilience to the stresses of modern medical practice.

Health Education England describes resilience as "the ability to bounce back—a capacity to absorb negative conditions, integrate them in meaningful ways, and move forward" (Oliver 2017). Medical schools are encouraged to select applicants who demonstrate resilience (Medical Schools Council 2018) and to provide resilience training to students (Minford and Manning 2017). Indeed, resilience training has been suggested as a solution to concerning numbers of medical student suicides (Munn 2017).

This focus on individual resilience through "self-care" as a reaction to burnout—and its threats to patient care and staff turnover—has culminated in a codified professional responsibility. Following his NHS Health and Well-Being report (Boorman 2009), lead reviewer Steven Boorman claimed that: "What we have got to do is raise awareness that [a doctor's] own individual health is your own responsibility. We are calling for basic training to include that—even for surgeons" (Eaton 2009). Others state that "doctors' well-being is integral to professionalism", proposing a "professional responsibility [for doctors] to be at their best" (Lemaire and Wallace 2017). One apparent implication of this approach is that burned-out doctors are not at their best, and that this is an individual failure of professionalism.

Doctors' well-being has been incorporated as a professional responsibility into the GMC's outcomes for newly qualified doctors. The GMC (2018) admonishes new doctors that they "must demonstrate awareness of the importance of their personal physical and mental wellbeing and incorporate compassionate self-care into their personal and professional life". Similarly, when the Royal College of Physicians (Tweedle et al. 2018) sought to outline what it means to be a health care professional in the twenty-first century, they included self-care as integral to modern professionalism: "patients and the public must believe that doctors put their interests first. To achieve this, doctors must care for themselves. *A basic building block of professionalism is to understand and manage oneself*" (emphasis added).

All this represents one possible approach to clinician welfare: first, that it is primarily the clinician's own responsibility, perhaps even a moral responsibility that they owe to their patients and colleagues. Second, that this entails a further responsibility to practise self-care as a way of avoiding burnout. An alternative view sees institutions as primarily responsible (Lancet 2019). However, such approaches tend to emphasise a broadly consequentialist case for an institutional focus, again centred primarily on patient welfare.

Finally, some organisations appear to endorse both approaches. We have noted that the GMC emphasises personal resilience as an outcome for graduates; but in the report *Caring for Doctors, Caring for Patients*, the organisation emphasises institutional responsibilities (West and Coia 2019). Still, these ideas are found in different documents and, as far as we are aware, no attempt has been made by the GMC to explain how these different responsibilities relate to one another.

This chapter explores the idea of self-care as a professional responsibility for doctors. We begin by charting the grounds of a professional or moral responsibility of self-care. Following this we suggest two concerns around the individual focus: instrumentalisation and demandingness. To put things another way, we suggest that even if there exist grounds for a responsibility of self-care, this must be balanced with attention to doctors' rights qua employees. Finally, we consider the potential scope of a responsibility of self-care given the limited control individual doctors have over their own mental health.

Our discussion is centred on the health care system that we know best, the NHS. We make two types of claims in this chapter: descriptive claims about the evidence for thinking that the NHS may make its medical employees perceive a professional duty to avoid burnout, and normative claims about whether such a duty exists, and should be communicated to employees. The former type of claim is restricted to the NHS; we do not suggest that other health care systems face similar issues, though they may do. The normative claims, we think, apply more widely.

We also focus our discussion on doctors, particularly training documents and guidance given to doctors. Again, this is a question of familiarity (one of us is a doctor in the NHS), but we believe that our normative claims apply beyond these specific documents (which are likely to change over time) and also beyond the doctor role. Indeed, they may be relevant outside the medical sector, including teachers, university staff, prisons and probation officers, and any other profession where there is a direct duty of care which may be compromised by burnout among professionals. We think, however, that there is a particular problem in medicine because of the extremely moralised nature of discussion about medical professionals' professional responsibilities. Additionally, an important aspect of doctors' identities is their professional identity. Professional norms become embedded within these identities and so, if becoming burned out were a violation of professional norms and thereby their professional identity, this would be especially significant for doctors. Although this is not unique to medicine, the serious impacts of medical work mean that every aspect of medical professionals' behaviour is ripe for moralisation since it could potentially have significant implications for patients.

Finally, while we focus on a responsibility of "self-care", where convenient we speak of a "responsibility to avoid burnout" and a "responsibility to be resilient". These terms are different, but we see their connection as follows: according to the individualistic view of responsibility, physicians have a responsibility to avoid burnout because this compromises patient care. In order to do this, physicians must develop resilience against the causes of burnout, and one primary way of doing so is to practise self-care. Thus, while these terms are not synonymous, and one might posit one of them without the others, they are sufficiently tightly linked in this case to be able to use them interchangeably.

11.2 Responsibilities of Self-Care

When reflecting on a professional responsibility to practise self-care, one way of understanding the idea of "responsibility" is as a form of role responsibility. To say that doctors have a role responsibility means that they are entrusted to exercise a degree of agency and judgement to meet the demands of the role that they occupy (Goodin 1987). As Cane (2016: 281) points out, role responsibility has received considerably less theoretical attention than "liability", a form of responsibility that depends on a person bearing, inter alia, a certain causal relationship to an outcome. In contrast, one can bear role responsibility for an outcome without being causally responsible for it. For instance, senior clinicians may have responsibility for the actions of their juniors. Doctors' role responsibilities are derived from the specific functions they perform, which are informed by the expertise they hold and the goals of medicine they serve. Some specific role responsibilities will depend on medical specialism, and include making diagnoses, suggesting appropriate treatments, performing surgeries and, most importantly for our discussion, caring for patients.

In the next section we consider whether doctors have a role responsibility to practise self-care and avoid burnout. We begin by considering the responsibilities engendered by doctors' formal professional codes and broader professional norms. We then consider a broader picture, on which general moral principles which apply to everyone might generate a responsibility for doctors to avoid burnout because of the specific tasks they are employed to do.

11.2.1 The Grounds of a Responsibility of Self-Care

This subsection outlines several ways in which one might ground doctors' responsibility to practise self-care. We do not come down on the side of any particular approach. Rather, we outline various approaches—of varying plausibility—which would establish *some* resilience-related responsibilities. Following this, our main focus is on the limits of any such responsibility, and on its communication, explored in Section 11.3.

"Professionalism" plays an increasing role in contemporary medicine. Despite this, the nature of professionalism and its significance are contested. Many see professionalism as growing out of something like a social contract (e.g., Cruess and Cruess 2008). Professionals have a monopoly over a body of knowledge and certain skills. They are granted autonomy to use these in the pursuit of good, and on the basis of social trust. To maintain this trust, professionals act in accordance with ethical standards and regulatory frameworks. Broadly, there are three main positions on the ethical standards to be upheld in order to meet standards of professionalism. One perspective is that professionalism entails following certain rules and principles that represent the applied morality of the profession (Cruess and Cruess 1997; Cruess et al. 2002), but which do not apply outside that profession. Thus, failures of professionalism are failures to follow the rules. A second view is that to be a professional is to embody a role-specific set of characteristics or virtues (Pellegrino and Thomasma 1993). On this view, the professional doctor is the good doctor; the good doctor is the virtuous doctor. Finally, a third view is that detailed, specific professional duties for doctors can be derived from the application of moral duties that apply universally, by virtue of their specification via doctors' proper professional tasks. As we suggest, these views are not equally plausible; nonetheless, we consider each in turn.

Turn first to the "rule-based" view of professionalism. Many documents refer to the importance of doctors' well-being and resilience (Richards and Lloyd 2017; Tweedle et al. 2018). As we have suggested above, some do seem to suggest explicit professional duties of self-care. Moreover, rules can be informal as well as formal. Employees in various professions may be subject to "soft" expectations from within their professional culture that are no less powerful, and no less subject to penalty, than formalised orders. There are sufficient references to the importance of resilience and its connection to quality patient care (in academic discussions and policy on doctors' professional duties) for there to exist an implicit rule that doctors should be at their best. Take the canonical statement of doctors' duties, the GMC's (2013) *Duties of A Doctor*, which outlines the professional duties under-pinning good medical practice. The GMC's instruction to "take part in activities that maintain and develop your competence and performance" might—especially when combined with statements from professional bodies about the role of self-care (GMC 2018; Tweedle et al. 2018)—reasonably be taken to indicate to doctors that if activities that build resilience will maintain their performance, then they have a responsibility to perform a sufficient number of those activities to become resilient, that is, a responsibility of self-care (GMC, "Good medical practice").

Doctors also operate in a professional culture that encourages over-work, self-sacrifice, perfectionism, and emotional detachment. While these are likely to fuel burnout, they may also underpin cultural norms that emphasise resilience as a personal virtue. In this way, expectations of doctors, semi-formal statements about professionalism, and doctors' own reasonable inferences about how formal rules might be applied create a culture in which doctors are informally assumed to have responsibilities of self-care and resilience. According to a slightly looser interpretation of the "follow the rules" approach, this may be sufficient to generate a professional responsibility of self-care.

However, a perspective on professional responsibility that is so radically divorced from broader morality is implausible. It may be true that professional roles can generate responsibilities to act, or refrain from acting, in ways that cut against broader moral principles. For instance, duties of confidentiality might speak against reporting even quite serious crimes if they were confessed in relevant circumstances. Yet this does not mean that a professional code should be wholly divorced from broader morality, and that a member of a particular profession's responsibilities must conform with all of, and only, the rules and informal expectations of that profession. At least, a standard of reasonableness applies; codes and cultures should be subject to intentional change in the realisation that they go awry or deviate too radically from broader morality. Such changes seem impossible or at best arbitrary on the rule-following view, since on that view professional obligations are entirely determined by the rules, and not by external standards. Thus, the rest of this section canvasses two approaches which we think are more plausible. We do not adjudicate between these views, but show that each can ground a limited responsibility of self-care.

An alternative perspective on professionalism is that it involves the "cultivation of professional virtues, essential to the ethical concept of medicine as a profession" (Doukas et al. 2013). This perspective can itself be understood in two ways. First, one might take a character-focused approach: to be virtuous is to possess or cultivate particular character traits, without reference to particular ends. While the virtues in question are debatable, typical suggestions include fidelity to trust; benevolence; compassion; intellectual honesty; courage; and truthfulness (Pellegrino 2002). None of these suggest a straightforward prospective responsibility of self-care to avoid burnout; but such a virtue could in principle be added to the list.

An alternative virtue-based view of professionalism involves first determining what is good, and then outlining the virtues that are necessary to, or increase the likelihood of, achieving those ends. Edward Pellegrino, following Aristotle's argument that excellence in certain activities is to pursue the good, argues that "the well-being of the patient is the good end of medicine and of the physician's art and action" (Pellegrino 2001). On this view, clinical excellence means having the virtues necessary to achieve medicine's ends, which are found by considering the intrinsic nature of medicine as a practice.

One might think, therefore, that the professional doctor is one who does what they reasonably believe is necessary to provide good care. It follows that if part of what is required for performing at this level is a degree of resilience or well-being then, adopting strategies of personal resilience may be required for virtuous doctors (Dunn 2016).

A final approach is to set professionalism aside and question whether more general considerations of ethics can justify a responsibility for doctors to care for themselves. On such a view, although we are all governed by broadly the same moral rules, those rules recommend different actions and generate more specific responsibilities depending on our individual, including professional, circumstances. A central tenet of morality is to refrain from harming others. In medical ethics, this is the principle of non-maleficence (Beauchamp and Childress 2001). One might think, therefore, that if there is a causal connection between burnout and harm to patients, then the general duty to avoid harm might give rise to a more specific responsibility for doctors to avoid burnout.

Discussions of a doctor's duty of non-maleficence rarely occur in the absence of reference to beneficence. This is because, taken in isolation, non-maleficence would imply therapeutic nihilism. While the two principles are distinct, in practice they are typically considered together, with harms assessed in light of potential benefits. If self-care averts harms to patients, a duty of non-maleficence suggests that doctors ought, *ceteris paribus*, to engage in self-care. If being burned out risks harm to patients, then doctors' professional responsibilities include taking reasonable steps to avoid it, including strategies of personal resilience. These responsibilities must be assessed in light of potential benefits. This suggests that doctors have a responsibility to reduce the risks that burnout poses to their patients, unless they can benefit the patient more by practicing while burned out, or if burnout did not significantly compromise care quality (e.g., because a doctor was able to continue functioning despite exhaustion or significantly increased cynicism).

Each of the three views on professional responsibilities that we have considered can, we think, support a responsibility to practise self-care.

11.3 Responsibilities and Rights

The previous section suggests that support can be found for self-care or resilience being a professional responsibility. However, this does not mean that doctors' self-care should be seen *primarily* as an individual professional responsibility, or that it is reasonable for health care institutions to frame their members' or employees' mental health as a professional responsibility. To challenge these further claims we raise three objections. Section 11.3.1 suggests that the current focus on doctors' responsibility for their own well-being, particularly if framed as an obligation to avoid burnout, risks instrumentalising doctors and their well-being. Section 11.3.2 suggests that a *sole* focus on doctors' responsibilities ignores important elements of the context in which burnout occurs, and obscures the responsibilities of *institutions* in protecting doctors' welfare. Finally, Section 11.3.3 argues that adding an additional responsibility to avoid burnout to doctors' existing professional obligations may be excessively demanding, and thus counterproductive.

Our discussion thus far has intentionally replicated some significant problems with professional discussions of doctors' well-being and its relationship to patient care. We have considered the question of whether doctors might reasonably be thought to have a specific moral responsibility without consideration of how this potential responsibility fits into the broader moral landscape of the doctor's professional role. In particular, any discussion of whether someone has a particular responsibility must be situated in the context of other responsibilities, and of rights. Each of these considerations gives rise to a serious concern around an individual responsibility of self-care.

Consideration of rights gives rise to two concerns. First, a focus on resilience as a professional responsibility risks instrumentalising doctors' well-being, viewing (or at least presenting) them as valuable only insofar as they can benefit patients. Second, a focus on individual responsibility ignores the fact that doctors are employees. Individual professional responsibilities held by specific doctors qua employees must sit alongside an account of the responsibilities of their employer, and thus the corresponding employee rights held by doctors.

Consideration of other professional responsibilities generates a further concern. The arguments outlined above suggest that, since burned-out doctors compromise patient care, and doctors' professional or broader moral responsibilities require that they do *not* avoidably compromise care, we can straightforwardly assume a responsibility to avoid burnout. But this ignores the fact that doctors are already working with numerous professional responsibilities, and that there are moral and practical costs to explicitly adding another. The worry is that by ignoring the broader context of professional responsibilities which doctors already navigate, an isolated consideration of a responsibility to be resilient risks being overdemanding.

11.3.1   Instrumentalisation

Theories of well-being attempt to answer the question of what is good for a person, or what makes their life go well. While there are various competing theories in this area (Parfit 1984), burnout seems unambiguously bad for the person experiencing it on any plausible theory. The negative psychological states involved in burnout erode subjective well-being, undermine one's ability to perform a socially valuable role, and can cause a significant amount of mental distress. Attempts to stave off burnout are thus also ways of maintaining or protecting one's own well-being. Conversely, efforts to bolster or improve well-being may also be ways of avoiding burnout. Insofar as burnout compromises one's ability to practice medicine effectively, doctors with higher well-being will typically be better doctors.

One popular analogy for exploring physician burnout has been that of an aircraft emergency, following the familiar instruction to "put on your own oxygen mask before assisting others" (Yates 2020). This is a powerful image, and it is easy to see why it is frequently referenced in discussions of physician well-being. If doctors don't look after themselves, how can they be expected to care for others? This is emblematic of approaches to physician well-being where the motivation to make doctors better is to make them better doctors. Its significance is the implication that the reason to improve doctors' well-being is instrumental, with no mention of any intrinsic reason to do so. That such intrinsic reasons are not mentioned does not, of course, mean that those who use this analogy do not also believe in them. But, as we will argue, there is an issue of priority in focusing on the instrumental value of doctors' well-being.

Our view is that the primary value of doctors' well-being is intrinsic. It is important to be clear that in saying this we mean to invoke two forms of value. The first is the prudential value of an individual's well-being for herself. A person's well-being just *is* how well their life goes and, as such, has intrinsic prudential value for that individual. This is consistent, of course, with individuals not caring about their own well-being except insofar as it makes them instrumentally valuable. Perhaps some value their own well-being primarily in instrumental terms and find this unproblematic. We do not wish to insist a priori that such an individual is unreasonable: perhaps they are a committed utilitarian, or a particularly saintly altruist. In practice, however, we worry that valuing oneself in this way—that is, in primarily instrumental terms—is *typically* a sign that something has gone wrong, for example, with one's self-esteem or self-conception. Additionally, measuring one's own value in primarily instrumental terms may (though need not necessarily) lead one to fall prey to one of the problems of instrumentalisation we outline below. These risks are sufficient to worry considerably about any person who sees themselves this way.

The second thing that we mean when we say that the primary value of doctors' well-being is intrinsic concerns how others—particularly those with some influence on how individual doctors are treated, such as government officials, institutional managers and employers, and, to a lesser extent, patients—see the doctor. Unlike the case of how doctors see themselves, we do not think that there is any plausible justification for *other people* seeing a doctor in solely, or even primarily, instrumental terms. So, one idea we mean to invoke when we say that a doctor's well-being is of primarily intrinsic value is this: even if relevant others see the (undeniable) instrumental value of doctors' well-being, it is important that they focus primarily on the intrinsic value of that well-being. The reason for this is that seeing the individual herself as being of intrinsic value implies seeing her well-being as having intrinsic value. It is, we think, incoherent to accept that another person has intrinsic moral value, and yet view her well-being as being solely or primarily of instrumental value.

Thus, one problem with a focus on avoiding burnout as a professional responsibility is that it risks focusing on doctors' well-being in primarily instrumental terms. In other words, it risks "instrumentalising" doctors and their personal welfare.

We do not wish to suggest that "using" doctors to benefit patients is immoral; such a suggestion would be troubling for the operation of a health care system. Rather, our claim is twofold. First, against the valuing of doctors' well-being solely as a means to promote patient care, there is a familiar Kantian objection: that we ought not treat one another as *mere* means. To fully intrumentalise another person—that is, to value them and their welfare solely for what it can deliver you or for someone else—is to treat them with a form of disrespect inconsistent with their moral value.

However, as well as canvassing instances where doctors are said to have a responsibility to avoid burnout, we have also mentioned some examples where doctors' welfare is discussed in ways that seem to involve intrinsic valuing. We discuss this in greater detail shortly. One idea one might draw from this, though, is that even if it is wrong to wholly instrumentalise doctors by treating them as a mere means to quality care, it is not wrong to partially instrumentalise them by treating them as *primarily* a conduit for patient care. For instance, it does not seem wrong for a patient to think of a doctor primarily in these instrumental terms, even if there are clear limits (informed in part by the doctor's own worth as a person) on what the patient may demand. This is, however, why we have separated out the claims of patients from those of others who interact with doctors. Where it may be permissible for patients to view a doctor in primarily instrumental terms, this is fundamentally different to a doctor being instrumentalised by their employer due to the differing natures of the relevant relationships and the power dynamics contained within. Whereas doctors typically have power over patients, they are generally vulnerable to the power of other relevant institutions and officials.

It is thus relevant to note that the primary source of the suggestion that doctors have a responsibility to avoid burnout is not in fact patients, but health care institutions such as professional bodies and employers. The concern about instrumentalisation becomes particularly acute in each of these contexts, for distinctive reasons. With respect to employers, it is important that individuals in senior positions in an organisation cultivate respectful attitudes towards those over whom they have power, and that organisations themselves encourage respectful attitudes towards employees. One reason for this is hierarchy: occupying a position of power and authority over someone else, as is the case in an employer–employee relationship, makes it particularly important to encourage respect. Those with control over working conditions should be encouraged to see their employees primarily as individual people, not merely as "useful"; not to do so risks exploitation, bullying, and other workplace issues.

With respect to professional bodies, things are a little different. One reason that it is particularly important for professional bodies to recognise their members as individuals with intrinsic value, rather than primarily as of instrumental value, is that a professional body is an organisation which represents its members. It seems odd that organisations like the BMA or doctors' Royal Colleges might value their members' well-being primarily because it allows them to function as better doctors. Their role is to support and represent doctors' interests. This is not to say that such organisations never do this. And the picture is complex, because these organisations often have secondary aims that involve improving health care more generally as well as patient care. Nevertheless, documents placing emphasis on individual responsibility to avoid burnout, no matter how well intentioned, may instrumentalise the well-being of their members, and thus undermine their claims of member support.

We now return to an issue raised briefly above. Some readers may be sceptical of our framing, suggesting that we are making a mountain out of a molehill. Even if the phrasing of certain documents is unfortunate, they might say, it is implausible to think that professional health care bodies and employers solely value doctors instrumentally. Suggesting that doctors have a responsibility of self-care does not entail that this is the *primary* value of doctors' welfare. Rather, it is simply a way of highlighting a further reason, aside from the clear intrinsic value of an individual's welfare. After all, other documents we have already cited note the intrinsic value of doctors' well-being; what's more, one might think, it is simply obvious that this is not the dominant view.

We think, however, that it is far from obvious. While talking about the intrinsic value of doctors' well-being is necessary, talk is not enough. As we discuss below, the dominant framing of doctors' well-being in practical terms (i.e., in terms of what is actually to be done to protect mental welfare) concerns individual attempts to build personal resilience, often in response to increased workloads. At the very least, employers have strong instrumental interests in promoting doctors' well-being. Most discussion of the issue (aside from platitudes about valuing doctors) is around self-help, perhaps with some minimal support. Even if doctors' well-being really is valued intrinsically by institutional representatives, there is a concern about the development of a broader workplace and professional culture. Whether or not those at the top are virtuous or vicious with respect to their personal beliefs, what they actually communicate matters in terms of forming a working culture. The messages coming from health care leaders also matter in terms of what is communicated to individual doctors.

A practical issue is thus whether this way of communicating the importance of self-care is best for actually achieving the end of promoting doctors' well-being, even for instrumental reasons. Being told, or having it implied, that your well-being matters only, or mostly, in instrumental ways will be demoralising for many. Even from a purely instrumental perspective, it is thus likely conducive to doctors' welfare—and to patient care—that doctors *feel* intrinsically valued. Of course, "feeling valued" may mean different things. There is undeniable psychological benefit to feeling instrumentally valuable. But we suspect that most doctors—just like most other people—would prefer not to think that they are valued in a primarily instrumental way. Thus, in communicating personal resilience primarily as a professional responsibility owed to patients, health care institutions may unintentionally undermine their own goals by demoralising doctors and making them feel unvalued, making burnout more likely for some.

Assume again that employers, managers, government officials, and those who develop professional guidance really do value doctors' well-being intrinsically, so that we can say that they "subjectively respect" doctors. Qua employees and members of professional bodies, doctors not only have a moral claim to be subjectively respected; they also have a claim to have that respect and value communicated to them by those in control of their professional lives and identities, and not to feel reduced to what they can deliver for others. As such, not only does communication matter from a pragmatic, motivational perspective; it also matters from a moral perspective.

This claim is not absolute; it is subject to a reasonability restriction. If those in charge of health care institutions, guidance, and training make reasonable efforts to make doctors feel valued, then even if some individuals still do not feel valued, no claim has been illegitimately ignored. The precise boundaries of what constitutes such a reasonable effort will be complex. But, we suggest, a focus on doctors' well-being in primarily instrumental terms falls well short of such a requirement.

11.3.2 Institutional Responsibility

We ended Section 11.2 by suggesting that institutions should ensure that a reasonable employee or member feels intrinsically valued and respected. In the context of burnout among doctors, one element of this is to consider institutional responsibility for doctors' well-being, and how this constrains any putative responsibility doctors themselves have. In this sense, the claim that doctors have a responsibility to practise self-care is, at best, incomplete. There may be actions which doctors can take to become more resilient. But by focusing on these rather than on the broader professional and social influences on mental health—which are not in doctors' personal control, but which are to some extent in institutions' control—claims about a responsibility to be resilient highlight only a small part of the solution, unreasonably distorting the role doctors can play. Moreover, if the efficacy of individual acts of self-care in turn rely on wider institutional factors this further curtails any putative individual responsibilities.

Our core focus thus far has been establishing a potential role responsibility for health care professionals to practice self-care. A further question is whether, if they fail to fulfil this responsibility, others may legitimately hold them accountable. The dominant way of understanding such accountability is by reference to *liability* responsibility: being answerable or accountable on the basis of previous actions. As we outlined above, this is distinct from role responsibility (Dworkin 1981). The generally accepted framework for determining liability responsibility remains relatively unchanged since Aristotle, who outlines two conditions: a control condition and an epistemic condition (Eshleman 2016). The control condition has been interpreted in different ways, but broadly speaking holds that individuals cannot be liable for an outcome if they could not exercise a relevant form of control over it, for example, the ability to act so as to prevent the outcome, or the ability to "guide" the outcome (Fischer 2012). The epistemic condition stipulates that (non-culpable) ignorance of relevant information can also undermine liability. Insofar as somebody understands the consequences and moral relevance of their actions, and they have control over what they do, they are responsible. On this view, to assess whether doctors are accountable for failures to avoid burnout, we would need to assess whether they (i) knew that they were at risk of burning out, and (ii) had achievable means to avoid doing so.

Before considering this, it is worth briefly acknowledging an alternative sense of accountability relative to a person's role, which is sometimes understood as not involving the epistemic or control conditions.[2] For instance, Cane (2016: 284) highlights the idea of "ministerial responsibility" in UK government, where "ministerial responsibility for performance of public functions by officials... does not depend on the capacity of the Minister personally to control performance of those functions". Similarly, in a health care setting, the role of consultant may mean being held accountable for the actions of your subordinates even though you could not have known or changed what they might do. Some might suggest that self-care is a role responsibility for doctors, and that they can be held accountable for burning out *even if* they lacked control over whether this occurred.

We reject this idea. Role responsibilities are constrained by institutional and social structures; for instance, Cane suggests that ministers' role responsibility derives from their position of *authority*. Thus, we should consider our accountability practices in light of not only the epistemic and control conditions when it comes to role responsibilities but also the wider structures they exist in, and which enable those operating within them to perform their roles. The idea of a role presupposes a certain kind of social or institutional structure; where that structure is lacking, either the role or crucial aspects of it do not exist. Thus, we think that we should conceive of accountability for failing to fulfil this responsibility in the

---

[2] A related idea is the legal concept of "strict liability", where people can be legally accountable for an outcome despite lacking control over it.

more standard terms of liability, that is, in terms of the epistemic and control conditions.

We will focus on the control condition. Whether doctors are in control of burning out can also only be decided by considering the broader institutional and social context. For one thing, to be in control of avoiding burnout, the self-care strategies recommended must be effective. The evidence on whether strategies to increase personal resilience and reduce burnout are effective is mixed. Some reviews conclude that interventions to increase personal resilience are ineffective (McKinley et al. 2019). Others find that individual-focused interventions do reduce burnout, but also that organisational interventions are just as effective and suggest that both are probably necessary, though that has not been studied (West et al. 2016).

What this latter point suggests is that, to the extent that doctors do have a responsibility to engage in self-care, this must be understood as sitting within broader responsibilities *to* doctors with respect to their well-being. These latter responsibilities are held, as we have suggested, by their employers and to some extent their professional representative bodies. But, at least when doctors work in the public sector, they are also held by society in general, represented in particular by local and national government.

This is significant because the causes of doctors' burnout are multifactorial (Wallace et al. 2009). Many reasons doctors become burned out are accounted for by the environment in which they work: long hours; poorly organised rotas; staffing issues; inefficiencies in the system in addition to changing cultures within medicine; shifting patient expectations and austerity measures that undermine their ability to provide good care (Lemaire and Wallace 2017; Montgomery et al. 2019). Of these various causes of doctors' burnout, there are few that are predominantly under doctors' individual control.

If we subtract the institutional causes of doctors' burnout, then we are left with a realm in which doctors are in control and could be held liable, while still recognising that a doctor might do everything they reasonably can and yet still become burned out. Yet, the list of systemic causes above are also potential limitations on doctors performing self-care. Systemic problems act not only to *cause* doctors' burnout, but also as obstacles to the very resilience-building interventions that they can undertake. By working long and unpredictable hours with little opportunity for them to have direct control over this, doctors' opportunities to enact the recommended measures are limited. Even if there are evidence-based self-care interventions that doctors could use, the system they work within prevents them doing this, further weakening liability.

This has two implications. First, doctors can only have a responsibility to engage in self-care in the context of appropriate institutional and social efforts to protect their well-being. Thus, cases where doctors have significant liability for burning out may be less common than initial appearances suggest.

Second, even when doctors do bear some liability, it is often inappropriate, and indeed morally offensive, for those who make even greater contributions to burnout to frame doctors' well-being and resilience in terms of personal responsibility, unless this is explicitly acknowledged as subsidiary to the broader responsibilities that are held to doctors qua workers. Philosophers sometimes speak of the idea of "standing to blame" (Scanlon 2008: 175–9; Wallace 2010; Coates and Tognazzini 2012: 19–22; Herstein 2017; Fritz and Miller 2018; Todd 2019): that there can be cases where even if S is blameworthy, T may lack the right or some other form of proper moral authority to blame S, due to some feature of T's own position or past behaviour (e.g., the fact that it would be hypocritical because T has engaged in the same blameworthy behaviour). Even if one is sceptical about the idea that people require something like "standing" to blame (Bell 2012; King 2019), there are still ethical problems with the positional aspects of (expressed or unexpressed) blame, including hypocritical blame.

A focus on doctors' individual responsibilities needn't strictly imply that they will be blamed if they fail in those responsibilities. But the idea of inappropriate blame can, we think, be translated to the related issue of explicating responsibilities. There are cases where, even if S has some responsibility, it is inappropriate for T to explicitly *communicate* that responsibility given her own failure to exercise similar or related responsibilities. Imagine that T has failed to pay her share of the rent on a flat she shares with S. It would be hypocritical to blame S for not paying her share of the rent. But it would also be hypocritical to *remind* S that she is responsible for paying her share of the rent, or to emphasise how important paying the rent is for both of them. Indeed, we may suspect that there are other reasons for T to offer such reminders, such as deflecting possible scrutiny or blame from themselves. In the context of health care, there are clear political incentives to frame burnout as an issue for doctors themselves to address, rather than one of institutions and government placing excessive demands on workers.[3]

In focusing on individual responsibility for burnout, we suggest, those in charge of health care and other political institutions engage in hypocritical behaviour. Even if it is true that individual doctors bear some responsibility for their mental health, there is something inappropriate about institutions (both professional and political) which are *even more liable* for damage to doctors' well-being having such a rhetorical focus, even if it does not go as far as blame. Individuals working for and representing these institutions do not, at least in the absence of acknowledging institutional responsibility, have "standing to hold accountable". At the very least, it is hypocritical of them to do so without both acknowledging, and making concrete commitments to tackle, the institutional and social influences on doctors' well-being.

---

[3] Thanks to Neil Levy for pressing us to include this point.

This analysis leaves it open that others—such as individual patients—might permissibly hold doctors accountable for failing to avoid burnout where this was in their control. A patient who received substandard care, or their family, from a burned-out doctor could, on our analysis, reasonably hold that doctor liable in some circumstances. Even here, though, it would be unfair (though understandable given the distress of mistreatment) to focus only on the individual's failures, and not on broader institutional contributions.

11.3.3 Demandingness

A third potential problem with the idea of a responsibility of self-care is that it may be too demanding. Before considering ways in which such a responsibility might plausibly be too demanding, we want to briefly note an interpretation of this complaint which, in our view, does not work.

On any theory of well-being, there are numerous activities that can promote it. One might worry that, as such, the range of activities which might constitute self-care becomes extremely broad, since almost any activity that doctors undertake which boosts their well-being seems to be contained within a personal moral responsibility of self-care. Thus, going for a run, spending time with friends, reading to one's children, or watching television all become moral responsibilities for doctors.

The version of the demandingness worry which is not ultimately plausible is that offering a wide range of ways to boost your well-being implies that you must take on them all. For example, the Royal College of General Practitioners (RCGP) ("GP well-being", No date) has a webpage on well-being that includes a list of "five ways to improve your well-being as a GP": connect with people; take notice; keep learning; be active; and give. Attempting to do all of this on top of a doctor's regular duties may prove too much.

In principle, however, it seems more reasonable to read the idea of a responsibility of self-care not as maximising but satisficing. While we do not think that this is adequately communicated in current discussions of doctors' welfare, we do in principle accept that health care organisations see a responsibility of self-care as having practical limits. In particular, if the aim of self-care is to avoid burnout, then doctor's responsibility is not to have as high a level of well-being as possible, but to have sufficient well-being to avoid burning out.

Still, we think that there is a more plausible version of this worry, namely when it is applied to those who do not find that their well-being improves despite trying various acts of self-care. Starting by running through each item on the list suggested by the RCGP but finding little success, one might next try them in various combinations and, finally, all together. Even if it would be more reasonable to interpret the items on the list disjunctively, a doctor who is already suffering from mental and emotional pressure, and who feels like they risk failing their patients, may be at risk of interpreting the RCGP's list conjunctively, and thus over-demandingly. It is also important to remember, again, that injunctions of self-care will often be received by doctors who are already overworked and struggling to meet what are presented to them as the basic demands of their job. Where doctors are *already* finding their work-related duties overwhelming, adding even one more obligation may be excessively demanding.

More generally, there is an issue of "mission-creep", where the line between the professional and the personal becomes blurred and previously personal pleasures become moralised professional responsibilities. The GMC (2018) is explicit in seeing no daylight between professional and personal duties when they state that newly qualified doctors must "incorporate compassionate self-care into their *personal and professional life*" (emphasis added). To some degree, doctors' duties clearly do extend to how they choose to spend their time (e.g., by studying new developments in medical journals). But they also have rights to set limits on those duties, and to have time for themselves which is not subsumed under their professional life.

Finally, this way of thinking about well-being—as an essentially moralised good that doctors need to protect for their patients' sakes—may have a pernicious effect on the very activities that previously boosted it. To see this, consider an attempt to place a positive spin on the moralisation of doctors' free time. Pre-moralisation, various activities are engaged in for purely personal reasons. Post-moralisation, these personal reasons still apply, yet by spending time with family, exercising, or enjoying hobbies, doctors are *also* fulfilling a moral responsibility. Thus, valued activities take on extra value at no additional cost, creating a "win-win" situation!

Yet the downside of this is that there are now also more ways to fail. In a context where many doctors are often forced to forego things that they value due to tiredness, overwork, and stress, this becomes not only a personal sacrifice but also a moral failure. In lacking time or energy to go for a run at the weekend, you have not only missed out on an enjoyable activity, you have also let your patients down. Other activities already have a moral element to them, but casting them as ways of avoiding burnout burdens them with an additional moral weight. Getting home too late to put your children to bed, you haven't just let them down, you've also (yet again) let your patients down. Yet so too would you let patients down by leaving work on time when there is still care work to be done so that you can see your kids or have enough energy to exercise the next day. The win-win situation becomes lose-lose.

11.4 Conclusion

There is increasing recognition that burnout is a serious problem among health care staff. Due to the impact of burnout on patients and health care systems, it has been recognised by those with authority and power in health care that something must be done. Increasing individual resilience to the challenges of modern medical practice has been one core strategy. Professionalism has been utilised as a tool to encourage compliance with strategies of self-care and in light of the effect of burnout on patient care. Here we have offered some grounds of plausible support for such a role responsibility.

However, we have also pointed to a major flaw in the discussion around doctors' well-being, namely that its primary framing of the value of doctors' well-being is instrumental. We suggested two important issues with a putative professional responsibility to self-care that is framed in this way: demandingness and instrumentalisation. As employees, doctors hold rights not to have their mental well-being excessively eroded by their workplace and to be supported to do their job effectively. Moreover, effective interventions to avoid burnout are not individual alone and must come within an appropriate institutional framework. The bulk of responsibility for ensuring that doctors are resilient falls to institutions tasked with employing or supporting doctors.

References

Azoulay E., De Waele J., Ferrer R., et al. (2020), "Symptoms of Burnout in Intensive Care Unit Specialists Facing the COVID-19 Outbreak", in *Annals of Intensive Care* 10: 1–8.

Beauchamp, T. and Childress, J. (2001), *Principles of Biomedical Ethics* (Oxford University Press).

Bell, M. (2012), "The Standing to Blame: A Critique", in: D. Coates and N. Tognazzini (ed.), *Blame: Its Nature and Norms* (Oxford University Press), 263–81.

Boorman, S. (2009), *NHS Health and Well-Being: Final Report* (Department of Health).

Bourne T., Shah, H., Falconieri, N., et al. (2019), "Burnout, Well-Being and Defensive Medical Practice Among Obstetricians and Gynaecologists in the UK: Cross-Sectional Survey Study", in *BMJ Open* 9/e030968: 1–11.

Brazeau C., Schroeder, R., Rovi, S., and Boyd, L. (2010), "Relationships Between Medical Student Burnout, Empathy, and Professionalism Climate", in *Academic Medicine* 85: S33–36.

British Medical Association. (2019), *Mental Health and Well-Being in the Profession.* https://www.bma.org.uk/collective-voice/policy-and-research/education-training-and-workforce/supporting-the-mental-health-of-doctors-in-the-workforce (accessed 14 March 2022).

Cane, P. (2016), "Role Responsibility", in *Journal of Ethics* 20: 279–98.

Coates, D., and Tognazzini, N. (2012), "The Contours of Blame", in: D. Coates and N. Tognazzini (ed.) *Blame: Its Nature and Norms* (Oxford University Press), 3–26.

Cruess, S., and Cruess, R. (1997), "Professionalism Must Be Taught", in *BMJ* 315: 1674–7.

Cruess, R., and Cruess, S. (2008), "Expectations and Obligations: Professionalism and Medicine's Social Contract with Society", in *Perspectives in Biology and Medicine* 51: 579–98.

Cruess S., Johnston, S., and Cruess, R. (2002), "Professionalism for Medicine: Opportunities and Obligations", in *Medical Journal of Australia* 177: 208–11.

Doukas, D., McCullough, L., Wear, S., et al. (2013), "The Challenge of Promoting Professionalism Through Medical Ethics and Humanities Education", in *Academic Medicine* 88: 1624–9.

Dunn, M. (2016), "On the Relationship Between Medical Ethics and Medical Professionalism", in *Journal of Medical Ethics* 42: 625–6.

Dworkin, G. (1981), "Taking Risks, Assessing Responsibility", in *The Hastings Center Report* 11: 26–31.

Eaton, L. (2009), "Take More Responsibility for Your Own Health, NHS Staff Told", in *BMJ* 339: b3406.

Eshleman, A. (2016), "Moral Responsibility", in E. Zalta (ed.) *The Stanford Encyclopedia of Philosophy.* https://plato.stanford.edu/archives/win2016/entries/moral-responsibility/ (accessed 14 Mar 2022).

Fahrenkopf, A., Sectish, T., Barger, L., et al. (2008), "Rates of Medication Errors Among Depressed and Burnt Out Residents: Prospective Cohort Study", *BMJ* 336: 488–91.

Fischer, J. (2012), *Deep Control: Essays on Free Will and Value* (Oxford University Press).

Fritz, K., and Miller, D. (2018), "Hypocrisy and the Standing to Blame", in *Pacific Philosophical Quarterly* 99: 118–39.

General Medical Council. (2013), *Good Medical Practice.* Domain 1: Knowledge skills and performance. https://www.gmc-uk.org/ethical-guidance/ethical-guidance-for-doctors/good-medical-practice/duties-of-a-doctor (accessed 14 March 2022).

General Medical Council. (2018), *Outcomes For Graduates 2018.* https://www.gmc-uk.org/education/standards-guidance-and-curricula/standards-and-outcomes/outcomes-for-graduates/outcomes-for-graduates (accessed 14 March 2022).

Goodin, R. (1987), "Apportioning Responsibilities", in *Law and Philosophy* 6: 167–85.

Herstein, O. (2017), "Understanding Standing: Permission to Deflect Reasons", in *Philosophical Studies* 174: 3109–32.

Iliffe, S., and Manthorpe, J. (2019), "Job Dissatisfaction, 'Burnout' and Alienation of Labour: Undercurrents in England's NHS", in *Journal of the Royal Society of Medicine* 112: 370–7.

King, M. (2019), "Skepticism about the Standing to Blame", in: D Shoemaker (ed.) *Oxford Studies in Agency and Responsibility Volume 6* (Oxford University Press), 265–88.

Lancet, T. (2019), "Physician Burnout: The Need to Rehumanise Health Systems", in *The Lancet* 394: 1591.

Lemaire, J., and Wallace, J. (2017), "Burnout Among Doctors", in *BMJ* 14/358: j3360.

Linzer, M. (2018), "Clinician Burnout and the Quality of Care", in *JAMA Internal Medicine* 178: 1331–2.

McKinley, N., Karayiannis, P., Convie, L., et al.. (2019), "Resilience in Medical Doctors: A Systematic Review", in *Postgraduate Medical Journal* 95/1121: 140–7.

Medical Schools Council. (2018), *Statement On the Core Values and Attributes Needed to Study Medicine*. www.medschools.ac.uk/media/2542/statement-on-core-values-to-study-medicine.pdf (accessed 14 March 2022).

Minford, E., and Manning, C. (2017), "Current Status and Attitudes to Self-Care Training in UK Medical Schools", in *Journal of Compassionate Health Care* 4/3: 1–6.

Montgomery, A., Panagopoulou, E., Esmail, A., Richards, T., and Maslach, C. (2019), "Burnout in Healthcare: The Case for Organisational Change", in *BMJ* 366: l4774.

Munn, F. (2017), "Medical Students and Suicide", in *BMJ* 357: j1460.

Office for National Statistics. (2018), "Sickness Absence Falls To the Lowest Rate On Record", www.ons.gov.uk/employmentandlabourmarket/peopleinwork/employmentand-employeetypes/articles/sicknessabsencefallstothelowestratein24years/2018-07-30 (accessed 14 March 2022).

Oliver, D. (2017), "When 'Resilience' Becomes a Dirty Word", in *BMJ* 358: j3604.

Parfit, D. (1984), *Reasons and Persons* (Oxford University Press).

Pellegrino, E. (2001), "The Internal Morality of Clinical Medicine: A Paradigm for the Ethics of the Helping and Healing Professions", in *Journal of Medicine and Philosophy* 26: 559–79.

Pellegrino, E. (2002), "Professionalism, Profession and the Virtues of the Good Physician", in *Mount Sinai Journal of Medicine* 69: 378–84.

Pellegrino, E., and Thomasma, D. (1993), *The Virtues in Medical Practice* (Oxford University Press).

Richards, V., and Lloyd, K. (2017), *Core Values for Psychiatrists* (Royal College of Psychiatrists).

Royal College of General Practitioners. (No date), *GP Wellbeing*. www.rcgp.org.uk/training-exams/practice/gp-wellbeing.aspx (accessed 14 March 2022).

Scanlon, T. (2008), *Moral Dimensions: Permissibility, Meaning and Blame* (Belknap).

Todd, P. (2019), "A Unified Account of the Moral Standing to Blame", in *Noûs* 53/2: 347–74.

Tweedle, J., Hordern, J., and Dacre, J. (2018), *Advancing Medical Professionalism* (Royal College of Physicians).

Vahey, D., Aiken, L., Sloane, D., Clarke, S., and Vargas, D. (2004), "Nurse Burnout and Patient Satisfaction", in *Medical Care* 42: II57–66.

Wallace, J, Lemaire, J., and Ghali, W. (2009), "Physician Wellness: A Missing Quality Indicator", in *The Lancet* 374: 1714–21.

Wallace, R. (2010), "Hypocrisy, Moral Address, and the Equal Standing of Persons", in *Philosophy & Public Affairs* 38/4: 307–41.

West, C., Dyrbye, L., Erwin, P., and Shanafelt, T. (2016), "Interventions to Prevent and Reduce Physician Burnout: A Systematic Review and Meta-Analysis", in *The Lancet* 388: 2272–81.

West. M., and Coia, D. (2019), *Caring For Doctors, Caring For Patients* (General Medical Council).

World Health Organisation. (2019), *Burn-Out An "Occupational Phenomenon", in International Classification of Diseases.* http://www.who.int/mental_health/evi dence/burn-out/en/ (accessed 14 March 2022).

Yates, S. (2020), "Physician Stress and Burnout", in *American Journal of Medicine* 133/2: 160–4.

PART V

BEYOND PATIENT
RESPONSIBILITY—SOCIETY

12

Fighting Vaccination Hesitancy

Improving the Exercise of Responsible Agency

Daniel J. Miller, Anne-Marie Nussberger, Nadira Faber,
and Andreas Kappes

12.1 Introduction

Addressing vaccine hesitancy—the refusal to receive safe and recommended available vaccines—is a major challenge in the fight against infectious disease such as measles or COVID-19. Theorists working in various fields focus on different questions about how to respond to vaccine hesitancy. Philosophers working on moral responsibility, for example, may concern themselves with whether vaccine sceptics or "anti-vaxxers" are deserving of blame or sanction. This concern, however, seems distinct from psychologists' focus on what might be the more pressing question: which responses to vaccine hesitancy would be most effective in improving vaccine uptake? Despite initial appearances, we aim to show that the answers to these questions are closely related. Vaccine uptake depends upon an individual's capacity to recognize and respond to reasons to vaccinate—capacities that are essential to philosophical theories of moral responsibility. Because of this, the factors that operate as obstacles to the exercise of morally responsible agency in this regard are the very ones that psychologists can identify (and aim to mitigate) in the effort to improve vaccine uptake.

We begin by introducing recent work from the philosophical literature on agency and responsibility, with a special focus on the rational capacities at work in the context of deciding whether to vaccinate. Drawing upon this literature, we present a framework for the exercise of ideally responsible agency in three stages that correspond to these "responsibility-relevant" capacities. This framework allows for the identification of various obstacles to vaccination as they pertain to one or more of these stages of responsible agency. We then show how this philosophical framework fits well with extant psychological theories of human behaviour and can also offer a more unified explanation of various determinants of vaccine uptake listed in recent psychological approaches to vaccine readiness. Finally, we flesh out the philosophical framework with lessons drawn from a variety of studies in psychology and behavioural economics on how individuals

Daniel J. Miller, Anne-Marie Nussberger, Nadira Faber, and Andreas Kappes, *Fighting Vaccination Hesitancy: Improving the Exercise of Responsible Agency* In: *Responsibility and Healthcare*. Edited by: Benjamin Davies, Gabriel De Marco, Neil Levy, Julian Savulescu, Oxford University Press. © Edited by Benjamin Davies, Gabriel De Marco, Neil Levy, Julian Savulescu 2024. DOI: 10.1093/oso/9780192872234.003.0013

reason, decide, and behave in the context of vaccination. These lessons help illuminate the extent to which individuals exercise responsibility-relevant capacities in such contexts while also pointing to strategies that can improve the exercise of those capacities.

12.2 The Rational Capacities of Responsible Agency

Responsible agency requires the possession of certain rational capacities.[1] This is perhaps best explained by the view that holding an individual responsible for their beliefs and behaviour is fair or reasonable only if that individual is capable of guiding their beliefs and behaviour in light of the relevant reasons.[2] If an individual lacks these capacities, or if the exercise of these capacities is inhibited in particular ways, then it is unreasonable to expect them to know better or to act as they ought to, and thus unfair to hold them responsible when they don't (Rosen 2004: 306; Widerker 2006: 63; FitzPatrick 2008: 603; Levy 2009: 729; Sher 2009: 100ff). The fact that we are ordinarily disinclined to hold very young children or individuals with severe cognitive disabilities responsible for their behaviour reflects tacit acceptance of this view. This view also explains why we sometimes excuse individuals who possess the relevant capacities but whose behaviour is temporarily characterized by lack of control (e.g., intoxication) or blameless ignorance.[3] In such cases, the exercise of the individual's capacities is inhibited or interfered with in a way that undermines responsibility for a particular belief or behaviour.

Notably, even divergent responses to individuals who fail to behave as they ought to may nevertheless reflect a shared underlying view about the relationship between rational capacities and responsible agency. In the context of COVID-19, for example, some have responded to individuals' refusal to vaccinate with outrage. One possible explanation for this response is the belief that "anti-vaxxers" *are* capable of responding to the reasons to vaccinate but refuse to do so (and are therefore culpable for this failure).[4] Others, however, have dubbed individuals

---

[1] Work on these capacities is well-developed in the philosophical literature on moral responsibility. As such, the discussion that follows draws upon this literature.

[2] It is widely accepted by philosophers working on moral responsibility that an individual is morally responsible for something if and only if it would be appropriate to hold the individual responsible (Wallace 1994: 91; Fischer and Ravizza: 6–7; McKenna 2012: 34ff).

[3] For example, I may be forcibly restrained (thus inhibiting the exercise of bodily control) or deceived (thus inhibiting the exercise of my capacity to become aware of the relevant facts).

[4] These "anti-vaxxers" may be aware that they should vaccinate but (contrary to their own best judgment) refrain from doing so. Alternatively, they may be culpably ignorant of the reasons in favor of vaccination (either due to a culpable failure to become aware of the relevant facts, or else due to a culpable failure to recognize that the relevant facts provide sufficient reason to vaccinate). If so, then such individuals are culpable for failing to exercise certain rational capacities (we discuss these capacities in more detail below).

who refuse to vaccinate as "covidiots" (Capurro et al. 2022), which may indicate the belief that at least some who refuse to vaccinate are deficient in their capacity to recognize and respond to reasons. If so, then some might respond to anti-vaxxers with condescension more so than anger.

We call the capacities required for responsible agency *responsibility-relevant capacities*. The philosophical literature provides the resources for categorizing responsibility-relevant capacities into three general kinds.[5] In what follows we briefly explain the nature of these capacities and their application to the context of vaccination.

Responsible agency first requires the capacity to become aware of the norma-tively relevant facts: facts that bear upon what one has reason to do (or to refrain from doing). The fact that an action (or omission of an action) risks serious harm to others is a paradigmatic example of a normatively relevant fact. One's capacity to become aware of a fact like this depends in turn upon a host of more particular sub-capacities to make observations, take steps to investigate matters, recall and reflect upon evidence, draw inferences, assess and compare various risks, and more (Rosen 2004: 301; Robichaud 2014: 147 Clarke 2017: 242; Miller 2017: 1569–71). In the case of COVID-19, a number of facts are normatively relevant to the decision to vaccinate: vaccination (for most individuals) poses negligible risks of adverse side-effects, decreases the risk of transmission to others, vastly decreases the risk of hospitalization and death for those who are vaccinated, and thus allows for hospitals to reserve capacity for individuals who require hospital-ization (Center for Disease Control). One's capacity to become aware of such facts depends upon at least some further sub-capacities, such as the capacity to under-stand public health messaging or investigate various sources of information concerning vaccines, the capacity to compare the relative risks of vaccination and non-vaccination, and the capacity to infer that the risks associated with vaccination are vastly lower than the risks associated with non-vaccination.

Second, responsible agency requires the capacity to recognize the normative significance of the relevant facts. This is essential because an individual might become aware (for example) of the risk of serious harm while nevertheless failing to recognize that fact as a *sufficient* reason to perform (or refrain from perform-ing) some action (where the reason is, in fact, sufficient).[6] It is possible that some

---

[5] The philosophical literature often groups these three kinds of capacities into two general categor-ies, where the first two we discuss here are treated as sub-categories of the more general capacity to recognize reasons. Fischer and Ravizza (1998) is the most influential work that draws this distinction. Fischer and Ravizza describe the general capacity for responsible agency as "reasons responsiveness", which they then divide into the more specific capacities of "reasons receptivity" (the capacity to recognize reasons for action) and "reasons reactivity" (the capacity to act on those reasons). Other prominent theorists who maintain a similar framework include Wallace (1994) and Nelkin (2011).

[6] As we use the term, a "sufficient reason" (to act or refrain from acting) is one that, in the particular circumstances, is weighty or important enough to make it so that the overall balance of reasons favours a certain course of action. In some cases a sufficient reason is a set of reasons (e.g., the fact that an action would harm others *and* constitute a violation of some prior commitment).

individuals are aware that the failure to vaccinate poses serious health risks to themselves and others while still failing to recognize this fact as a sufficient reason to vaccinate. Importantly, the capacity to recognize that these risks provide a sufficient reason to vaccinate seems to require, minimally, the belief that the interests of others matter, a belief which may in turn depend upon emotional capacities involved for example in empathy (Russell 2004: 293; Nelkin 2011: 22ff; Rudy-Hiller 2017: 415, fn.48). Those who are at a lower risk of serious illness may judge that it's not worth being vaccinated because they fail to care about the risks that their failure to vaccinate may pose to others who are more vulnerable. The capacity to recognize some fact (or set of facts) as a sufficient reason to act in a certain way may also require the more particular sub-capacity to weigh competing reasons against each other and assess where the balance of reasons lies. This more particular capacity enables one to recognize that, even though the slight risk of adverse effects posed by vaccination provides a minor reason to refrain from vaccination, such a reason is clearly outweighed by the reasons to vaccinate. In sum, one must be capable of recognizing that the overall balance of reasons favours vaccination.

Third, responsible agency requires the capacity to respond to one's recognition of reasons by being motivated to behave in the relevant way.[7] An individual who judges that they have sufficient reason to vaccinate must be capable of deciding and behaving in accordance with this judgement. However, a number of factors can inhibit the exercise of this capacity: an individual might judge that they have sufficient reason to perform (or omit from performing) some action and yet nevertheless fail to behave accordingly due to compulsion, fear, or some other psychological obstacle (Fischer and Ravizza 1998: 41–2; Mele 2006: 150ff). One may also have the capacity to respond to one's recognition of reasons but fail to exercise it due to weakness of will (Fischer and Ravizza 1998: 42–3; Mele 2012). In the case at hand, an individual who judges that it's best to vaccinate may nevertheless be prevented from doing so out of fear of adverse side-effects.

The successful exercise of these capacities results in corresponding outcomes: (1) the awareness of relevant facts (e.g., an accurate assessment of various risks and benefits of vaccination), (2) the recognition of their normative significance (i.e., since the benefits of vaccination far outweigh the risks, these facts provide sufficient reason to vaccinate), and (3) behaviour motivated by this recognition (i.e., vaccination).[8] To structure the following discussion, we can frame the exercise of the capacities of responsible agency in these three stages.

---

[7] This capacity could be further sub-divided into the capacity to *choose* on the basis of the reasons and the capacity to *act* (i.e. perform overt bodily actions) on the basis of these choices (Fischer and Ravizza 1998: 69). We use "the capacity to respond to reasons" to refer to one or both of these (depending upon the context).

[8] In what follows we sometimes refer to an individual's actions or behaviour, intending this language to cover both cases of action and omission.

The successful exercise of capacities across these three stages provides a picture of ideal responsible agency. An individual who fulfils this ideal would recognize reasons to minimize harm and, all else being equal, be motivated to act accordingly. In the context of vaccination, the ideally responsible individual would (1) recognize that the risks of harm (to oneself and others) posed by vaccination is significantly lower than the risks of harm posed by non-vaccination, (2) judge that this fact is a sufficient reason to vaccinate (rather than to not vaccinate), and (3) vaccinate on this basis.

12.3 Psychological Foundation of Responsible Agency

In considering the usefulness of the outlined philosophical framework for responsibility-level capacities, one might wonder how it relates to the psychological understanding of vaccine hesitancy. While the philosophical framework does not claim to provide a descriptive account of why people might or might not decide to receive safe and recommended vaccines, it might be more credible if it can capture the key psychological drivers of health-related behaviour and vaccine hesitancy.

In response, we suggest that this framework provides a way of highlighting intersections between two distinct literatures that focus on human agency—the philosophical literature and the psychological literature—while also applying the insights from both literatures to pressing matters in public health. In particular, while the philosophical framework provides a theory of why rational capacities are required for the exercise of responsible agency, the psychological literature can flesh out how the possession (or lack) of these capacities in fact plays a role in human behaviour as it pertains to vaccine-uptake.

Most psychological theories that attempt to explain health-related behaviour share a common set of variables that are associated with the intention to perform a certain behaviour such as getting a vaccine, variables that can be related to the capacities underlying responsible agency. Prominent examples of such theories are the health belief model (Rosenstock 1974), protection motivation theory (Rogers 1975), theory of reasoned action (Fishbein and Ajzen 1975), and theory of planned behaviour (Ajzen 1985). These models assume that the intention to perform a behaviour is the best predictor of behaviour. Intentions in turn are best predicted by attitudes, norms, and expectations. For instance, the most influential example of these theories, the theory of planned behaviour, suggests that people are more likely to form an intention to vaccinate if they feel that getting vaccinated is overall positive (attitude), that people who are important to the person approve of getting vaccinated (subjective norm), and when people feel confident that they can access the vaccine (behavioural control beliefs). The variables described in the theory of planned behaviour are crucial determinants for vaccination uptake

(Larson et al. 2014; Tickner, Leman, and Woodcock 2010; Liau and Zimet 2000). The health belief model additionally suggests that people need to feel at risk for getting the disease and a vulnerability to the disease, perceptions that also seem important to consider when predicting vaccine intentions (Rosental and Shmueli 2021). Finally, extensive research has shown that people often fail to act on their intentions (intention–behaviour gap) (Sheeran and Webb 2016), suggesting that people lack the self-control to act in line with their intentions. One might believe that one should get vaccinated, or want to do so, but then fail to find the time in one's schedule or be stopped by an unwillingness to endure the needle injection.

The outlined psychological drivers of health behaviour map well onto the components of the framework for responsible agency. First, perceptions of risk as well as perceived vulnerability seem to result from exercises of the capacity to become aware of the normatively relevant facts. Second, insofar as a positive attitude suggests that people feel that the advantages of getting the vaccine outweigh the disadvantages, and that the associated risks are perceived to be minor (attitude reference), this attitude may reflect the capacity to recognize the normative significance of those facts (in particular, it may reflect a belief that the reasons in favour of vaccination outweigh the reasons against). Subjective norms are also related to the capacity to recognize the normative significance of the relevant facts, and in some cases may present obstacles to the successful exercise of this capacity. For example, some individuals may believe that they aren't obligated to sacrifice more than others around them. If so, then even individuals that have a positive attitude towards vaccination may nevertheless fail to judge that they have an obligation to vaccinate if others in their community are not doing so. Finally, behavioural control beliefs and a lack of self-control relate to the capacity to respond in accordance with one's recognition of reasons. For example, feeling that one is not capable of getting the vaccine seems to be an obstacle to being motivated to act in line with one's convictions about what one has sufficient reason to do.

Other descriptive approaches in psychology focus exclusively on vaccine hesitancy, trying to capture the determinants of getting a vaccine, though with less consideration of a specific theoretical framework. The 5Cs approach, recently extended to the 7Cs, is one of the most widely used (Betsch et al. 2018; Geiger et al. 2021). The philosophical framework we offer provides a way to unify the various determinants listed in these approaches under the more fundamental categories of the three rational capacities required for responsible agency.

To illustrate, the 7Cs approach suggests that people need to trust in the safety of vaccines (*confidence*), see themselves at high risk for the disease (*complacency*), and give little credence to fake news and conspiracy theories around vaccines (*conspiracy*). These three components of the 7Cs approach are related to the capacity to become aware of the normatively relevant facts—facts that provide individuals reasons to behave in certain ways. Whether one has sufficient reason to vaccinate depends upon facts about whether vaccines are safe and whether one is at high risk for the disease. Credence in conspiracy theories can operate as obstacles to recognition of these facts.

The 7Cs approach also emphasizes that individuals must have a willingness to protect others by getting vaccinated (*collective responsibility*) and must positively compare the benefits of being vaccinated against the personal costs of doing so (*calculation*). These two components of the 7Cs approach are related to the capacity to recognize the normative significance of the relevant facts. An individual might, for example, recognize that vaccines are safe and effective, but lack the sense of community or social obligation required for the willingness to protect others. Furthermore, the recognition that one has sufficient reason to vaccinate will depend upon one's capacity to weigh the reasons for and against vaccination—in particular, the benefits of vaccination (including benefits to others) against the minor personal costs—and to recognize that the former outweigh the latter.

Lastly, the 7Cs approach also recognizes that even those who judge that it would be best to vaccinate may nevertheless be prevented from doing so due to various psychological or everyday structural obstacles such as an inconveniently located vaccination clinic. For this reason, some individuals may judge that they have sufficient reason to vaccinate but fail to be sufficiently motivated to act in accordance with this judgement due to lack of self-control or weakness of will.[9] Thus, individuals must perceive few hurdles in their everyday life hindering them from accessing the vaccine (*constraints*). This component of the 7Cs approach fits well with the capacity to respond to one's recognition of reasons by being motivated to act in the relevant way. In sum, research indicates that the components outlined above explain a large percentage of variance in adults' intentions to vaccinate against COVID-19 (Geiger et al. 2021).[10]

The normative philosophical framework of personal responsibility, then, maps well onto descriptive accounts of why people form the intention to get a vaccine and act on this intention. However, neither of these approaches offers a detailed or sufficient treatment of the factors that explain why people might fail to do so, and in particular the psychological obstacles that can prevent individuals from successfully exercising the capacities associated with each of the stages of responsible agency. In what follows we supplement the philosophical framework we have offered by drawing upon a variety of studies in psychology and behavioural economics to identify a number of these obstacles as they pertain to vaccine hesitancy.

---

[9] For an account of "weakness of will" see Mele (2012).

[10] We leave out one of the components of the 7Cs approach, which concerns whether an individual is disposed to support monitoring to control adherence to vaccination regulations (*compliance*). While this component may be related to the effectiveness of public health policies, it is not clearly related to an *individual's* decision to vaccinate.

12.4 Obstacles to Becoming Aware of
Normatively Relevant Facts

People are more likely to vaccinate when they perceive a high risk associated with an infectious disease (Norman, Boer, and Seydel 2005) but they are less likely to vaccinate when they perceive a high risk associated with a vaccination (Mills et al. 2005; Betsch and Schmid 2013). Such risk perceptions drive attitudes towards vaccination, including current vaccine hesitancy with regard to COVID-19 (Machingaidze and Wiysonge 2021). Risk (mis)perceptions misconstrue minor safety concerns as more dangerous than evidence would suggest (Brown et al. 2010; Ropeik 2013). Such risk perceptions present a key obstacle to responsible agency. In particular, misperceptions of risks associated with vaccination indicates an obstacle to the exercise of the capacity to become aware of the relevant facts. The mistaken belief that vaccines are linked with autism is one such obstacle (Brown et al. 2010). One obvious culprit of misperceptions of risk is misinformation, but sometimes the way information about risk is communicated might also lead people to form a false impression of the risks associated with vaccines.

Misinformation is a key problem for the formation of accurate risk perceptions and has negatively impacted the uptake of safe vaccinations during the COVID-19 pandemic (Levy 2021; Roozenbeek et al. 2020). It may sometimes seem inevitable that people will fall for misinformation, especially with how quickly it spreads on the Internet in general and social media in particular, as well as the tailoring of the misleading messages to the psychological preferences of Internet users. The latter threat is often discussed in research on echo bubbles– closed information systems that reinforce the same messages without giving space to counter arguments (Levy 2021). However, research on the impact of such echo bubbles on people's attitudes revealed a complicated picture (Ecker et al. 2022). Some research suggests that people who tend to favour intuitive over deliberative reasoning are more likely to share fake news (Mosleh et al. 2021) while other lines of research add that the key component might not be that people are inherently biased or incapable of processing risk perceptions accurately, but rather too lazy to activate critical reasoning when needed (Pennycook et al. 2021). Importantly, an extensive line of research shows that accuracy prompts delivered on social media help people to effectively identify misinformation and stop sharing it (Pennycook and Rand 2021). It seems that often, when people form misperceptions of risks about vaccines it is not that they are too senseless to develop accurate perception ("covidiots"), but rather failed to activate the dormant capacity to assess the accuracy of information.

Ironically, the way experts attempt to correct misinformation might sometimes increase the misperception or increase confidence in false beliefs. For instance, strong negations of vaccine-adverse events—relative to weak negations—increase perceived risks associated with vaccination (Betsch and Sachse 2013). Hence, experts might feel that they must powerfully reject misleading information, and thereby might make people more suspicious about the accuracy of the experts' claims. However, it is important to note that authorities should also not downplay risks, as this can undermine trust dramatically (Betsch and Sachse 2013; Brown et al. 2010; Larson et al. 2011).

Furthermore, experts might tend to speak about the risks of a disease by focusing exclusively on numbers and statistics as well as medical terms such as herd immunity. However, statistics and medical terms depend on analytic risk perceptions and are not likely to evoke emotional reactions, the latter playing a crucial role in how people build risk perceptions (Slovic and Slovic 2015). The power of emotional risk perceptions also implies that a child dying due to a severe reaction to a vaccine could dramatically increase risk perceptions of vaccines, even when the chances of such a reaction are literally one in a million. Communicating that those chances are low would be unlikely to have a significant impact on vaccination acceptance, given the emotion-centred nature on risks.[11] Embedding emotional language such as "feeling protected" or "preventing regret" alongside medical evidence would help people to enhance their capacity to form accurate risk perceptions (Chapman and Coups 2006).

Uncertainty attached to predictions about the risk associated with a disease and the vaccination against it also enhance biases in information processing that might prevent people from forming or holding accurate beliefs about normatively relevant facts. People exhibit two different effects of uncertainty on prosocial behaviour: when people are uncertain about *whether* a self-serving choice will harm others, they behave selfishly (Dana et al. 2006). However, when uncertain about the *extent* of the harmful impact self-serving choices might have on others' welfare, people tend to err on the side of caution and do whatever minimizes harm to those others (Kappes et al. 2018, 2019). For instance, people might feel that it is uncertain if not being vaccinated for a disease like COVID-19 might be harmful to others, leading them to form self-serving beliefs (e.g., "I don't need to get vaccinated to protect others, they will be fine"). However, when people feel uncertain about the *extent* to which one's failure to vaccinate might potentially harm someone (especially when considering worst-case scenarios), they form pessimistic beliefs (e.g., "I need to get vaccinated, otherwise I might infect a vulnerable person and thereby endanger their life").[12] Self-serving and pessimistic beliefs are themselves not accurate perceptions of the situation, but the former increases

---

[11] Notably, the emotions associated with risk perceptions may also present an obstacle to an accurate recognition of the normative significance of the relevant facts. An individual may, for example, be aware that a risk of adverse reaction is low, but nevertheless place too much normative significance upon it, seeing a small risk of an adverse reaction as a significant reason to avoid vaccination.

[12] This is perhaps another instance where some factor can present an obstacle to the successful exercise of more than one of the three capacities of responsible agency. Self-serving beliefs, for example, may also prevent someone from seeing the potential for harm to *someone else* as a significant reason to vaccinate.

vaccination hesitancy, while the latter decreases it. In line with this idea, communicating the fate of a specific individual who severely suffered from others' decision not to vaccinate (as opposed to speaking about a number of nondescript people) may increase vaccination acceptance (Li et al. 2016).

An additional obstacle to the awareness of the normatively significant facts is a sense of unrealistic optimism, the widespread belief that negative events like getting a disease are likely to happen to others, but not oneself (Weinstein 1980). Similarly, theoretical approaches to health behaviour assume that without a personal feeling of vulnerability, people are unlikely to form the intention to get vaccinated (Weinstein et al. 2007). For a while, research suggested that people might form and defend such unrealistic optimistic beliefs by deliberately constructing scenarios about how relevant evidence might not apply to them (Lovallo and Kahneman 2003). Accordingly, a person might feel that it is likely that most people will get COVID-19 if they do not get vaccinated, but that they themselves are unlikely to get the virus since they rarely get the flu, either. However, more recent research indicates that people might automatically reject information that suggests they need to increase their risk perception without the need for deliberate thought (Kappes and Sharot 2019). The latter finding suggests that it might be difficult to change beliefs of unrealistic optimism. However, even if this is so, drawing attention to the risks that non-vaccination poses to others may be enough to elicit the judgement that one has sufficient reason to vaccinate. We discuss this further in the following section.

To summarize, the inevitable uncertainty attached to the risks of diseases, vaccinations, and the consequences for oneself and others, along with unrealistic optimism, often make it difficult for people to form accurate presentations of normatively relevant facts, but different communication strategies might be able to counteract these effects.

12.5 Obstacles to Recognizing the Normative Significance of the Relevant Facts

Whether an individual decides to vaccinate depends not only upon their awareness of the relevant facts (e.g., concerning various risks), but also upon whether they are able to recognize that those facts provide sufficient reason to do so. There are a number of different obstacles to the successful exercise of this capacity. One such obstacle is the tendency to focus on one's own interests while neglecting the interests of others.

One strategy to overcome this obstacle is to draw attention to the potential of negative consequences to others, rather than to oneself. When considering behaviour that entails risk to self and others, people must balance their individual aversion to risk with their inclination to do what is good for others (Gross et al.

2021). In general, people are less self-interested than many suspect, and can be highly motivated when they become aware that their behaviour (e.g., getting vaccinated) can help their community. For example, framings that highlight social rather than individual benefits of becoming vaccinated such as herd-immunity, the protection for vulnerable groups (e.g., the very young or very old, or those who cannot be vaccinated on medical grounds) have been found to increase intentions to get vaccinated in experimental settings (Betsch et al. 2013; see also Li et al. 2016). Similarly, parents often seem to be willing to vaccinate their children for the benefit of others, rather than the benefits to the child (Quadri-Sheriff et al. 2012). These findings suggest that people seem to be responsive to the needs of others and social benefits when contemplating whether they have sufficient reason to vaccinate. Furthermore, when people appear to ignore the costs that not getting vaccinated poses to others, this may simply indicate that they have not considered them, rather than being incapable of feeling concern for others.

Vaccination rates often rely on parents' compliance with vaccination schedules for their children. From a parent's perspective there are undeniably more desirable things to do with one's child than taking them to the doctor and having them injected with a vaccine, which may be slightly painful and cause harmless (though unpleasant) secondary effects. On the other hand, omitting to do so could cause one's child much greater suffering. Despite the fatal consequences an omission to vaccinate could have for one's child, research has shown that parents perceive harm which could occur because of deciding to immunize their child (commission/action) as less acceptable than harm that could occur as a result of deciding not to immunize their child (omission/inaction; Meszaros et al. 1996; Wroe et al. 2005). The tendency to favour omissions over commissions—commonly labelled the "omission bias"—is expressed in parental statements such as "I feel that if I vaccinated my kid and he died I would be more responsible for his death than if I hadn't vaccinated him and he died—sounds strange, I know. So I would not be willing to take as high a risk with the vaccine as I would with the flu" (Ritov and Baron 1990). Strategies aimed to counteract the omission bias could start by stressing that one is as much responsible for omissions than for commissions, or by framing the failure to vaccinate itself as a commission.

12.6 Obstacles to Implementing the Behaviour

Finally, people may form a judgement about what is best (e.g., that they ought to become vaccinated) and yet be prevented from acting on their better judgement because of a lack of self-control or fear of the procedure itself. Extensive research on health behaviour shows a substantial gap between people's intention and their behaviour (Sheeran 2002). A review including more than forty experimental studies found that a medium-to-large change in intention caused only a small-to-medium change in behaviour (Webb and Sheeran 2006). People often strongly feel that it would be the right thing to do to eat healthier foods, exercise more, practice safe sex, or get vaccinated, but fail to implement their intentions. Reasons for why people might fail to implement their intentions is that they might fail to get started (e.g., finding the initiative to book a vaccinate appointment), get derailed by distractions (e.g., postponing the appointment to finish writing an important book chapter), or because they are already stretched too thin (e.g., fail to find time in their schedule to go to an appointment because of obligations towards work, family, friends, and oneself) (Gollwitzer and Oettingen 2019; Gollwitzer and Sheeran 2006). While some research suggests that such failures of self-control reflect potentially innate, stable individual differences (Metcalfe and Mischel 1999), other research shows that teaching people strategies to boost their self-control significantly increases the relation between intentions and behaviour (Gollwitzer and Oettingen 2019). For instance, research by Liao and colleagues (2011) suggests that it is more efficient to facilitate vaccination planning—an effective strategy to increase self-control (Gollwitzer and Sheeran 2006) than provide more information to boost vaccination intentions. If people are already convinced that getting vaccinated is the right thing to do, then providing them with more information relating to significant normative beliefs will have little impact on their behaviour, but helping them to find a convenient time to get the vaccine will.

12.7 Conclusion

We began by setting out a philosophical framework concerning the rational capacities required for responsible agency. Our subsequent discussion provided insights from psychology and social science as they apply to these capacities concerning vaccination uptake. Research in these fields identify risk misperceptions, affective components of risk perceptions, and systemic biases as relevant factors in vaccine uptake. Although these factors present potential obstacles to vaccination, reflection upon these factors also yields promising strategies for reducing their deleterious effects, thus promoting the successful exercise of responsible agency in this regard.

References

Ajzen, I. (1985), "From Intentions to Actions: A Theory of Planned Behavior", in J. Kuhl and J. Beckman (ed.), *Action Control: From Cognitions to Behaviors* (Springer), 61–85.

Betsch, C., Böhm, R., and Korn, L. (2013), "Inviting Free-Riders or Appealing to Prosocial Behavior? Game-Theoretical Reflections on Communicating Herd Immunity in Vaccine Advocacy", in *Health Psychology* 32: 978–85.

Betsch, C., and Sachse, K. (2013), "Debunking Vaccination Myths: Strong Risk Negations Can Increase Perceived Vaccination Risks", *Health Psychology* 32: 146–55.

Betsch, C., and Schmid, P. (2013), "Does Fear Affect the Willingness to Be Vaccinated? The Influence of Cognitive and Affective Aspects of Risk Perception During Outbreaks", *Bundesgesundheitsblatt, Gesundheitsforschung, Gesundheitsschutz* 56: 124–30.

Betsch, C., Schmid, P., Heinemeier, D., et al. (2018), "Beyond Confidence: Development of a Measure Assessing the 5C Psychological Antecedents of Vaccination", *PloS One* 13/12: e0208601.

Brewer, N., Chapman, G., Gibbons, F., et al. (2007), "Meta-Analysis of the Relationship Between Risk Perception and Health Behavior: The Example of Vaccination", *Health Psychology* 26: 136–45.

Brown, K., Kroll, J., Hudson, M., et al. (2010), "Factors Underlying Parental Decisions about Combination Childhood Vaccinations Including MMR: A Systematic Review", *Vaccine* 28: 4235–48.

Capurro, G., Jardine, C., Tustin, J., and Driedger, M. (2022), "Moral Panic sbout 'Covidiots' in Canadian Newspaper Coverage of COVID-19", *PloS One* 17/1: e0261942.

Center for Disease Control. Date accessed: 31 January 31 2022. https://www.cdc.gov/coronavirus/2019-ncov/vaccines/effectiveness/index.html

Chapman, G., and Coups, E. (2006), "Emotions and Preventive Health Behavior: Worry, Regret, and Influenza Vaccination", *Health Psychology* 25: 82–90.

Clarke, R. (2017), "Ignorance, Revision, and Commonsense", in P. Robichaud and J. Wieland (ed.) *Responsibility: The Epistemic Condition* (Oxford University Press), 233–51.

Dana, J., Weber, R., and Kuang, J. (2006), "Exploiting Moral Wiggle Room: Experiments Demonstrating an Illusory Preference for Fairness", *Economic Theory* 33: 67–80.

Ecker, U., Lewandowsky, S., Cook, J., et al. (2022), "The Psychological Drivers of Misinformation Belief and Its Resistance to Correction", *Nature Reviews Psychology* 1/1: 13–29.

Fischer, J., and Ravizza, M. (1998), *Responsibility and Control: A Theory of Moral Responsibility* (Cambridge University Press).

Fishbein, M., and Ajzen, I. (1975), *Belief, Attitude, Intention, and Behavior: An Introduction to Theory and Research* (Addison-Wesley).

FitzPatrick, W. (2008), "Moral Responsibility and Normative Ignorance: Answering a New Skeptical Challenge", *Ethics* 118/4: 589–613.

Geiger M., Rees, F., Lilleholt, L., et al. (2021), "Measuring the 7Cs of Vaccination Readiness", *European Journal of Psychological Assessment*: 1–9. DOI: 10.1027/1015-5759/a000663.

Gollwitzer, P., and Oettingen, G. (2019), "Goal Attainment", R. Ryan (ed.) *The Oxford Handbook of Human Motivation* (Oxford University Press), 247–68.

Gollwitzer, P., and Sheeran, P. (2006), "Implementation Intentions and Goal Achievement: A Meta-Analysis of Effects and Processes", *Advances in Experimental Social Psychology* 38: 69–119.

Gross, J., Faber, N., Kappes, A., et al. (2021), "When Helping is Risky: The Behavioral and Neurobiological Tradeoff of Social and Risk Preferences", *Psychological Science* 32/11: 1842–55.

Kappes, A., Nussberger, A.-M., Faber, N., Kahane, G., Savulescu, J., and Crockett, M. (2018), "Uncertainty About the Impact of Social Decisions Increases Prosocial Behaviour", *Nature Human Behavior* 2: 573–80.

Kappes, A., Nussberger, A.-M., Siegel, J., et al. (2019), "Social Uncertainty is Heterogeneous and Sometimes Valuable", *Nature Human Behaviour* 3: 764.

Kappes, A., and Sharot, T. (2019), "The Automatic Nature of Motivated Belief Updating", *Behavioral Public Policy* 3/1: 87–103.

Larson, H., Cooper, L., Eskola, J., Katz, S., and Ratzan, S. (2011), "Addressing the Vaccine Confidence Gap", *The Lancet* 378: 526–35.

Larson, H., Jarrett, C., Eckersberger, E., Smith, D., and Paterson, P. (2014), "Understanding Vaccine Hesitancy around Vaccines and Vaccination from a Global Perspective: A Systematic Review of Published Literature, 2007–2012", *Vaccine* 32: 2150–9.

Levy, N. (2009), "Culpable Ignorance and Moral Responsibility: A Reply to FitzPatrick", *Ethics* 119/4: 729–41.

Levy, N. (2021), "Echoes of COVID Misinformation", *Philosophical Psychology*: 1–17. DOI: 10.1080/09515089.2021.2009452

Li, M., Taylor, E., Atkins, K., Chapman, G., and Galvani, A. (2016), "Stimulating Influenza Vaccination Via Prosocial Motives", *PloS One* 11: e0159780.

Liao, Q., Cowling, B., Lam, W., and Fielding, R. (2011), "Factors Affecting Intention to Receive and Self-Reported Receipt of 2009 Pandemic (H1N1) Vaccine in Hong Kong: A Longitudinal Study", *PloS One* 6: e17713.

Liau, A., and Zimet, G. (2000), "Undergraduates' Perception of HIV Immunization: Attitudes and Behaviours as Determining Factors", *International Journal of STD and AIDS* 11: 445–50.

Lovallo, D., and Kahneman, D. (2003), "Delusions of Success", *Harvard Business Review* 81/7: 56–63.

Machingaidze, S., and Wiysonge, C. (2021), "Understanding COVID-19 Vaccine Hesitancy", *Nature Medicine* 27/8: 1338–9.

McKenna, M. (2012), *Conversation and Responsibility* (Oxford University Press).

Mele, A. (2006), *Free Will and Luck* (Oxford University Press).

Mele, A. (2012), *Backsliding: Understanding Weakness of Will* (Oxford University Press).

Meszaros, J., Asch, D., Baron, J., et al. (1996), "Cognitive Processes and the Decisions of Some Parents to Forego Pertussis Vaccination for Their Children", *Journal of Clinical Epidemiology* 49: 697–703.

Metcalfe, J., and Mischel, W. (1999), "A Hot/Cool-System Analysis of Delay of Gratification: Dynamics of Willpower", *Psychological Review* 106/1: 3–19.

Mosleh, M., Pennycook, G., Arechar, A., and Rand, D. (2021), "Cognitive Reflection Correlates with Behavior on Twitter", *Nature Communications* 12/1: 1–10.

Miller, D. (2017), "Reasonable Foreseeability and Blameless Ignorance", *Philosophical Studies* 174/6: 1561–81.

Mills, E., Jadad, A., Ross, C., and Wilson, K. (2005), "Systematic Review of Qualitative Studies Exploring Parental Beliefs and Attitudes Toward Childhood Vaccination Identifies Common Barriers to Vaccination", *Journal of Clinical Epidemiology* 58: 1081–8.

Nelkin, D. (2011), *Making Sense of Freedom and Responsibility* (Oxford University Press).

Norman, P., Boer, H., and Seydel, E. (2005), "Protection Motivation Theory", M. Conner and P. Norman (ed.), *Predicting Health Behaviour: Research and Practice with Social Cognition Models* (Open University Press), 81–126.

Pennycook, G., Epstein, Z., Mosleh, M., et al. (2021), "Shifting Attention to Accuracy Can Reduce Misinformation Online", *Nature* 592/7855: 590–5.

Pennycook, G., and Rand, D. (2021), "Reducing the Spread of Fake News by Shifting Attention to Accuracy: Meta-Analytic Evidence of Replicability and Generalizability". https://psyarxiv.com/v8ruj

Ritov, I., and Baron, J. (1990), "Reluctance to Vaccinate: Omission Bias and Ambiguity", *Journal of Behavioral Decision Making* 3: 263–77.

Robichaud, P. (2014), "On Culpable Ignorance and Akrasia", *Ethics* 125/1: 137–51.

Rogers, R. (1975), "A Protection Motivation Theory of Fear Appeals and Attitude Change", *Journal of Psychology* 91: 93–114.

Roozenbeek, J., Schneider, C., Dryhurst, S., et al. (2020), "Susceptibility to Misinformation about COVID-19 around the World", *Royal Society Open Science* 7/10: 201199.

Ropeik, D. (2013), "How Society Should Respond to the Risk of Vaccine Rejection", *Human Vaccines and Immunotherapeutics* 9: 1815–8.

Rosen, G. (2004), "Skepticism about Moral Responsibility", *Philosophical Perspectives* 18/1: 295–313.

Rosental, H., and Shmueli, L. (2021), "Integrating Health Behavior Theories to Predict COVID-19 Vaccine Acceptance: Differences between Medical Students and Nursing Students", *Vaccines* 9/7: 783–796.

Rosenstock, I. (1974), "The Health Belief Model and Preventive Health Behavior", *Health Education Monographs* 2: 354–86.

Rudy-Hiller, F. (2017), "A Capacitarian Account of Culpable Ignorance", *Pacific Philosophical Quarterly* 98: 398–426.

Russell, P. (2004), "Responsibility and the Condition of Moral Sense", *Philosophical Topics* 32: 287–305.

Sheeran, P. (2002), "Intention–Behavior Relations: A Conceptual and Empirical Review", *European Review of Social Psychology* 12/1: 1–36.

Sheeran, P., and Webb, T. (2016), "The Intention–Behavior Gap", *Social and Personality Psychology Compass* 10/9: 503–18.

Sher, G. (2009), *Who Knew?: Responsibility Without Awareness* (Oxford University Press).

Slovic, P., and Slovic, S. (2015), *Numbers and Nerves: Information, Emotion, and Meaning in a World of Data (1 ed.)* (Oregon State University Press).

Tickner, S., Leman, P., and Woodcock, A. (2010), "The Immunisation Beliefs and Intentions Measure (IBIM): Predicting Parents' Intentions to Immunize Preschool Children", *Vaccine* 28: 3350–62.

Wallace, R. (1994), *Responsibility and the Moral Sentiments* (Harvard University Press).

Webb, T., and Sheeran, P. (2006), "Does Changing Behavioral Intentions Engender Behavior Change? A Meta-Analysis of the Experimental Evidence", *Psychological Bulletin* 132/2: 249–68.

Weinstein, N. (1980), "Unrealistic Optimism About Future Life Events", *Journal of Personality and Social Psychology* 39/5: 806–20.

Weinstein, N., Kwitel, A., McCaul, K., et al. (2007), "Risk Perceptions: Assessment and Relationship to Influenza Vaccination", *Health Psychology* 26/2: 146–51.

Widerker, D. (2006), "Blameworthiness and Frankfurt's Argument", D. Widerker and M. McKenna (ed.) *Moral Responsibility and Alternative Possibilities* (Ashgate), 53–73.

Wroe, A., Bhan, A., Salkovskis, P., and Bedford, H. (2005), "Feeling Bad about Immunising Our Children", *Vaccine* 23: 1428–33.

13

Progressive Reciprocal Responsibility

A Pre-emptive Framework for Future Pandemics

Julian Savulescu and Peter Marber

13.1 Introduction

Why were some countries able to implement quarantines and roll out vaccines effectively during the COVID-19 pandemic, while others suffered with poor wide scale public support and compliance? Part of the reason is because of a lack of fair, well-articulated government policies to deal with the costs of necessary lockdowns and other protocols, along with uneven citizen compliance and behaviour.

We argue for a "progressive reciprocal responsibility" framework, as an ethical and economic template for governments and citizens responding to pandemics. It provides incentives for greater public quarantine and vaccine compliance which can end future health crises faster, thereby reducing overall societal harm. We ask individuals to take responsibility during a pandemic, but this is a reasonable demand only if states themselves take responsibility for the ongoing welfare of their citizens, including those who bear the costs of government strategy to address pandemic threats, such as lockdowns and vaccination efforts. Our proposal also addresses the long-term economic damage when such crises have not been contained quickly enough. Such an accord should become a universal governing principle, perhaps under the auspices of the World Health Organization.

13.2 Background: The Need for Quarantine and Its Costs

As COVID-19 has shown us, the ill effects of prolonged pandemics can be devastating; hopefully, many lessons have been learned. One is the need for speed. While countries wait and debate quarantine, the risks of spread and deep economic contraction increase exponentially.

If a pandemic is not curbed quickly, the fallout looks remarkably like those of world wars, with both losers and winners (as we will later show). There are millions of fatalities and health-compromised victims, with billions displaced

Julian Savulescu and Peter Marber, *Progressive Reciprocal Responsibility: A Pre-emptive Framework for Future Pandemics* In: *Responsibility and Healthcare*. Edited by: Benjamin Davies, Gabriel De Marco, Neil Levy, Julian Savulescu, Oxford University Press. © Edited by Benjamin Davies, Gabriel De Marco, Neil Levy, Julian Savulescu 2024.
DOI: 10.1093/oso/9780192872234.003.0014

from jobs and school. Some industries, companies, and their employees are decimated while others profit like never before.

This has been a particular problem in democracies in Europe and the USA. Not only have these nations struggled with public compliance to quarantines, but the situation has been aggravated by governments' inconsistent messaging, slapdash start-and-stop lockdown policies, and an overall failure to acknowledge their reciprocal obligations of mandated quarantines. The result: a more than 4.5 per cent contraction in global economic activity in 2020 according to the World Bank, one of the deepest in decades, which has impacted all countries—even those that have curbed quickly or eliminated the spread of COVID-19. Some countries, such as Spain, saw their output fall by more than 10 per cent.

Global GDP losses from the travel industry alone have been estimated at over $4 trillion for 2020 and 2021. Developing countries experienced the brunt of the impact, accounting for nearly 60 per cent of the global GDP losses. Continued travel restrictions prevented tourism from picking up, as well as insufficient containment of the virus. Although tourism has recovered to a significant extent, international travel was still at 80% of pre-pandemic levels in the first quarter of 2023 (UNWTO 2023).

While the late 2020 news of vaccines caused many to believe the global economy would rebound sharply, research suggests that still only just over 70% of the global population has been vaccinated in late 2023. Limited vaccine rollout in developing countries slowed the global vaccination rate—and economic growth—and continues to impact poorer parts of the world more significantly than in developed countries (Economist Intelligence Unit 2021).

It is also worth considering the declining birth rates that can accompany an economic downturn, especially when the downturn is worldwide. One study has already shown that southern Europe, in particular, has been heavily affected by declining fertility linked to COVID-19, with birth rates in Italy, Spain, and Portugal seeing −9.1 per cent, −8.4 per cent, −6.6 per cent declines respectively (Aassve et al. 2021). Declining birth rates can have lasting impacts on a country's economy, with a shrinking workforce leading to slower economic growth and country-wide stagnation in development and innovation.

The flipside to these individual and collective losses is that there are also individual and corporate financial winners from the COVID-19 pandemic. Sectors such as personal protective equipment manufacturing, e-commerce and home delivery, online entertainment streaming, banks, and social media have certainly made excess profits due to prolonged lockdowns. The already wealthy have also prospered as government-engineered low interest rates have propelled markets to record levels.

With a better understanding of the potential costs of quarantines to modern economies, governments need to develop plans not only to effectively prevent and contain future pandemics, but also to create ethical and economic frameworks to prepare for, recover from, and finance their fallout. We propose a pre-emptive and curative framework for pandemics. Such tenets are to be enacted before future crises to promote speedier public compliance to reduce potential economic and health costs to societies, as well as to support those who bear the costs of quarantine and other measures. A multilateral accord of such principles might garner greater trust from citizens worldwide, as well as reinforce the notion that pandemics, like climate change, are global risks shared by all countries. Such principles have the potential to remove the localized politicization of pandemic policies, particularly as witnessed during the 2020 US presidential campaign season and many elections worldwide since then.

13.3 Government, Citizens, and Quarantines

Quarantines have been a human response to pandemics for millennia. There are references to them in the Old Testament for leprosy and in Greek writings by Hippocrates. They have been accepted public health policies for centuries, their use based on a fundamental responsibility of government, as well as pragmatic and moral arguments, to protect society against harm. In more recent times, quarantines were used during the Spanish flu of 1918, the Hong Kong flu of 1968, SARS in 2003, and Ebola in 2014–16.

Four ethical principles should underpin government pandemic calls for autonomy-limiting quarantines, including: (1) justifying by evidence the clear and measurable harm to others without such policies; (2) designing the least restrictive policies to bring viral spread under control; (3) clear justification for the decisions and the democratic right to appeal such actions; and (4) insuring responsible "reciprocity", meaning that if governments require individuals to make economic, welfare, or other sacrifices for the good of others, they have the obligation to reciprocate and assist with the costs and harms of such quarantine compliance (Upshur 2003).

Amid COVID-19 shelter-in-place orders, this last concept of reciprocity warrants further debate. Its ethos is embedded in many democratic constitutions such as America's 15th Amendment, which notes that private property shall not be taken for public use without "just compensation", along with similar concepts in Canada, most of Western Europe, and many other countries such as Chile, Brazil, and India. It is also rooted in social contract theory whereby citizens agree to cooperate for broader social benefits which sometimes requires the sacrifice of some individual freedoms.

As witnessed in 2020 and 2021, government-imposed quarantines created economic hardship worldwide. The unfortunate lost jobs, and underwent bankruptcies and evictions, while deteriorating health and education from prolonged lockdowns may have affected significantly more people than those hurt directly by the virus. Similarly to a world war, a pandemic also creates a responsibility for governments to reintegrate millions of economic casualties and refugees in their respective countries. But it also requires responsibilities on the parts of citizens who need to heed required quarantines, isolation, and other measures such as vaccination as required for the common good.

Under such circumstances, we address the responsibilities of each party—government and citizens—and questions of who should be justly compensated. We propose a framework to achieve this while also strengthening economies. We argue for "progressive reciprocal responsibility", not only to fairly compensate citizens who fully comply with pandemic orders, but also with a larger share of costs borne by beneficiary populations, some of whom may receive windfall profits due to government responses to the pandemic. Without such redress, quarantine policies may unfairly socialize the costs while privatizing the gains from pandemics. The proposed framework provides for greater long-term socio-economic prosperity—while fostering responsible citizenship and building greater national resilience against future health crises.

13.4 Responsibility, Reciprocity, and Just Compensation

While debatable, most governments possess inherent sovereign powers to quarantine populations amid a crisis like Ebola or COVID-19, or to enact eminent domain—the ability not only to expropriate facilities for use as health care centres, but also food, medicine, and other supplies as is often done during wartime. It also extends to the restriction of freedom of movement. This might be seen as a responsibility of government under most social contract theories. But such wide-ranging power does not eliminate statutory protections such as "just compensation" to restore the economic value had citizens' property and liberties not been taken away by quarantine measures.

Many COVID-19 policies that prohibited people from working or attending school are similar to eminent-domain actions which appropriate property: they hurt current and future livelihoods, and so must be fairly compensated. Again, this situation parallels wartime conditions.

While many governments responded to COVID-19 lockdowns with ad hoc policies such as temporary financial support for citizens and corporations, they were largely "act first, think later" decisions with minimal considerations of compensation amounts, outcomes, or fairness. This is to be expected during crises when time is of the essence, like pandemics. However, the practical reality is that some of the world's most vulnerable populations absorbed quarantine costs while some of the most fortunate actually gained. The recent experience with COVID-19 provides an opportunity to rethink and reaffirm reciprocal social contract responsibilities between the state and its citizens.

13.5 Government Responsibilities During Quarantines

To be fair, dozens of governments worldwide extended trillions of dollars in economic and social assistance during the COVID-19 pandemic. As new waves of the virus spread, some adapted programmes to help specific industries and populations particularly hard hit. However, it would be fair to categorize many actions as "just enough" and "let's wait and see" responses. Few countries answered comprehensively or systematically, and few contemplated or planned for the economic impacts of a global pandemic that would keep billions quarantined for a year or more.

Policy failures began early on, with preparation. In the USA, for example, dozens of plans had been drawn up by previous Federal administrations to address a potential pandemic, and yet initial action was largely left up to state and local governments. In addition, America's Strategic National Stockpile was short on personal protective equipment and ventilators before the pandemic began, and those shortages were not addressed in time to respond to COVID-19. In 2015, projections suggested that if a flu-like pandemic were to occur, the USA would need between 1 billion and 7 billion masks. Only 10 million had been stockpiled when COVID-19 began (Wallach and Myers 2020).

Testing was also a point of failure as early as January and February of 2020, just as COVID-19 was beginning to spread to other countries. COVID-19 tests were available by mid-January, and the World Health Organization ensured that these tests were distributed worldwide not long after they were developed. Despite early availability, the Center for Disease Control (CDC) decided to develop its own tests, but they proved to be faulty and unusable. The problem went unaddressed for two weeks or more—an eternity in a fast-moving pandemic (Wallach and Myers 2020).

While COVID-19 spread, and testing fell behind desired levels in many countries, certain governments actively discouraged people from doing anything differently in light of the developing situation. Most local and state governments in the USA, for example, were reluctant to disturb business as usual, and only once the situation was critical were quarantines and other stricter measures put in place suddenly and without any compensation to the public.

To some extent, well-defined and well-communicated frameworks for quarantine compensation would have a pre-emptive, reflexive effect: compliance would be higher, which would end the crisis quicker and curb economic fallout—in essence, reducing the need for such a framework. Indeed, providing people with assurances about their livelihoods during quarantine is an important component of public health compliance, which should help end the crisis faster and reduce damage. A 2020 study highlighted a 94 per cent compliance rate when compensation was offered versus less than 57 per cent without (Bodas and Peleg 2020). Most epidemiologists would agree improving quarantine compliance to 94 per cent from 57 per cent—37 percentage-point gain—could be the difference between a crisis ending in a few weeks versus lingering on for years. Although the virus was not eradicated, the swift, draconian clampdowns by the Chinese government resulted in relatively low infection rates with its "Zero COVID" policy.

13.6 Our Proposed Progressive Reciprocal Responsibility Framework

Our framework offers ten broad rules to address social sacrifice during pandemics. However, given that every country's economy may be affected idiosyncratically by a health crisis like COVID-19, we advocate that each sovereign state tailor such principles to their given situations and resources similar to recent global accords on climate change.

1. Compensation for Appropriating Facilities

Should government require use of hospitals, hotels, or other facilities, just compensation would be required as a result of such actions.

In the USA, the Fifth Amendment stipulates as much, stating that, "nor shall private property be taken without just compensation". Many US state and local governments have similar guidelines addressing the seizure of property in emergency circumstances. The California Emergency Services Act, for example, allows the governor to use private property if necessary in a state of emergency, but the reasonable value of the private property must be paid (Madden et al. 2020).

2. Lost Wage Assistance

If certain economic sectors are effectively closed because of a government-determined quarantine, compensation for lost wages and other financial assistance should be provided by government.

During COVID-19, many countries expanded their existing job retention schemes and developed new ones to address the loss of jobs due to the pandemic. France's Activité Partielle allows companies to declare *force majeure* in times of health crises, ensuring employees receive 70 per cent of their gross wages while relieving employers of the burden of paid-but-unworked hours. During the pandemic, France lengthened the duration of the programme from six months to twelve months. Germany lowered the threshold for support from its Kurzarbeit programme; before COVID-19 a firm could only request assistance if hours were cut for 30 per cent of its workforce, but this was lowered to 10 per cent in the early months of the pandemic. Under Kurzarbeit, employers pay employees 60 per cent of their lost earnings due to hours cut and are reimbursed for those costs.

These job retention schemes are aimed at employers to prevent large-scale layoffs and ensure employees not only stay in their jobs during a health crisis, but also receive assistance for both hours worked and those lost.

3. Hazard Pay for At-Risk Workers

Health care and other essential workers should receive extra hazard pay during pandemics. Not only are they asked to do risky work, but often it is at the expense of caring for families. This is similar to combat pay during wartime, as well as hazard pay in construction work or deep-sea diving.

While the adoption of hazard pay has largely been a voluntary decision for companies, some US states and cities have pushed for mandates that require companies to offer essential workers hazard pay during a pandemic. California and Seattle successfully passed mandates that require grocery, food retail, and pharmacy companies to provide hazard pay, and Pennsylvania and Vermont used federal funding from the "Coronavirus Aid, Relief, and. Economic Security" (CARES) Act to create hazard pay programmes. Whether state mandates or government funding prove more effective in getting hazard pay to workers remains to be seen.

4. Medical Assistance

If quarantining results in any physical or mental issues requiring professional care, such support needs to be provided and paid for by government. In Australia, for example, citizens experiencing dramatic or lasting mental health impacts from COVID-19 and mobility restrictions were eligible for ten extra Medicare-subsidized psychological therapy sessions each calendar year until 30 June 2022 (Australian Government Department of Health 2021). In England, the government launched programmes to help ensure people stuck to self-isolation including alternative accommodation for overcrowded households, "buddy" programmes for those whose mental health has been affected by lockdowns, along with support for individuals whose native language was not English (Department of Health and Social Care 2021).

5. Family Support Assistance

For those who must work while their children or other dependent family members (such as parents or grandparents) are sheltering-in-place, some financial assistance should be provided by government to enable effective quarantine.

When Romania closed schools to stop the spread of the COVID-19 virus, many parents who were required to work were left without a way to care for their children. In response, the country passed Law no. 19/2020, which required employers to provide paid leave to parents of children 12 years and under (one parent per household) if schools were closed. This law only applied to closures that were enacted under "extreme situations" (Eurofound 2020).

6.   Education Assistance

If schools are closed due to quarantine, government support must be made available for affected students which may include technology support (computers, WIFI, etc.) or remedial help.

   For example, Croatia's government-funded Croatian Academic and Research Network, or CARNET, partnered with the EU to provide laptops, tablets, high-speed Internet connectivity, and other tech and learning equipment to 920 teachers and more than 6000 students when the country's schools switched to distance learning. An additional 700 students were due to receive equipment before the end of 2022 (European Commission 2021).

   Argentina's Ministry of Education kicked off an educational media programme in April of 2020, which broadcast fourteen hours of educational television content and seven hours of radio content daily. Students who lacked televisions or radios, or access to the broadcasts, received book learning resources instead (World Bank Group 2021).

7.   Free Personal Protective Equipment (PPE)/Testing and Tracing

Should policies require PPE, it should be provided free of charge by government. Similarly, if testing and tracing equipment is required, government should supply them free of charge.

   While PPE shortages left many countries scrambling for supplies in the early months of the pandemic, governments tried to distribute what little PPE they did have to at-risk workers, primarily in the health care sector. In America, the United States' Families First Coronavirus Response Act ensured that everyone in the USA, even those without insurance, would be able to receive COVID-19 testing services for free. In the UK, lateral flow testing kits were provided free of charge to anyone who requested them.

8.   Vaccine and Treatment Trials

To the extent that trials are required to reduce pandemic harm, volunteers should be compensated by government for the risks taken for the broader society.

   Payments for research trial participation typically function as reimbursement and/or compensation. Reimbursement enables participation regardless of financial status by covering participation-related expenses, while compensation aims to prevent the exploitation of research participants who may receive no benefits in exchange for their time and contributions. Funds for both reimbursement and compensation typically come from limited budgets, but placing the burden of payment on government could lead to greater participation if payments are increased to levels more fitting to the risks. This applies even to risky challenge studies (Grimwade et al. 2020).

9. Free and Incentivized Vaccines and Treatments

Once effective vaccines and treatments have been developed, government should supply them free of charge. Moreover, governments should determine a fair, but not excessive, profit from such a vaccine. Reciprocity demands that the broad population not be burdened by the cost to enable the economy and country to continue to function. Moreover, it may be that governments also offer citizens cash incentives to receive vaccinations in an effort to bolster national immunity or consider mandatory policies (Savulescu 2021).

Lotteries are also a promising avenue to encourage vaccination, as participants tend to focus more on the large sum of money on offer than their chances of winning. A vaccine lottery in the city of Philadelphia used unique tactics to encourage vaccination: citizens are auto-enrolled based on their address, but the money up for grabs, $50,000, would only be awarded if the winner was vaccinated (City of Philadelphia 2021). This method takes advantage of what is known as "anticipated regret", or the sense that if you were to win but not be able to accept the prize due to not being vaccinated, you would feel regret.

How much these incentives have pushed vaccination rates is up for debate. The US state of Ohio successfully boosted rates with lotteries whose prizes included $1 million and more than one dozen full scholarships to university (Sehgal 2021). A study in Germany found that while cash incentives motivated vaccination particularly in those who were undecided, the payment had to be significant to have a promising impact (Klüver et al. 2021). Still, for those whose barrier to vaccination is financial (considering the time and cost associated with receiving even a free vaccine), cash incentives can decrease the financial burden of vaccination not felt by those who are more well-off.

It is essential that countries have vaccine compensation schemes, such as that of the UK. In the case of limited supply, governments should allocate vaccines according to ethical principles, balancing the requirement to bring about the most good with offering all citizens an equal opportunity to be vaccinated (Giubilini et al. 2021).

10. Reciprocal Tax Adjustments

The heavy financial costs of pandemics should be paid for by progressive taxation that considers not only those hurt by policies, but also the inadvertent economic beneficiaries. Possible strategies are detailed in the next section.

During COVID-19, countries worldwide adjusted individual and business taxes as well as offering extensions as part of their economic relief plans. Some of these adjustments included payment deferrals (Canada), early tax refunds for small businesses (Chile), VAT reductions (several countries), and reduced interest fees on late payments (Netherlands). Many of these programmes applied to all tax-payers and businesses, however, with few taking into account the lopsided impact of the pandemic and adjusting accordingly (Enache et al. 2020).

Responsible compliance is a key part of our proposed principles. It would hardly be fair to offer compensation to those who refuse to comply with quarantine regulations along with those who do. In order to be compensated by government, people must do their part, and that includes adhering to quarantine guidelines and getting vaccinated.

The proposed compensation and incentive framework serves not only as a philosophy for fair, reciprocal national sacrifices during pandemics, but also to strengthen national economic resiliency and curb trends towards inequality. In a perfect world, such schemes would boost compliance and end crises quickly, making their costs smaller than one might otherwise imagine.

13.7 The Progressive Principle

Most recent health crises that required lockdowns—including SARS, MERS, and Ebola—were relatively isolated with small ringfenced populations; COVID-19 is unique in its prolonged quarantines of billions of people around the globe. Unemployment and bankruptcy rates in some countries hit levels not seen since the Great Depression of the 1930s.

One unintended consequence of quarantines witnessed recently is that certain populations may suffer disproportionately as a pandemic persists. In many countries, knowledge and office-sector employees may be able to work from home with little or no interruption to their incomes. At the same time, many more who toil in lower-paying sectors such as travel, retail, hospitality, and day-labour may either lose their jobs or be furloughed for long periods. In the USA, this is roughly 25 per cent of the American workforce, and as the US Bureau of Labor Statistics reported during COVID-19, "occupations with lower wages are more common in the shutdown sectors than elsewhere in the economy and higher paying jobs are less common in those sectors. Consequently, shutdown policies disproportionately affect workers in lower paying jobs" (Dey and Loewenstein 2020).

In the UK, two-thirds of the population living below the poverty line lost their job or experienced a change in hours and consequently in earnings, while only one-third of the population above the poverty line experienced negative employment changes. In addition, 60 per cent of low-income families had to borrow money through credit cards or payday loans (Soodeen 2020).

Unfortunately, essential workers suffer from potentially hazardous conditions and their children often are left at home unattended. Single-parent households may also be at risk. For example, 23 per cent of US children under the age of 18 live with one parent and no other adults (Kramer 2019). Quarantines hit these vulnerable populations excessively hard; they tend to be the least educated and financially capable, often uninsured, with fewer resources to deal with long interruptions to income and family neglect. In the USA, studies before COVID-19 highlighted how 40 per cent of Americans could not afford a $400 emergency cost (Chen et al. 2019). In short, under quarantine, socioeconomic disparities can worsen for those already struggling.

These economic results have been called "K-Recoveries", with some corporations and better resourced workers normalizing, and even advancing, while others stagnate or fall further behind. E-commerce companies, for example, flourished during COVID-19 quarantines both from increased revenue and rising stock prices at the expense of traditional brick-and-mortar operators. Restaurants, movie theatres, and live entertainment venues suffered while grocery, video streaming, and social media sites gained. These winners, surprisingly, have relatively small workforces compared to physical stores, restaurants, and theatres.

Financial markets, too, generally are boosted from government monetary measures during a pandemic, including lower interest rates and quantitative easing. However, stock market ownership is relatively narrow in most countries. In the USA, for example, only 10 per cent of Americans control more than 84 per cent of stock market capitalization, with 50 per cent owned by the top 1 per cent (Wigglesworth 2020). Meanwhile, as a result of pandemic policies, the world's billionaires have seen their already huge fortunes grow to a record high of more than $10 trillion. The wealth of Jeff Bezos, one of the world's richest people, rose by more than a startling $76.3 billion during the pandemic—which is more than the individual GDPs of 120 countries (Smart 2021).

In short, the unintended consequences of quarantine policies cumulatively can accelerate inequality: a small slice of wealthy, better educated people can quickly pull further ahead of less-resourced, more imperilled citizens.

A further issue we cannot fully address here is the responsibilities governments accrue to other countries and their citizens. The most publicized example is the distribution of vaccines to poorer countries. Clearly, governments will have to balance responsibilities to their own citizens and those other countries (Emanuel et al. 2020). However, it is also clear that pandemics often cross borders, and as a result it can be argued that fair vaccine distribution is a global responsibility. Several countries certainly think so: the USA pledged to donate more than one billion doses of the COVID-19 vaccine to poorer countries, while the UK, Japan, Canada, and the European Union intend to contribute tens of millions. Most of these commitments are being routed through COVAX, the scheme run by organizations like WHO and UNICEF dedicated to providing low-income countries with the vaccine (BBC 2021).

Beyond the global responsibility to poorer nations is the economic concern associated with low worldwide vaccination rates. According to one study, if the worst-case scenario were to take place, in which wealthier nations achieve full vaccination while middle- and low-income countries lag far behind, the cost to the global economy could reach $9 trillion as a result of global supply chain disruption. The world's economies are interdependent to the point where full vaccination in wealthy countries would not stop them from taking on 49 per cent of the global costs of the pandemic if poorer economies are not helped. For both morally relevant and economically relevant reasons, then, it falls on wealthier nations to ensure fair vaccine distribution worldwide (Çakmaklı et al. 2021).

13.8 Funding Progressive Reciprocal Responsibility: Tax the Winners for Win-Wins?

What if such a framework cannot pre-empt or curb a pandemic? What about the costs to an economy if a crisis lingers for months, or even years such as COVID-19? Such scenarios should be considered, similar to the costs of wars, with a fair and reciprocal philosophy.

Given the mutual responsibilities defined earlier, quarantine costs and economic consequences should be absorbed by governments. However, this often requires borrowing given reduced tax collection during periods of lower economic activity, saddling future generations with the pandemic costs.

Interestingly, when examining subsidies during the recent COVID-19 pandemic, many countries saw assistance go disproportionately to large businesses and publicly traded companies rather than individuals directly. To the extent that the broader population underwrites these costs, the debt is levied disproportionately on vulnerable and future citizens.

Under utilitarian principles, a country's tax philosophy should be structured to maximize broad social welfare. The nineteenth-century economist Francis Edgeworth highlighted the fact that marginal financial gains are valued less as an individual's income rises, particularly in the wealthiest segments. As a result, a country's social welfare is maximized through progressive taxation. At certain income levels, tax rates rise marginally higher to proportionally shoulder public costs since equal rates cause more hardship to lower income populations.

There is some precedent for taxing "profiteers" during national crises, often at times when countries assume large deficits to keep their economies afloat. During World War I, the UK levied an Excess Profits Duty on companies that were profiting excessively at a time when the government ran huge annual deficits. This tax on war-time profits in excess of pre-war averages fluctuated between 40 per cent and 80 per cent and remained in effect through 1921 (Rutterford 2020).

This progressive principle should be an even higher priority during pandemics and lockdowns which impose severe costs and harms to some of the most vulnerable citizens. A temporary, fair, and transparent tax regime should be established to prevent effective exploitation and achieve reciprocity. In this progressive vein, several tax adjustments could be considered, including:

(1)   Temporary Changes in Base Income Tax Rates

Progressive taxation during a national crisis might involve reducing or eliminating rates on lower income populations while raising them on higher income populations. Interestingly, after World Wars I and II, US personal income rates were adjusted with higher marginal rates for the top quintile of earners to help pay down government debts accumulated during the wars.

This also could take the form of higher tax rates on larger companies with some revenue threshold. During the COVID-19 pandemic, several countries announced such moves to raise government revenue. Such adjustments could be tailored for each country's specific situation. In Malaysia, for example, the government raised taxes on companies with income of more than 100 million ringgit (approximately $25 million) from 24 per cent to 33 per cent (Lee 2021).

(2)   Temporary Changes in Sales Taxes and VATs

Sales taxes and value added tax (VAT) are considered regressive because they hit a larger percentage of income from lower-income taxpayers than from wealthier ones. During pandemics, countries could opt to reduce them on most goods, which acts as an immediate break for most citizens. To make such taxes less regressive, basic household items could be exempt up to a certain level, say $200. Luxury items might still be taxed at high rates since they can be better afforded by higher income groups. Indeed, many governments around the world made such adjustments during COVID-19 (Deloitte 2021).

(3)   Temporary Changes in Capital Gains Tax Rates

In many countries, profits made on investments held for a period of time are often taxed at a rate lower than half that paid on wages. This inherently favours wealthy populations who derive much of their income from investments versus labour.[1] Amid a pandemic in which securities may be inflated by monetary and fiscal

---

[1] This inequity was highlighted by billionaire Warren Buffett in 2016 who remarked that his effective tax rate in the USA was much lower than that paid by his secretary.

policies, not only do windfall gains go to relatively small populations; they are taxed, ironically, at lower rates. This is regressive and unfair under normal circumstances, and even more so during national crises.

Several studies highlight that nearly all capital gains benefits flow to the highest income brackets. In the USA, this costs the government more than $100 billion per annum with more than 75 per cent of the benefits going to those earning over $1 million, and a remarkable 50 per cent captured by only one in 1000 taxpayers (Lenzner 2011). Fairer treatment under quarantine could be achieved by raising the low rate or abolishing this tax break for a period of time. Such rates could be restored or adjusted when the economy has normalized after quarantine measures are fully lifted.

(4)   Temporary Windfall Taxes

Just as the UK raised taxes on profiteers during World War I, there are progressive arguments to levy taxes on specific populations, sectors, or companies that that have benefited the most from the economic windfall during pandemics. There are many examples of windfall policies in more recent periods, which have often been associated with energy price spikes (NPR 2008; Power 2009). During the COVID-19 pandemic, there were several calls for windfall taxes on pandemic profiteers such as pharmaceutical companies or online retailers (Gneiting et al. 2020; Topham 2021). Like the other tax suggestions above, these rates could and should be temporary and reset when the economy has normalized and would be tailored for countries' respective experiences.

In total, these progressive tax suggestions, while certainly controversial, could help to promote a more equitable distribution of gains, losses, and costs from pandemics within countries. In the long run, such adjustments help to counter trends in inequality that may otherwise result from extensive lockdowns during pandemics. It also attempts to relieve future generations of taxpayers the burden of paying for past health crises.

13.9   Conclusion

Quarantine measures which restrict liberty and harm citizens economically are only justified under conditions of comprehensive reciprocal compensation. Moreover, as COVID-19 highlighted, such lockdown policies have the potential to create a few winners and many losers. Similarly, requiring vaccination, even through mandatory policies, requires just compensation for harm. Morality dictates that no one should benefit greatly during such crises at the expense of those harmed and weakened. Indeed, it is exploitation to profit from vulnerable populations.

Such health crises require a thorough, wartime policy response versus piecemeal approaches which often lead to further unfair socioeconomic damage. Our proposed framework ethically aims to: (1) end pandemics faster with incentives for compliance, wherever they take hold, thereby reducing total costs of lives and livelihoods; (2) support the weakest people in their respective societies versus destabilizing them; (3) help reduce the chance of spread beyond a country's borders; and (4) improve a country's ability to normalize economically and recover even after lengthy quarantines. Fewer people would be set back by such an approach, for shorter periods of time, with fewer winners profiting excessively.

On a global level, a multilateral accord ratifying such a philosophy of progressive reciprocal responsibility acts as collective security against future health crises. Agreeing to such principles helps to guard against spreads across borders, which benefits all countries. The plan reminds us that pandemics, like climate change or an invasion from outer space, are everyone's business everywhere. It is a small world, after all. As COVID-19 highlighted, pathogens know no geographic boundaries. The increase in the amount of food, manufactured goods, services, and people that have crossed borders over the last century—largely through cooperative, multilateral-set rules—has created more wealth and increased human wellness than in the previous ten millennia. Pandemics have been with humans in the past, and we will see them undoubtedly in the future. However, countries need not respond in piecemeal. Our progressive reciprocal responsibility framework could go a long way towards solving our next pandemic and possibly reinvigorate multilateral problem-solving in the twenty-first century.

References

Aassve, A., Cavalli, N., Mencarini, L., and Sanders, S. (2021), "Early Assessment of the Relationship between the COVID-19 Pandemic and Births in High-Income Countries", in *PNAS* 18/36: e2105709177. https://doi.org/10.1073/pnas.2105709118

Australian Government Department of Health (2021), "Mental Health Support", 23 December 2021. https://www.health.gov.au/health-alerts/covid-19/support/mental-health

BBC (2021), "Covax: How Many Covid Vaccines Have the US and the Other G7 Countries Pledged?", *BBC News*, 23 September 2021. https://www.bbc.com/news/world-55795297

Bodas, M., and Peleg, K. (2020), "Self-Isolation Compliance in the COVID-19 Era Influenced by Compensation: Findings from a Recent Survey in Israel", in *Health Affairs* 39/6: 936–41.

Çakmaklı, C., Demiralp, S., Kalemli-Özcan, S., Yesiltas, Ş., and Yıldırım, M. (2021), "The Economic Case for Global Vaccinations: An Epidemiological Model with

International Production Networks", *National Bureau of Economic Research Working Paper 28395*. https://doi.org/10.3386/w28395

Chen, L., Duchan, C., Durante, A., et al. (2019), "Report on the Economic Well-Being of US Households in 2018", *Board of Governors of the Federal Reserve System*. https://www.federalreserve.gov/publications/files/2018-report-economic-well-being-us-households-201905.pdf

City of Philadelphia (2021), "Mayor Kenney Announces 'Philly Vax Sweepstakes' Giving Vaccinated Philadelphians Dozens of Chances to Win Up to $50,000", 7 June 2021. https://www.phila.gov/2021-06-07-mayor-kenney-announces-philly-vax-sweepstakes-giving-vaccinated-philadelphians-dozens-of-chances-to-win-up-to-50000/

Deloitte (2021), *Covid-19 Related VAT and Sales Tax Measures: Global Summary*, 1 September 2021. https://www2.deloitte.com/content/dam/Deloitte/be/Documents/tax/COVID-19_TaxSurvey.pdf

Department of Health and Social Care (2021), "Government Launches New Pilots to Further Support People to Self-Isolate", 23 May 2021. https://www.gov.uk/government/news/government-launches-new-pilots-to-further-support-people-to-self-isolate.

Dey, M., and Loewenstein, M. (2020), "How Many Workers Are Employed in Sectors Directly Affected by COVID-19 Shutdowns, Where Do They Work, and How Much Do They Earn?", *U.S. Bureau of Labor Statistics*, April 2020. https://www.bls.gov/opub/mlr/2020/article/covid-19-shutdowns.htm

Economist Intelligence Unit (2021), "More than 85 Poor Countries Will Not Have Widespread Access to Coronavirus Vaccines before 2023", 27 January 2021. https://www.eiu.com/n/85-poor-countries-will-not-have-access-to-coronavirus-vaccines/

Emanuel, E., Persad, G., Kern, A., et al. (2020), "An Ethical Framework for Global Vaccine Allocation", in *Science* 369/6509: 1309–12.

Enache, C., Asen, E., and Bunn, D. (2020), "Covid-19 Economic Relief Plans Around the World", *Tax Foundation* 8 May 2020. https://taxfoundation.org/coronavirus-country-by-country-responses/#table

Eurofound (2020), "Free Paid Days to Parents for the Purpose of Childcare", http://eurofound.link/covid19eupolicywatch

European Commission (2021), "E-Learning through Lockdown", 24 March, 2021. https://ec.europa.eu/info/strategy/recovery-plan-europe/recovery-coronavirus-success-stories/digital/e-learning-through-lockdown_en

Giubilini, A., Savulescu, J., and Wilkinson, D. (2021), "Queue Questions: Ethics of COVID-19 Vaccine Prioritization", in *Bioethics* 35: 348–55.

Gneiting, U., Lusiani, N., and Tamir, I. (2020), "Power, Profits and the Pandemic: From Corporate Extraction for the Few to an Economy That Works for All", *Oxfam*. https://oxfamilibrary.openrepository.com/handle/10546/621044

Grimwade, O., Savulescu, J., Giubilini, A., et al. (2020), "Payment in Challenge Studies: Ethics, Attitudes and a New Payment for Risk Model", in *Journal of Medical Ethics* 46: 815–26.

Klüver, H., Hartmann, F., Humphreys, M., Geissler, F., and Giesecke, J. (2021), "Incentives Can Spur COVID-19 Vaccination Uptake", in *Proceedings of the National Academy of Sciences* 118/36: e2109543118.

Kramer, S. (2019), "US Has World''s Highest Rate of Children Living in Single-Parent Households", *Pew Research Center*, 12 December 2019. https://www.pewresearch.org/fact-tank/2019/12/12/u-s-children-more-likely-than-children-in-other-countries-to-live-with-just-one-parent/

Lee, Y. (2021), "Malaysia Stocks Fall 2 per cent as Government Announces 'Windfall' Tax on Companies", *CNBC*, 1 November 2021. https://www.cnbc.com/2021/11/01/malaysia-stocks-fall-as-government-announces-windfall-tax-on-companies.html

Lenzner, Robert (2011), "The Top 0.1 per cent of the Nation Earn Half of All Capital Gains", *Forbes* 20 November 2011. https://www.forbes.com/sites/robertlenzner/2011/11/20/the-top-0-1-of-the-nation-earn-half-of-all-capital-gains/?sh=6ebe3d767893

Madden, K., Mercer, S., and White, C. (2020), "Can State and Local Government Seize Your Private Property During a Global Pandemic?", *Brownstein Hyatt Farber Schreck*, 26 March 2020. https://www.bhfs.com/insights/alerts-articles/2020/can-state-and-local-government-seize-your-private-property-during-a-global-pandemic-

NPR (2008), "Historian Details Roots of Windfall Profit Tax", 2 May 2008. https://www.npr.org/templates/story/story.php?storyId=90142714

Our World In Data (2023), "Coronavirus (Covid-19) Vaccinations". https://ourworldindata.org/covid-vaccinations

Power (2009), "Finland to Tax Nuclear, Hydropower to Cut 'Windfall' Utility Profits", 8 April 2009. https://www.powermag.com/finland-to-tax-nuclear-hydropower-to-cut-windfall-utility-profits/

Rutterford, J. (2020), "Taxing Financial Winners from Coronavirus to Pay for the Crisis: Lessons from WW1", in *The Conversation*, 13 October 2020. https://theconversation.com/taxing-financial-winners-from-coronavirus-to-pay-for-the-crisis-lessons-from-ww1-147790

Savulescu, J. (2021), "Good Reasons to Vaccinate: Mandatory or Payment for Risk?", in *Journal of Medical Ethics* 47: 78–85.

Sehgal, N. (2021), "Impact of Vax-a-Million Lottery on COVID-19 Vaccination Rates in Ohio", in *American Journal of Medicine* 134/11: 1424–6.

Smart, T. (2021), "The Pandemic Has Been a Windfall for Billionaires", *U.S. News*, 24 February 2021. https://www.usnews.com/news/economy/articles/2021-02-24/the-pandemic-has-been-a-windfall-for-billionaires

Soodeen, F. (2020), "Emerging Evidence on Covid-19's Impact on Money and Resources", *The Health Foundation*. 17 September 2020. https://www.health.org.uk/news-and-comment/blogs/emerging-evidence-on-covid-19s-impact-on-money-and-resources

Topham, G. (2021), "Fresh Calls for Windfall Tax on Companies That Prospered During Covid", in *Guardian*, 20 September 2021. https://www.theguardian.com/politics/2021/sep/20/fresh-calls-for-windfall-tax-on-companies-that-prospered-during-covid

UNWTO (2023), "Tourism on Track for Full Recovery as New Data Shows Strong Start to 2023". https://www.unwto.org/news/tourism-on-track-for-full-recovery-as-new-data-shows-strong-start-to-2023

Upshur, R. (2003), "The Ethics of Quarantine", in *AMA Journal of Ethics* 5/11: 393–5.

Wallach, P, and Myers, J. (2020), "The Federal Government's Coronavirus Actions and Failures", *Brookings*, 1 April 2020. https://www.brookings.edu/research/the-federal-governments-coronavirus-actions-and-failures-timeline-and-themes/

Wigglesworth, R. (2020), "How America's 1 per cent Came to Dominate Stock Ownership", *Financial Times*, 11 February 2020. https://www.ft.com/content/2501e154-4789-11ea-aeb3-955839e06441

World Bank Group (2021), "Remote Learning during COVID-19: Lessons from Today, Principles for Tomorrow", 10 December 2021. https://www.worldbank.org/en/topic/edutech/brief/how-countries-are-using-edtech-to-support-remote-learning-during-the-covid-19-pandemic

14

Inequalities in Prospective Life Expectancy

Should Luck Egalitarians Care?

Shlomi Segall

When you're alive, you're alive; but when you're dead, you're dead.
An old Jewish saying

14.1 Introduction

According to recent research by the UK Office for National Statistics (2018), boys born in England in 2043 could expect to lead lives that are almost two years longer than those born in Scotland that year (ONS 2018).[1] The result is a disparity in life expectancy between two possible persons, potentially being born on each side of the border. Is there something unfair about this disparity? In a recent paper, Alex Voorhoeve (2021) argues that there is. On Voorhoeve's reading, we (society) have responsibility to address such disparities (we shall soon see the details of his proposal). I want to try to argue that this case presents no concern for egalitarians, and suggest an alternative way for handling distributions across persons who are merely possible.

On the face of it, it seems that egalitarians, and perhaps especially, luck egalitarians, should find disparities across all persons, including possible ones, to be objectionable. Luck egalitarians after all hold that all inequalities in health whose origins cannot be traced to the patient's own responsibility are thereby rendered unjust. It would follow that we have a prima facie responsibility to compensate such patients. Some luck egalitarians have even gone so far as to say that inequalities in life span between men and women, say, are therefore also unjust (Segall 2010). Shouldn't the same be true also for unequal *prospective* lives? Shouldn't we be exercised by inequalities (in life expectancy) between possible

[1] I am grateful to the audience at the Rutgers Center for Population-Level Bioethics Seminar, and to Ben Davies, Julia Mosquera, and Alex Voorhoeve for their extensive written feedback.

Shlomi Segall, *Inequalities in Prospective Life Expectancy: Should Luck Egalitarians Care?* In: *Responsibility and Healthcare.* Edited by: Benjamin Davies, Gabriel De Marco, Neil Levy, Julian Savulescu, Oxford University Press. © Edited by Benjamin Davies, Gabriel De Marco, Neil Levy, Julian Savulescu 2024. DOI: 10.1093/oso/9780192872234.003.0015

persons? Some (e.g., Voorhoeve) think that we should. I will try and argue in this paper that we should not.

According to what Voorhoeve calls *Equality for Prospective People* (henceforth EPP):

> It is important to ensure equality in people's life prospects, not merely between actual individuals, but also between all individuals who, given our choices, have a chance of coming into existence.   (2021: 304)

The challenge for the egalitarian is obvious. In the case presented, as indeed the case of persons who are merely possible, there is no certainty of *outcome* inequality across persons' life span. Instead we have an inequality in prospective lives, between individuals who will not co-exist. Standard, *ex post*, egalitarianism, of course, detects no badness when there is no inequality of outcome. (*Ex post* egalitarianism is the view that the badness of inequality is a function solely of the inequality in outcome between individuals.) To pass judgement according to which something is still amiss would require a novel addition to egalitarian thinking. Voorhoeve offers us precisely that. He, moreover, suggests that such a concern is in line with brute luck egalitarianism. I want to say, in reply, that in this particular context luck egalitarianism is out of place, and that our intuitions with regard to such cases are better explained by *ex post* prioritarianism. (By '*ex post* prioritarianism' I mean that the value of a prospect is determined by the expected value of the outcome, where benefits to individuals matter more the worse off, in absolute terms, the individuals are.)

14.2 Equality for Possible Persons

There is something troublesome about the disparity in life expectancy north and south of Hadrian's Wall. That is obvious. What is less obvious is why, exactly, it is troubling.

Since the aforementioned gap is between boys of different localities rather than between the sexes, it is obvious that some social determinants of health operate behind the scene.[2] And on any straightforward reading of justice (say, luck egalitarianism), it is not the fault of the future Scottish boys that they are born into these disadvantageous circumstances. This is pretty straightforward. Moreover, not only don't we need an elaborate ethical theory in this case in order to know that the state of affairs is wrong; we also know what the remedy ought to be. Since for the foreseeable future humans will probably continue to

---

[2] This is designed to get rid of the type of concerns raised by Hausman (2006).

inhabit both England and Scotland, and since these humans are likely to continue to procreate, it is pretty clear that we ought to ameliorate the state of affairs. We ought to try to equalize the prospects of good living between those born north and south of the border.

That, however, is *not* the question before us here. I am not asking whether it is wrong that in the future English men will live longer than Scottish men with whom they *will* co-exist. It is obvious that this *is* unjust. Rather, I ask a narrower question. Is it bad because unfair that *a* boy who might be born in England in 2043 can expect to live two more years than an *alternative* boy who might be born in Scotland at that same time? In other words, are inequalities in prospective lives between persons who will not co-exist unfair? Put differently, do we have a duty to prevent inequalities between alternative possible persons, as well as between future actual persons?

To flesh this out, consider the following case, adapted from one introduced by Voorhoeve (2021: 306). Suppose that through *in vitro* fertilization a woman who is a stranger to you will have a child (call him Chris). The prospective mother will develop a condition that will not affect her well-being but will affect the well-being of the child. The condition if untreated will cause the child to have a merely tolerable level of well-being (say at 30). The mother's condition comes in two equiprobable types, Type 1 and Type 2, the occurrence of which is independent of your choice. We can however improve the well-being of the child in one of two ways. Option A will unfortunately be ineffective in the event that the mother has Type 1. However, that treatment will fully cure the child if the mother has Type 2 of the condition, allowing him to have a very good life (at 90 say). Option B involves a treatment that is equally partially effective, ensuring that the child will have a moderate level of well-being (60 minus [c], where c is a small but positive amount), independently of which type of the condition the mother has. The case has an intra-personal variant (Table 14.1), and an inter-personal one (Table 14.2).

In the intra-personal variant alternative A is the course of action that maximizes the sole person's expected utility. Suppose therefore that you think this is the only intuitive course of action (I don't, but leave that aside for now). Let us consider the inter-personal variant first (Table 14.2), and let us bracket any potential inequalities concerning third parties (assume for the sake of argument that Dominic and Evan are the only persons in existence). Notice that whatever

Table 14.1 Case One: Future one child: Intra-personal

Alternative		Type 1 (0.5)	Type 2 (0.5)
A	Chris	30	90
B	Chris	60−c	60−c

Table 14.2 Inter-personal

Alternative	Type 1 (0.5)	Type 2 (0.5)
Dominic A	—	90
Evan	30	—
Dominic B	—	60−c
Evan	60−c	—

Table 14.3 Case Two: Concurrent inequalities between possible persons

Alternative	Type 1 (0.5)	Type 2 (0.5)
Dominic A	90	—
Evan	30	—
Dominic B	60	—
Evan	60	—

happens only one person will come into being, and therefore, there is never going to be outcome inequality. Suppose, consequently, that you have the intuition that we ought to opt for B. Standard, *ex post* egalitarianism, would not give us this desired result. That view is silent here, because whatever happens there is not going to be outcome inequality.

It is easy to see that EPP reaches attractive judgements in both variants of Case One. In the intra-personal scenario, since both prospects are Chris's, Voorhoeve (2021: 307) says that we should simply maximize his expected utility, which implies opting for A. In the inter-personal scenario, in contrast, EPP implies that we ought to opt for B:

> A makes the world a much more welcoming place for one of the two children who may arrive, whereas B makes the world just as welcoming for any child who might arrive, at modest cost. Equal concern therefore requires that we choose B. (2021: 310)

EPP seems to get us the correct result in Case One. But to see that it has a distinct input, contrast Case One with Case Two (Table 14.3).

Here Dominic and Evan are possible persons who, if they come about, *will* co-exist. Here as well, I presume, most of us have the intuition that we ought to opt for B. But unlike the previous case, Case Two can be easily explained by standard, *ex post* egalitarianism. That view seeks to maximize the *expected egalitarian value*,

Table 14.4 Case Three

Alternative	Type 1 (0.5)	Type 2 (0.5)
Howard A	—	90 + e
Idris	30 + e	—
Howard B	—	—
Idris	30	90

or in other words, the chance of bringing about the less unequal outcome. Therefore in the choice between A and B it opts for the prospect that maximizes the egalitarian prospect, namely B.

As I said, it is cases such as Case One that interest us in this paper, not Case Two. Notice that in Cases One and Two, even though they differ in their reasoning, Voorhoeve's EPP and *ex post* prioritarianism would offer the same verdict. (Although I should qualify this: on Voorhoeve's account, outcome inequality between co-existents is weightier than prospect inequality. It follows that the 'cost' for which equality must be achieved in Case One is lower than the cost for which it is to be achieved in Case Two.)[3] To see where these two views may disagree consider a third case, where (e) can be anything equal to or larger than zero (Table 14.4).

Here *ex post* prioritarianism has a clear preference for A. But a view committed to equalizing life *prospects* between possible persons, the way EPP is, may well favour B (for a small enough [e]). A contains large inequalities in life expectancy between possible persons, whereas B contains no such inequality. (I leave until the last section the question of how plausible the EPP verdict is in this particular case.)

As mentioned, I agree that in Case One Inter-personal we should opt for B. But I disagree that this is so for egalitarian reasons. In the rest of this paper I shall try to convince you of that. Egalitarianism, I maintain, is silent on Case One.

14.3 The EPP: Preliminary Difficulties

One, perhaps chief, motivation behind the EPP is that it not only explains our intuition in the inter-personal case (concerning possible persons); it also explains well the so-called 'shift' between the inter-personal variant and the intra-personal one. Voorhoeve writes that the EPP only kicks in in inter-personal scenarios, when competing claims of different individuals obtain. In contrast, in intra-

---

[3] I am grateful to Alex for pointing this out to me.

personal cases, when there are no competing claims of different individuals, we should opt to maximize the sole person's expected utility, conditional-on-existence. 'Expected utility conditional-on-existence' refers to taking into account only the prospect where that person exists, rather than taking into account (say, by factoring in zero well-being) also the prospects where that person does not exist. In Case One, intra-personal, we should opt for A. 'This option maximizes [the child's] expected well-being. Given our measure of well-being, it is the option that is most choiceworthy on [the child's] behalf' (2021: 307).

Let me first say something about the intra-personal variant of Case One. Opting for A in the intra-personal variant seems very sensible. But as Voorhoeve himself hints, as a rule, maximizing the expected utility, conditional-on-existence, of the sole person in question may not always be the best rule. In the following I will rehearse an objection (put forward by Toby Handfield (2018: 608)) to Voorhoeve's principle for intra-personal trade-offs. Voorhoeve actually anticipates this objection, so after presenting it I will spend time trying to deflect his reply to it.

But first here is Handfield's objection. Suppose you can bring a child, Henry, into the world but he is destined to have a horrible life, indeed one worth ending on his behalf. If you were Henry's guardian that is what you ought to do on his behalf (end his life). In Prospect X, there is a 50 per cent chance of Henry having a life worth *not* living at −10, and another 50 per cent chance of him leading a slightly better life, although still worth not living, at −2. You may, alternatively, opt for Prospect Y, thereby giving Henry a 50 per cent chance of the same miserable life at −10, and a 50 per cent of not coming into being at all. The options are summarized in Table 14.5, amended from Handfield's (2018) example.

Maximizing the sole person's expected utility conditional-on-existence, as Voorhoeve's EPP mandates, would lead us here to opt for X. But surely that would be wrong.

Notice that the problem is unique to views reliant on well-being conditional-on-existence. For if we took non-existence to equal zero well-being the problem would not arise. In that case, Henry's expected well-being in Y would be −5, and so a view committed to non-existence as equaling zero welfare would opt for that. That, however, is not the route taken by Voorhoeve (and for good reason, but we cannot go into that here).

Voorhoeve, I said, anticipates this objection, and in fact his EPP has two mechanisms meant precisely for averting this forceful objection. The first is that he brackets

Table 14.5 Case Four: the sadistic conclusion

	S_1 (0.5)	S_2 (0.5)
X	−2	−10
Y		−10

negative welfare, that is, lives worth not living, from the purview of the EPP.[4] In doing so Voorhoeve alludes to discussions by Jeff McMahan and others, to say that we typically think that different distributive rules govern benefits rather than harms, and bringing about good lives as opposed to lives not worth living. But notice that McMahan's point concerns the strength of our *reasons* for bringing people into existence.[5] He, very sensibly, goes on to say that different principles govern these reasons when it comes to good lives as opposed to our reasons for bringing (or refraining from bringing) lives worth ending. But notice that in the present context our concern is with something else, namely *distributive value* (say, of welfare). In that context, it would be rather arbitrary to delimit the discussion to positive levels only. Let me stress the point. It is perfectly plausible to say that we have one set of reasons to guide us in deciding to bring about good lives, and a different set of reasons to guide us in thinking about lives worth not living. One may disagree with this, but there is at least nothing prima facie incoherent about it. But when we discuss the value of distribution of X (again, think of welfare), it *would* be arbitrary to delimit your distributive principle to certain levels and not others. Let me give two illustrations for why this restriction is problematic. First, notice that some distributive views have specific input about the different value of different levels of well-being. For example, sufficientarianism and critical-level utilitarianism (and prioritarianism) are distributive views, and ones that treat different levels of well-being differently. You may happen not to find these views particularly attractive (neither do I), but that's beside the point. The point rather is that a distributive view can have different things to say about different levels of well-being. What it cannot do is simply ignore a whole range of well-being, just because it happens to yield uncomfortable verdicts. Here is a second illustration. Certain levels of welfare, and especially negative welfare, can teach us a lot about a distributive principle. Ingmar Persson for example has advanced the view that it is not equality that is valuable, but rather inequality that is disvaluable. He argues that if equality were good this would have given us a reason to bring into existence equally situated individuals with lives worth not living, which is counterintuitive. But if inequality is bad, we lack such reason (Persson: 2001). This is an example of how negative welfare teaches us something about the value of distribution. The details of the argument are less important here; what is important is that different welfare levels have an important input into the very distributive principles we ought to adopt. Delimiting a distributive view to a certain range is therefore at high risk of being ad hoc.

---

[4] 'Different principles apply to the distribution of benefits than to the distribution of harms'. Voorhoeve (2021: 305).

[5] See for example, McMahan (1981), especially p. 105 onwards. McMahan, in critiquing Narveson, discusses not distributive value, but rather the asymmetry in the strength of our *reasons* to bring a good life as opposed to causing a life worth not living to exist.

Table 14.6 Case Five: Henry's positive predicament

	S_1 (0.5)	S_2 (0.5)
X	80	100
Y	—	100

Still, suppose we accept Voorhoeve's delimitation of his *distributive* principle to positive levels only. In the following I switch the example to one containing only positive values. Suppose that whatever you do, if S_2 occurs, Henry would live until the ripe age of 100. If S_1 occurs, and you opt for X, he would have the additional 0.5 chance of leading a slightly less long life, at 80. If you opt for Y, in contrast, he will have an additional 0.5 chance of not coming into being (Table 14.6).

Maximizing Henry's expected well-being conditional on existence would lead us here to opt for Y (100>90), which is also (if not equally as) counterintuitive.

Now to this objection Voorhoeve has a different ready reply, namely yet another restriction of the purview of his argument. He explicitly writes that his view is restricted to 'same actual number … cases' (2021: 305), which Case Four and Five are manifestly not.[6] 'Same actual number cases' refers to comparing two or more populations that have the exact same number of persons in them, even when the identity of these individuals may vary. Now, this further restriction is not, I want to quickly concede, as problematic as the previous restriction (to positive welfare). Still, I want to say that it is not costless. Let me make two points to that effect, one extremely straightforward, the other less so. First, suppose the restriction to same actual number is indeed non-problematic. But suppose also that the rival principle that I will propose, *ex post* prioritarianism, delivers the correct result in cases Four and Five (which it in fact does). At minimum, then, my proposed view has at least one advantage over Voorhoeve's EPP. Here is the second point. It is true that it is quite common in population ethics to 'quarantine' distributive principles from applying to Different Numbers Cases.[7] That is, it is often agreed that distributive principles are exempt from applying to comparisons of cases that have different numbers of individuals in them. Now, it would be useful to conduct a little detour to remind ourselves why exactly it is that some in the literature (myself excluded) agree that distributive principles are exempt from comparing different number cases. The major test case is of course the Repugnant Conclusion (Parfit 1984: 388). What that objection shows is how aggregative principles struggle with a case that has the following two elements.

---

[6] See Voorhoeve (2021: 305, fn. 4).
[7] For recent discussions of this 'quarantining' strategy see Otsuka (2022), and Segall (2022).

The Repugnant Conclusion compares populations that have a large disparity in number of individuals on the one hand, and a large disparity of welfare (between these two or more sub-populations). A typical case would compare ten million people, say, all living equally good life at level 100, say, with a billion people, say, all living equally wretched lives at level 1, say. If the aggregate utility in the second population exceeds that of the former, we have encountered a repugnant conclusion. As a consequence, many people in the field think it is appropriate to not apply distributive principles to such Different Numbers cases. So far so good.

Now, Case Four and Five are, strictly speaking, Different Numbers cases. But they lack the two features of the repugnant conclusion mentioned above. They involve neither huge disparities in welfare, nor huge disparities in numbers of individuals. This is significant because whatever problems the EPP encounters in Cases Four and Five, these problems do *not* originate from (or merely from) aggregation, the original sin that motivates quarantining distributive views from Different Numbers cases. The fatal problem it encounters stems not from aggregation but from its tendency to *average*. In short, whatever problem the EPP has is not a problem unique to different numbers. It is a problem unique to views that rely on averaging. This, in turn, is significant because unlike the much less controversial quarantining of distributive views from (vastly) different numbers cases, the EPP's quarantining from applying to Case Four and Five is much more ad hoc and controversial.

I conclude that whatever we might think of the EPP's handling of inequalities in life expectancy *between* possible persons (our main concern in this paper), we can already see that the intra-personal variant of this view is not as attractive as it may initially appear. This is worth bearing in mind for its own sake, but also in so far as one (perhaps the main) underlying motivation behind the EPP is its ability to account for the so-called moral shift between inter- and intra- personal variants of cases concerning possible persons. Whatever we might find out about the EPP's handling of inter-personal cases, we can already register that despite appearances, it is not obvious that it handles successfully the intra-personal variants of such cases.

14.4 The Standing Objection

Suppose, however, we were to stick, as stipulated by Voorhoeve, to cases involving trade-offs between populations of same number of possible persons. Here, there is something quite appealing about the EPP. After all, we know that our actions can affect either future child, and so equal concern arguably mandates that we adopt an attitude that is equally accommodating to both. A rather similar sentiment has been put forward by Sophia Reibetanz Moreau (1998: 304):

As long as we know that acceptance of a principle will affect *someone* in a certain way, we should assign that person a complaint that is based upon the full magnitude of the harm or benefit, even if we cannot identify the person in advance.

This seems appealing, at least initially. After all, why should it matter, for egalitarians especially, that you cannot identify the person in advance? Indeed, this view seems to be shared also by non-egalitarians. According to Derek Parfit's 'No Difference View', if some people are less well off, it makes no difference whether they are less well off than *they* themselves could have been, or less well off than *different people* might have been, if they had existed instead (Parfit 2011: 219, 2017: 123). Once we have a distributive principle in place (be it egalitarian or prioritarian), it does not seem to matter who the bearer of the complaint (or benefit) is. The idea that we might have a responsibility to make the world as equally welcoming as possible to all possible persons therefore sounds attractive and a plausible addition to an egalitarian view. In the following, I would try and convince you that despite this initial appeal, egalitarians should resist it. To do so, I want to mount a *standing objection* to the value of equality between possible persons. We shall shortly get into the details of my claim but by a 'standing objection' I refer to the claim that a particular individual (in my case, possible persons) do not have a moral standing to complain about X.

To mount my particular standing objection, let me first distinguish *shortfall complaints* from *comparative complaints*. The former concerns complaints that individuals might have about the gap between their actual well-being, and a counterfactual well-being that they could have had. Comparative complaints pertain to the gaps between one's well-being (whether actual or possible) and the well-being of others who would co-exist with her. My claim in a nutshell is that possible persons have moral standing to complain about *shortfalls* to their absolute levels of being, but that they do not have moral standing to complain about relative shortfalls, including to complain about inequality. This sounds rather strange and arbitrary, so I have some work to do justifying this.

Let us start by noticing that saying that possible persons have no standing to complain about inequality does not mean they have no interests that we are bound by. That possible persons have interests that we are bound by should actually be quite easy to see. Consider the following simple example, amended from Otsuka (2018: 198), in Table 14.7.

Table 14.7 Case Six

	Bea
D_1	
D_2	50
D_3	70

Suppose you think, as most non-total-utilitarians do, that D_1 is a permissible option. That is, you believe we are under no obligation to bring Bea into existence. Even so, we think that whatever you do, you must not opt for D_2. In other words, if you do opt to bring Bea into existence, you ought to bring her at the highest level of well-being, other things being equal.[8] This shows that possible persons do have interests that we are bound by. I should now turn to say why those interests do not translate into complaints about relative standing, including inequalities.

I want to start by rehearsing Otsuka's claim that possible persons have no moral standing to complain about inequality, and differentiate his version of the Standing Objection from mine. To illustrate the Standing Objection, think of an elaboration on Case One (Table 14.8).

Case One included alternatives A and B, and here I have added alternative C. Recall that the EPP says that 'it is important to ensure equality in people's life prospects, not merely between actual individuals, but also between all individuals who, given our choices, have a chance of coming into existence' (Voorhoeve 2021: 304). On this reading we have a strong (although not necessarily overwhelming) reason to opt for B, which in this case implies levelling down. Now notice, this is *not* your ordinary case of levelling down, for here there isn't even the redeeming effect of avoiding inequality in final well-being. Recall, whatever you do, only one person will come about. Instead, this is an incidence of what I have elsewhere called Gratuitous Levelling Down (Segall 2019). Let me put this differently. In comparing A and B it is easy to see why B is preferable to A even though it involves some loss of aggregate well-being. Option A causes Evan to have a worse prospect than he could have otherwise had. But if equality of prospects is valuable (even if it is defeasible) then the same should be true for B in comparison to C. It

Table 14.8 Case Seven: gratuitous levelling down

Alternative	Type 1 (0.5)	Type 2 (0.5)
Dominic A	—	90
Evan	30	—
Dominic B	—	60−c
Evan	60−c	—
Dominic C	—	70
Evan	65	—

[8] I say 'other things being equal', because of course there may be scenarios under which D_2 is permissible or even mandatory, say when there are other persons around, and they are all at 50. (I am grateful to Larry Temkin for bringing this to my attention.) But most would agree that, say, if Bea is the only person in the world, D_2 is impermissible.

follows that proponents of EPP are committed to the idea that B is better in at least one respect compared to C. Otherwise there would be no value to equality of prospects between possible persons. But is that true? In what sense could B be better than C? Obviously, it is not better in terms of equality in final well-being. Is it better in terms of some unfairness? That seems implausible. Evan has no complaint against our choice in C. If Type 1 occurs he will fare better than if B was chosen. And if Type 2 occurs he has no complaint because he in any case would not exist. In short, in choosing C we have not disadvantaged Evan either compared to himself nor compared to Dominic. To judge from this example, it is very hard to see why inequality in life expectancy between possible persons has any disvalue.

The case of Gratuitous Levelling Down bears out a strong intuition, namely scepticism about any disvalue of inequalities across possible persons. But to back up that intuition I want to draw on Otsuka's Standing Objection to such inequalities and throw in my twist on it. To set this up, notice that Voorhoeve's EPP mandates equality not only between possible persons, but also between possible and necessary persons. 'It is important to ensure equality in people's life prospects, not merely between actual individuals, but also between all individuals who, given our choices, have a chance of coming into existence' (Voorhoeve 2021: 304). Think now of an example from Otsuka (2018: 195), which I have slightly amended (Table 14.9).

Making the world equally welcome across necessary and possible persons would imply opting for O_1. But according to Otsuka (2018: 195) we should prioritize the well-being of persons who are going to exist for sure:

> [Bea's] coming into existence is a fait accompli. [Alice's] existence, by contrast, is dependent on your choice regarding whether or not to benefit [Bea]. The fact that [Alice's] existence is dependent on your choice is morally significant for the following reason. You can ensure, simply by opting to benefit [Bea], that [Alice] will never exist. In the event that you opt to benefit [Bea], [Alice] will be nothing more than a possible person, and never an actual person. It would, moreover, be a mistake to maintain that such a merely possible person might have any standing to complain about your failure to bring her into existence.

Notice, it is not that possible persons have no standing to lodge complaints, according to Otsuka. Instead, possible persons do have complaints, but those

Table 14.9 Case Eight: the standing objection

	Bea	Alice
O_1	70	70
O_2	85	

complaints are weaker than those held by necessary persons. Moreover, they do not include complaints about a failure to bring oneself into existence (2018: 199).

As mentioned, I agree with Otsuka that merely possible persons have no egalitarian complaints. But I also think his particular defence of the Standing Objection is problematic. This has been brought out in a recent critique by Matthew Adler (2022), which will be useful for my purposes here to examine. In an illuminating discussion, Adler argues that the Standing Objection, as presented by Otsuka, is arbitrary in at least three respects. It is *agent-relative, time-relative, and discriminates against those unable to voice a complaint.* Let me take those in turn.

Otsuka, we saw, writes that it would be a 'mistake to maintain that such a merely possible person might have any standing to complain about *your* failure to bring her into existence' (2018: 195). I agree with Adler (2022: 19) that this violates agent-neutrality: '[T]he independence or dependence of someone's existence on an actor's choice cannot be reflected in an agent-neutral outcome ranking.' Second, our assessment of the badness of inequality must not violate time-neutrality. The fact that at the point of decision somebody already exists is irrelevant. 'A time-relative ranking is at odds with a plausible axiom of rational choice, namely time-consistency' (Adler 2022: 19). Third and finally, says Adler, a moral assessment of the unfairness towards possible persons must not be hijacked by their inability to voice a complaint:

> It is obscure why a claim-holders' ability to lay claim to his or her claim should affect its weight. The claims framework, let's remember, is a precisification of the Person-Affecting Idea. Do we believe that the moral weight of a given benefit or harm to some individual—moral weight, specifically, as an ingredient in the moral goodness of outcomes—is somehow increased by the individual's ability to use language to describe that benefit or harm? (2022: 20)

If Adler is correct, there is no Standing Objection to the force of possible person's complaints, including complaints concerning inequality in life expectancy. In short, if Adler is correct that would be detrimental to my case against Voorhoeve's EPP.

I now want to say why I think the Standing objection could yet be amended and rescued. To see this think of the following thought experiment. Suppose that you could either benefit an existing extremely well-off person, say Donald Trump, or a poor thirteenth-century Inca peasant. (To keep the thought experiment clean, assume that changes to that particular Inca individual do not alter the course of history in any way that would affect Mr Trump's well-being, including the fact of his very coming into existence.) Both egalitarians and prioritarians, I venture, would say that there is more value to be had by benefitting the Inca compared to benefitting Trump. Now, there is of course the small technical issue of how to deliver the benefit to the Inca peasant. But most would agree that *if* we could somehow travel back in time, we should definitely prioritize the well-being of the Inca over that of Trump. So far so good.

Comparing the Inca to future possible persons helps us, I think, meet all three of Adler's concerns. Suppose you agree that the Incas do have a standing to complain about inequalities, and suppose we further stipulate that possible persons, because they are merely possible, do not. *That* version of the standing objection (to such a disvalue) need not violate agent-neutrality. On this version, a possible person has no complaint, but not because *you* chose not to bring her about. Rather she lacks a standing to complain because her existence is not certain. (Notice, I am not, yet, trying to convincing you that she lacks standing; I am merely showing that such a claim does not violate agent-neutrality.) She is not certain to exist, independently of the control that some agent has over the decision to bring her about. Second, saying that inequality across possible persons lacks egalitarian disvalue need not violate time-neutrality. Indeed as the Inca case shows, egalitarians could and should be time-neutral. If we could travel back in time, no egalitarian I know would dispute that there is a (strong) egalitarian reason to rectify the inequality. (With the risk of stating the obvious, my discussion is exclusive to telic, not deontic views.) The same goes for forward time-travel. It is not the fact that possible persons occupy the strange land of the future that nullifies any egalitarian disvalue their position may give rise to. It is rather their lack of (for sure) existence. But 'the Incas also do not exist!' you might say. True, they don't. But they have existed, that is, they have a fixed identity, as evidenced by the realization that if we did travel back in time we would find there a distinct person, with a for-sure existence, and a fixed identity, and therefore with a complaint that we could address. Third and finally, denying that inequalities across and between possible persons are the business of fairness does not imply discrimination against the voiceless. After all, the Incas are equally voiceless and yet the egalitarian is quick and willing to recognize her complaint. The Standing Objection to the disvalue of inequalities between possible persons does not violate agent-neutrality, time-neutrality, or discriminates against the voiceless.

I conclude that the Standing Objection to equality across possible persons stands, but that it stands for reasons different from those invoked by Otsuka. Egalitarians need not care about inequalities involving merely possible persons. Merely possible persons do not have complaints of unfairness. This is not because they lack a voice but rather because they lack, well, existence. We can therefore formulate the following Standing Objection. Inequalities between persons always matter; be it between past, present, and future persons. But 'possible persons' are not, or not yet, persons; they are, as Otsuka (more) accurately says, a *possibility* of a person. Inequalities between persons matter, but inequalities between a person and a mere possibility of a person don't.

Let us see where we have gotten to. I have shown that it *is* possible to mount a Standing Objection to the disvalue of inequalities across possible persons *without* this objection violating core tenets of rational choice.

14.5 Prioritarianism, Possible Persons, and the Separateness of Persons

While persons can complain about inequalities, merely possible persons cannot. But possible persons, I have conceded, do have shortfall 'complaints'. In this section I want to look at two ways in which a prioritarian may account for these shortfall complaints. I want to argue against a prioritarian version of the complaint model used by Voorhoeve. And I want to argue in favour of a more traditional *ex post* prioritarian model to account for possible persons and our responsibilities towards them.

First then, would a *prioritarian* version of the idea underlying EPP, that is, a prioritarian complaint model, arrive at better judgements in Case One? As Voorhoeve nicely shows, this does not seem to be the case. Consider Jake Nebel's Person-Affecting Prioritarianism (PAP), according to which, in brief, a person has a complaint when we fail to maximize her *expected* well-being, and that complaint is stronger, the worse off she is. When no such competing complaints obtain, we ought to simply do whatever would maximize the expected value of the outcome (Nebel 2017: 908). In Case One this view would perform well, on both variants. In the intra-personal variant it says that, since there are no competing claims, we should maximize Chris's expected well-being, opting for A. And the inter-personal variant PAP says that if the prospective mother develops Type 1, Evan's complaint against A is stronger than Dominic's complaint against B (in case the mother develops Type 2). And that is because Evan starts off from a worse-off position. We should therefore choose B, which is the desired result.

While Nebel's Person-Affecting Prioritarianism handles Case One well, it struggles with other cases, as demonstrated by Voorhoeve. Consider Case Nine (Table 14.10), where (c) is equal to or greater than zero (Voorhoeve 2021: 318).

Table 14.10 Case Nine

Alternative	Type 1 (0.5)	Type 2 (0.5)
Dominic A	—	90
Evan	30	—
Dominic B	60−c	—
Evan	—	60−c

Here, there are no competing claims. If we choose A and Evan exists, it means that Type 1 obtains, which means that he would not have existed had we chosen B. And the same goes for Dominic. Since there are no competing claims, in this case, Person-Affecting Prioritarianism tells us to maximize the expected value, which would be A. Notice, Person-Affecting Prioritarianism does *not* tell us (that is, when there are no competing claims) to maximize the expected *prioritarian* value (that is, maximize the chance of the greater prioritarian value, where that value is greater the worse off the recipient is); it tells us to maximize value, which would lead to A. But I submit that for a small enough (c) that does not seem like the right result for a prioritarian. This is easier to see when we set (c) to equal zero, in which case PAP would be indifferent. That, however, would be a very strange, and implausible, verdict for a prioritarian to entertain. Put differently, such a case shows that PAP can be insensitive to what we earlier called *shortfall* complaints.

Person-Affecting Prioritarianism is not an attractive alternative when dealing with possible persons. Think now of standard *ex post* prioritarianism (as an alternative to Voorhoeve's EPP and Nebel's PAP). In Cases One (inter-personal) and Case Nine it would straightforwardly opt for B, since that is the option that maximizes the expected *prioritarian* value. *Ex post* prioritarianism therefore does better than Person-Affecting Prioritarianism with respect to promoting the prioritarian value. But of course standard *ex post* prioritarianism suffers from an obvious shortcoming when it comes to handling cases involving possible persons, namely the intra-personal variants (of such cases). I have already mentioned the Separateness of Persons (SOP) Objection, and the charge that *ex post* prioritarianism yields the wrong result in intra-personal cases. In Case One Intra-personal, *ex post* prioritarianism reaches the same verdict it reaches in the inter-personal scenario, and thus recommends B, which many people deem to be counterintuitive. We saw that the view is the best at handling the tough question of distribution between possible persons, but so far only of the inter-personal variant. Can its handling of the intra-personal variant be somehow defended?

I think the answer to that question is affirmative. Elsewhere I have suggested that proponents of *ex post* prioritarianism should bite the bullet and insist on the *impersonal* nature of their view (Segall 2015). They should maintain that option B makes things better *in one respect*, namely in preventing a scenario in which someone is made badly off. This may not be in the best interests of the sole person in question (Chris). But preventing the risk of the scenario in which he leads a life of 30 years, as opposed to a life of 60, is better *in one, impersonal, respect.*[9]

Many people remain unconvinced by this reply. In the next and final section I want to show that the current debate about possible persons may lend that reply

---

[9] I am hardly the first to suggest something of this nature. For example, in his reply to critics, Derek Parfit (2012) has made a point somewhat similar to this, even while not going all the way in embracing the impersonal point of view.

some added support. And that is because, I want to argue, *ex post* prioritarianism does no worse than Voorhoeve's account.

14.6 The Separateness of Persons Objection and Impersonal Values

My contention is that *ex post* prioritarianism does no worse, with respect to the *unity of the person*, than does Voorhoeve's account. Let me make two points to that effect.

Think of the intra-personal version of Case One again (Table 14.11).

And now consider a variant of it (Table 14.12).

Case Ten is a two-person case, but the trade-off in it is, at least *ex ante*, an intra-personal one. In choosing between option A and B you are disadvantaging neither person in comparison to one another. In a case that is structurally identical to this, Voorhoeve (with Fleurbaey) says that, for a small enough positive (c), we should opt for B. They note that, *ex ante,* the trade-off in the two-person case is a purely intra-personal one (and as such identical to the one in the one-person case). Still they reason that a commitment to what would be justifiable, *with full information,* to both Hiroshi and Iwao, should stir us to favour B. So even though *ex ante,* it is A which is in the best interests of each person taken separately, taken together and with full information, we should opt for B (Fleurbaey and Voorhoeve 2013, see especially 117–20).

Fair enough. Now, there are two ways VF could defend opting for B here. However, neither of them spells good news for EPP, I want to say.

Table 14.11 Case One: single person intra-personal

Alternative		Type 1 (0.5)	Type 2 (0.5)
A	Chris	30	90
B	Chris	60−c	60−c

Table 14.12 Case Ten: Two-person intra-personal

Alternative	Type 1 (0.5)	Type 2 (0.5)
Hiroshi A	90	30
Iwao	30	90
Hiroshi B	60−c	60−c
Iwao	60−c	60−c

Table 14.13 Case Three

Alternative	Type 1 (0.5)	Type 2 (0.5)
Howard	—	90 + e
A		
Idris	30 + e	—
Howard	—	—
B		
Idris	30	90

First, the most straightforward way an egalitarian could justify the choice of B over A (in Case Ten) is simply to refer to the impersonal, or non-person affecting, value of fairness (or equality) in final well-being. Voorhoeve in fact does explicitly that in the paper in which the EPP is laid out. He invokes the impersonal value of fairness when offering a response to the charge that his EPP violates State-Wise Dominance.[10] (State-Wise Dominance is a condition of rationality according to which one lottery is better than another if it generates a better outcome *in each* of [or at least as good as, in some of] the possible states of nature.) Now consider Case Three again (Voorhoeve 2021: 314; Table 14.13).

Suppose you have the intuition, like I do, that A is better. This could be explained by State-Wise Dominance, certainly if we take final well-being as the determinant of the goodness of options. Whatever the state of the world, option A offers the higher well-being. A, however, involves large inequalities in expected well-being, whereas B involves none. Therefore for a small enough (e), EPP would favour B. I have already indicated that Voorhoeve has no qualms endorsing this result. What is of interest here is how, exactly, he defends this somewhat counter-intuitive verdict. His view, Voorhoeve says, does not violate State-Wise Dominance because the value of an outcome 'is not wholly determined by the distribution of final well-being'. Instead, it is determined also 'by the prospects enjoyed by both potential individuals at the time of decision. These prospects contribute to the value of the outcome, since they determine how potential individuals' prospective interests were served at the moment of decision, *and therefore how fair this decision was*' (2021: 314, emphasis added).

As an egalitarian I am sympathetic to such non-person affecting concerns. But then the question of course presents itself: Why are impersonal considerations such as fairness so acceptable in Case Three (where, recall, the trade-off, at least

---

[10] In discussing the merits, *ex post*, of holding a lottery over a kidney to simply giving it straightaway to one of the patients, he writes. 'If the second patient were to lose out in the random draw, in terms of final wellbeing, both patients may be worse off than if the kidney had been assigned directly to the first patient. [. . .] Nonetheless, the outcome of the random assignment is clearly better in one respect, and therefore possibly better overall, because it results from a fairer distribution of chances' (Voorhoeve 2021: 315).

ex ante, is entirely intra-personal), but so allegedly absurd in the equally intra-personal variant of Case One? Recall that in Case One Intra-personal the *ex post* prioritarian alluded to the impersonal greater value of raising a person from a worse-off absolute level, even when this conflicts with person-affecting considerations. If impersonal (non person-affecting) considerations can trump prudential ones in the *ex ante* intra-personal Case Three, why should that be precluded in the equally intra-personal Case One? To stress: I am not asking the anti-prioritarian egalitarian to convert and believe in the prioritarian value; I am only asking her to accept that in so far as there is such a value, there is nothing preposterous about deferring to that impersonal value over a prudential one (in intra-personal cases).[11]

Voorhoeve, then, allows himself recourse to non-person affecting considerations in defence of his egalitarian handling of *ex ante* intra-personal cases, recourse that he does not allow prioritarians in their handling of intra-personal cases. Could, perhaps, the justification for this be that while the prioritarian dilemma pertains to a case that is purely intra-personal, the egalitarian one, in Case Ten, pertains to one that is only *ex ante* intra-personal? Let us look at this contrast without using the term 'intra-personal'. In describing his objection to the prioritarian's handling of Case One Voorhoeve (2021: 308) writes that for the prioritarian to decree that option A is better (in one, impersonal respect) 'would be to take a perspective on moral value that is curiously divorced from the interests of the only person whose fate is under consideration'. But how different is his judgement that in Case Ten we should opt for B? Doesn't a choice of B amount to taking (to paraphrase) 'a perspective on moral value that is curiously divorced from the interests of the only *two* persons whose fate is under consideration'? I doubt that it does not.

I said that Voorhoeve could defend the choice in B (in Case Ten) in one of the two ways and that neither of them spells good news for the EPP. Here is the second strategy mentioned. Suppose, instead, that egalitarians were to justify the choice in B (in Case Ten) *not* from an impersonal perspective, but rather from the (very much person-affecting) *competing claims* framework. They would then argue that Teresa, the person making the decision, ought to operate as if she had full information. She should thus eliminate the situation by which one of the two, Hiroshi or Iwao, ends up with a complaint. That would imply opting for B, for only then, *ex post,* none of the claimants would have a complaint.[12] Recall the quote from Reibetanz Moreau (1998: 304) that we invoked earlier:

---

[11] VF seem to accept this. In a footnote they write: 'our argument against ex ante Pareto is equally effective if one instead gives priority to the worse off because one is a prioritarian [...]'. Fleurbaey and Voorhoeve (2013: 116, fn. 8).

[12] As in fact they do. See Fleurbaey and Voorhoeve (2013: 116–17). Otsuka (2012: 374) holds the same view.

As long as we know that acceptance of a principle will affect *someone* in a certain way, we should assign that person a complaint that is based upon the full magnitude of the harm or benefit, even if we cannot identify the person in advance.

But notice now what it would imply for Voorhoeve and Fleurbaey to rely on something like the logic invoked here by Reibetanz Moreau (as in fact they do). Johann Frick has rightly noted, I think, that to be guided by the fact that there is bound to be a loser in such cases is to treat Hiroshi and Iwao as 'placeholders' for losses and gains (Frick 2013: 141). I want to take this one step further and say that if we were to take that second route, in opting for B, it would amount to looking at the gains and losses not of Hiroshi and Iwao, taken separately, but of 'Hiroshi and Iwao', taken as a collective. It is true of 'Hiroshi and Iwao' that a choice in A is unjustifiable to them, because we know that one of them is bound to lose. But to look at things this way is not to respect the individuality of Hiroshi and Iwao. In other words, if Voorhoeve were to justify B by means other than the impersonal value of equality he would end up violating something that is even dearer to the EPP, namely the Separateness of Persons.

To sum up these last two sections, *ex post* prioritarianism handles well the inter-personal version of possible persons, and does this much better than its rivals, both egalitarian and prioritarian (the person-affecting variant). To defend its handling of the intra-personal variant, its proponents could invoke the impersonal nature of its value. That, I said, could not be easily dismissed, especially by proponents of EPP such as Voorhoeve. That is the case because the latter are committed either to invoking impersonal values in defence of their view, or are themselves at risk of violating the much-touted Separateness of Persons Objection (when attempting to produce the correct verdict in multi-person *intra*-personal cases such as Case Three). It would therefore be wrong to dismiss *ex post* prioritarianism in this context solely on the grounds that it invokes a non-person affecting value.

14.7 Conclusion

It *is* bad that boys born in Scotland in 2043 could expect a lower life span compared to boys born in England. In contrast, if only one of these boys will come into being, it is not bad in itself that one had worse prospects *compared to another*. It is not at all obvious that egalitarians should be exercised by inequalities in life expectancy between merely possible persons. I agree with Voorhoeve that we need to make the world as welcoming as possible to whomever may arrive. But I disagree that we have a responsibility to make it *equally* welcoming to possible, alternate individuals.

References

Adler, M. (2022), 'Claims Across Outcomes and Population Ethics', in G. Arrhenius, K. Bykvist, T. Campbell, and E. Finneron-Burns (ed.) *The Oxford Handbook of Population Ethics* (Oxford University Press).

Fleurbaey, M. and Voorhoeve, A. (2013), 'Decide as You Would with Full Information! An Argument Against Ex Ante Pareto', in N. Eyal, S. Hurst, O.F. Norheim, and D. Wikler (ed.), *Inequalities in Health: Concepts, Measures, and Ethics* (Oxford University Press): 113–28.

Frick, J. (2013), 'Uncertainty and Justifiability to Each Person: Response to Fleurbaey and Voorhoeve', in N. Eyal, S. Hurst, O.F. Norheim, and D. Wikler (ed.), *Inequalities in Health: Concepts, Measures, and Ethics* (Oxford University Press): 129–46.

Handfield, T. (2018), 'Egalitarianism About Expected Utility', in *Ethics* 128/3: 603–11.

Hausman, D. (2006), 'What's Wrong with Health Inequalities?', in *Journal of Political Philosophy*, 15: 46–66.

McMahan, J. (1981), 'Problems of Population Theory', in *Ethics* 92: 96–127.

Moreau, S.R. (1998), 'Contractualism and Aggregation', in *Ethics* 108/2: 296–311.

Nebel, J. (2017), 'Priority, Not Equality for Possible People', in *Ethics* 127/4: 896–911.

Office for National Statistics (ONS). (2018), www.ons.gov.uk/peoplepopulationand-community/birthsdeathsandmarriages/lifeexpectancies/bulletins/pastandproject-eddatafromtheperiodandcohortlifetables/1981to2068

Otsuka, M. (2012), 'Prioritarianism and the Separateness of Persons', in *Utilitas* 24/3: 365–80.

Otsuka, M. (2018), 'How it Makes a Moral Difference That One is Worse Off Than One Could Have Been', in *Politics, Philosophy, Economics*, 17/2: 192–215.

Otsuka, M. (2022), 'Prioritarianism, Population Ethics, and Competing Claims', in J. McMahan, T. Campbell, J. Goodrich, and K. Ramakrishnan (ed.), *Ethics and Existence: The Legacy of Derek Parfit* (Oxford University Press): 527–51.

Parfit, D. (1984), *Reasons and Persons* (Oxford University Press).

Parfit, D. (2011), *On What Matters (Vol. 2)* (Oxford University Press)

Parfit, D. (2012), 'Another Defense of the Priority View', in *Utilitas*, 24: 399–440.

Parfit, D. (2017), 'Future People, the Non-Identity Problem, and Person-Affecting Principles', in *Philosophy & Public Affairs*, 45/2: 118–57.

Persson, I. (2001), 'Equality, Priority, and Person-Affecting Value', in *Ethical Theory & Moral Practice*, 4: 23–39.

Segall, S. (2010), *Health, Luck, and Justice* (Princeton University Press).

Segall, S. (2015), 'In Defense of Priority (and Equality)', in *Politics, Philosophy, & Economics*, 14: 343–64.

Segall, S. (2019), 'Why We Should be Negative about Positive Egalitarianism', in *Utilitas*, 31: 414–30.

Segall, S. (2022), 'Quarantining Prioritarianism', in J. McMahan, T. Campbell, J. Goodrich, and K. Ramakrishnan (ed.), *Ethics and Existence: The Legacy of Derek Parfit* (Oxford University Press): 552–76.

Voorhoeve, A. (2021), 'Equality for Prospective People: A Novel Statement and Defense', in *Utilitas*, 33/3: 304–20.

Index

For the benefit of digital users, indexed terms that span two pages (e.g., 52–53) may, on occasion, appear on only one of those pages.